MICROBIAL GROWTH ON C$_1$ COMPOUNDS

MICROBIAL GROWTH ON C$_1$ COMPOUNDS

Proceedings of the 8th International Symposium on
Microbial Growth on C$_1$ Compounds,
held in San Diego, U.S.A., 27 August – 1 September 1995

Edited by

MARY E. LIDSTROM
California Institute of Technology,
Pasadena, California, U.S.A.

and

F. ROBERT TABITA
The Ohio State University,
Columbus, Ohio, U.S.A.

KLUWER ACADEMIC PUBLISHERS
DORDRECHT / BOSTON / LONDON

A C.I.P. Catalogue record for this book is available from the Library of Congress

ISBN 0-7923-3938-X

Published by Kluwer Academic Publishers,
P.O. Box 17, 3300 AA Dordrecht, The Netherlands.

Kluwer Academic Publishers incorporates
the publishing programmes of
D. Reidel, Martinus Nijhoff, Dr W. Junk and MTP Press.

Sold and distributed in the U.S.A. and Canada
by Kluwer Academic Publishers,
101 Philip Drive, Norwell, MA 02061, U.S.A.

In all other countries, sold and distributed
by Kluwer Academic Publishers Group,
P.O. Box 322, 3300 AH Dordrecht, The Netherlands.

Printed on acid-free paper

Printed in the Netherlands

Contents

Preface

The 8th International Symposium on Microbial Growth on C_1 Compounds was held at the Bahia Resort Hotel, San Diego, CA 27 August - 1 September, 1995. A total of 160 participants from 18 countries were registered. Eight Scientific Sessions were held in which 45 papers were presented, and 114 posters were displayed and discussed in 3 separate poster sessions.

The Symposium covered a broad range of topics, including acetogenesis, methanogenesis, CO_2 fixation, lithoautotrophy, carboxidotrophy, methylotrophy and methanotrophy. The theme of the Symposium was mechanistic, and under this rubric the physiology, biochemistry, molecular biology, and both applied and environmental aspects of microbial growth on C_1 compounds were addressed.

This Symposium Volume contains 46 chapters, including the text of an Opening Address delivered at the Symposium by J.R. Quayle. This chapter elegantly presents an historical perspective on the past 7 Symposia, in the context of major breakthroughs in the field and of what is termed "giant" topics. The reader of this Volume will be pleased to see that the tradition of the past Symposia Volumes is upheld, and that both familiar and new "giant" topics are covered. This Volume presents a cutting edge view of the broad field of microbial one-carbon metabolism, and provides a valuable resource for researcher and student alike.

ACKNOWLEDGMENTS

The 8th Symposium was achieved through the combined efforts of a number of dedicated persons, including the Organizing Committee (Mary Lidstrom, Chair and Symposium Editor; Bob Tabita, Symposium Editor; Sunney Chan; and Dick Hanson); Fran Matzen, the Symposium Secretary; and the members of the Applied Microbiology group at the California Institute of Technology: Olga Berson, Ludmila Chistoserdova, Andria Costello, Kelly Goodwin, Kelly Smith, Amy Springer, Sergei Stoliar, Hirohide Toyama, and Edye Udell.

The Symposium could not have been held without the generous financial support of the following organizations: Bayer Corporation, Environment Now, Pharmacia Corporation and the following U.S. agencies, the Department of Energy, the National Aeronautics and Space Administration, and the National Science Foundation, and their assistance is gratefully acknowledged.

October 1995

MARY E. LIDSTROM
F. ROBERT TABITA

Opening Address

J.R. Quayle
Compton Dando, Bristol BS18 4LA, United Kingdom

It gives me the greatest pleasure to welcome you all to the Eighth International C_1 Symposium. These symposia have so far always been held at some location in the world where there is a centre of excellence in the study of microbial growth on C_1 compounds. The emphasis has been on the proximity to the research centre rather than to some exotic island paradise. But in 1995 these separate academic and sybaritic strands have been neatly ligated here in San Diego because, besides being on a world famous stretch of the Pacific Ocean, it is close to Pasadena and Caltech where Mary Lidstrom has since 1987 led an illustrious C_1 research group.

I need hardly say how honoured I feel in being asked to open this symposium and in order to gain some perspective on the progress of the C_1 field since the symposia started 22 years ago in Edinburgh, I took the opportunity to look through all the past proceedings of each symposium. I should say not quite all, for despite writing to several founding members I could not turn up the proceedings of Symposium Zero held at Edinburgh in 1973. However, I have a clear memory of what excited us then, and some of the topics are still very much with us in 1995.

This brings me to my main theme that when you view the proceedings of all these conferences, you get a vivid and sustained picture of how a focused field of science progresses. It is as if the viewer's microscope is a magic one in which the field of view is continually widening while at the same time it has an ever-increasing resolving power. To illustrate this I have picked a few well defined areas and followed their development over these 22 years. I have deliberately avoided naming the principal players as there is no time for me to ensure that justice is done to all of them.

In the 1970's, covered by the Edinburgh, Tokyo and Puschino meetings there was excitement about the elucidation of the ribulose monophosphate pathway of formaldehyde fixation and the icl^+ serine pathway, although crucial steps of the icl^- pathway remained unsolved (as they still do today, as far as I am aware). There was excitement over the newly discovered methylotrophic yeasts with evidence accumulating for some kind of assimilation pathway operating which involved sugar phosphate intermediates. The Tokyo meeting in 1974 saw the first astonishing electron microscope picture of the cube-like peroxisomes in cells of methanol-grown *Hansenula polymorpha*. The methanol dehydrogenase discovered by Zatman and his research student, Chris Anthony, in 1964 was already a constant feature on all the programmes, as indeed it is today. The first

1

M. E. Lidstrom and F. R. Tabita (eds.), Microbial Growth on C₁ Compounds, 1–3.
© 1996 *Kluwer Academic Publishers. Printed in the Netherlands.*

purifications of methane monooxygenase were announced—another enzyme giant destined to be with us in all subsequent symposia. Applied aspects were mainly concerned with use of methylotrophs for production of amino acids and extracellular polysaccharides. ICI announced the isolation of *Methylophilus methylotrophus*, a methanol utilizer with a formidably fast growth rate which was to spearhead their attempt to make bacterial SCP from methanol.

Symposium 3 in Sheffield in 1980 opened up the 1980's with some major advances into the biochemistry of the 3-component, soluble methane monooxygenase from *Methylococcus capsulatus* and the particulate methane monooxygenase from *Methylosinus trichosporium*. The biochemical and physiological interrelation of these two systems has been, and still is, a major topic in all subsequent C_1 symposia. One of the highlights of the conference was the announcement of the long-awaited structure of the coenzyme of methanol dehydrogenase—pyrrolo-quinoline quinone, which introduced a new coenzyme to biochemistry, not only confined to the methylotrophs. The continued study of the new quinoproteins has resulted since 1988 in the budding off of its own series of triennial symposia. Other enzymes reached giant status and have remained so ever since, e.g., ribulose bisphosphate carboxylase. Serious interest began in the CO utilizers and the new carbon assimilation pathway of methanol utilizing yeasts—the xylulose monophosphate pathway or dihydroxyacetone pathway was firmly established. Much patient and detailed biochemical work on characterisation of methylated amine dehydrogenases and of cytochromes of methylotrophs was reported and one cannot help but reflect on how important all this detail has become as the full power of genetic analysis and X-ray crystallography has been applied to many of these systems in the 1980's and 90's. Sheffield also saw the apogee of the ICI single cell protein project at Billingham with the successful construction of the world's largest air lift fermenter. That such an Everest of applied microbiology had to be abandoned later due to the onslaught of the soya bean farmers was a disappointment to many of us. Genetics of methylotrophs, for long a no-go area for all but the brave, began to stir and bubble. The other great landmark was the assault on the biochemistry of the methanogens, mainly by teams of American, German and Dutch scientists. I have never ceased to marvel at the way in which the black box of conversion of C_1 substrates and of acetate into methane and carbon dioxide, together with the modes of energy coupling and carbon assimilation pathways have exploded in the 15 years since the Sheffield meeting in a welter of new coenzymes and enzymes. How many times have microbial biochemists sketched out putative linear C_1 assimilation sequences which, alas, have never seen the aerobic light of day? Whereas here in the anaerobes, including not only the methanogenic archaebacteria but also Gram-positive and Gram-negative acetogenic bacteria, sulphur-dependent archaebacteria, and possibly a few others, is the elegantly simple sequence of two C_1 making C_2 which then adds another C_1 to make C_3.

The next six years were covered by the meetings in Minnesota (1983) and Haren (1986). During these six years, the importance of Cu^{++} in the interrelation of the soluble and particulate methane monooxygenases had been recognised and much progress had been made on the biochemistry and enzymology of CO utilization by the carboxydotrophs. There was now great interest in the intracellular organisation of the periplasmic methanol dehydrogenase and its coupling with cytochromes c_L and c_H. In these six years the genetics of C_1 utilizers had really got under way. Recombinant DNA technology to a large extent had overcome the traditional difficulties of doing genetics with the methylotrophs, e.g., lack of mutants, rarity and low efficiency of chromosome-mobilising plasmids, lack of transducing phages. The full extent of this acceleration was exemplified by the work of Mary Lidstrom's group who provided the first comprehensive genetic study of methanol dehydrogenase activity by analysing the activity in *Methylobacterium extorquens* AM1 into 10 complementation groups each characterised for function.

And so to the last two symposia: Göttingen in 1989 and Warwick in 1992. The torrent of discoveries of new enzymes and coenzymes in the methanogenic and acetogenic field continued. Molecular genetic studies were now being extended into several anaerobic C_1 utilizers, while the full battery of 10 enzymes and 13 reactions of the Calvin cycle was now being studied in photosynthetic and hydrogen bacteria. The soluble methane monooxygenase had been cloned and expressed in *Escherichia coli*. These two meetings saw a substantial increase in ecological studies with new techniques for detecting specific groups of methylotrophs *in vivo* based on the use of oligonucleotide probes, an increased interest in the biogeochemical cycling of sulphur, in particular dimethyl sulphide and its oxidation product dimethyl sulphonic acid, a welcome renewal of interest in the global methane budget and the still unresolved question of anaerobic oxidation of methane in marine sediments. Surprises were sprung with methanol dehydrogenase where a new small enzyme subunit was found. This changed the long held view of the molecule as being a dimer to that of an $\alpha_2\beta_2$ tetramer. A Ca^{++} ion was fitted into the interface region of the two dimers. A picture was drawn up of the way in which the cytochromes c_L and c_H might be docked with the active enzyme. A preliminary X-ray analysis at 3Å resolution was shown, a foretaste of the very beautiful picture at 1.94 Å which no doubt we shall be admiring later in this meeting. Two new quinone coenzymes appeared: tryptophan tryptophylquinone in bacterial methylamine dehydrogenases and 6-hydroxydopa quinone in copper-containing amine oxidases and a full scale genetic analysis of bacterial methylamine dehydrogenase was reported.

So we come to 1995 in San Diego where we can confidently expect that the hypothetical microscope that I mentioned at the beginning will once again show its field further widened and its resolving power further increased.

The Ribulose Monophosphate (Quayle) Cycle: News and Views

Y.A.Trotsenko, V.N.Khmelenina and A.P.Beschastny
Institute of Biochemistry and Physiology of Microorganisms, Russian
Academy of Sciences, Pushchino, Moscow region, 142292 Russia

INTRODUCTION
Over 30 years ago the elegant radioisotopic experiments by Quayle and his colleagues
have led to a discovery of the novel metabolic pathway, the ribulose monophosphate
(RuMP) cycle. Further enzymic studies confirmed its operation in many obligately
and facultatively methylotrophic bacteria of different taxonomic position. A detailed
presentation of four possible variants of this cycle has been reviewed elsewhere
(Quayle, 1980). This paper briefly describes a new variant of the RuMP cycle coupled
with pyrophosphate-dependent 6-phosphofructokinase (PPi-PFK) and its operation in
Type (obligate methanotrophs. Current state and future prospects for our studies on
the metabolic role of inorganic pyrophosphate (PPi) in aerobic methylotrophs are also
discussed.

RESULTS AND DISCUSSION
According to Quayle (1980), an essentially irreversible phosphorylation of fructose 6-
phosphate to fructose 1,6-bisphosphate (FBP) being catalyzed by ATP -dependent 6-
phosphofructokinase (ATP-PFK):

$$Mg^{2+}$$
$$\text{Fructose-6-P + ATP} \longrightarrow \text{Fructose-1,6-P}_2 + \text{ADP}$$

is a key regulatory step in methane- and methanol - utilizing bacteria with glycolytic
variant of the RuMP cycle. However, extremely low (if any) activity of ATP-PFK
was detected by us in extracts of Type I obligate methanotrophs that correlated with
their lowered intracellular pools of ATP (Trotsenko, Shishkina, 1990). Instead, we
found in these bacteria enhanced levels of PPi and PPi-PFK (EC 2.7.1.90) catalyzing
a readily reversible reaction:

$$Mg^{2+}$$
$$\text{Fructose-6-P + PPi} \longrightarrow \text{Fructose-1,6-P}_2 + \text{Pi}$$

Following these findings we have proposed that PPi serves in obligate
methanotrophs as a phosphoryl and energy donor (Trotsenko, Shishkina, 1990;
Shishkina, Trotsenko, 1991). As will be discussed here, the use of PPi as both
phosphoryl and energy donors raises some important questions on the properties of
PPi-PFK and its role in primary C_1-metabolism of obligate methanotrophs. To answer
these questions a highly active PPi-PFK from the Type I obligate methanotroph
Methylomonas methanica 12 was first purified to electrophoretic homogeneity and
characterized (Beschastny et al., 1993). The properties of the PPi-PFKs purified from

4

M. E. Lidstrom and F. R. Tabita (eds.), Microbial Growth on C₁ Compounds, 4–8.
© 1996 Kluwer Academic Publishers. Printed in the Netherlands.

M.methanica and other methylotrophic and non- methylotrophic microorganisms are summarized in Table 1. In principle, their major properties such as molecular weights of the native enzyme and subunits, as well as the absence of effectors are rather similar with exception of the *Amycolatopsis methanolica* PPi-PFK having a tetrameric structure and rather low affinity for PPi.

Since the PPi-PFK discovery up to date there has been a question raised: in which direction (glycolytic or gluconeogenic) this enzyme operates in vivo? Although no clear answer has yet been obtained it is believed that PPi-PFK to be involved in glycolysis in microorganisms and our findings agree with this suggestion (Mertens, 1991). Furthermore, as the phosphorylation of fructose-6P in *M.methanica* is involved in FBP -variant of the RuMP cycle, PPi-PFK is obviously the vital and energy-saving enzyme in primary assimilation of formaldehyde. This should lead to reappraisal of the basic equation of the RuMP cycle and role of PPi in primary metabolism of Type I obligate methanotrophs. In this case we have to modify the general equation of the RuMP cycle (FBP-variant) postulated by Quayle (1980):

$$3HCHO + ATP ----> Triose-P + ADP,$$

to reflect the involvement of PPi instead of ATP:

$$3HCHO + PPi ----> Triose-P + Pi;$$

However, this basic stoichiometry does not consider the possibility for simultaneous operation of the FBP- and Entner-Doudoroff (ED) pathways in Type I methanotrophs that have been first postulated and questioned by Ström et al. (1974). Indeed, we have no enough experimental data to evaluate adequately their ratio in vivo and contribution to the regeneration of ribulose-5P (Ru-5P), the primary acceptor of formaldehyde. As judged by the high activity of PPi-PFK, the FBP cleavage dominates in *M.methanica* thus providing phosphotrioses for Ru-5P regeneration and phosphoenolpyruvate (PEP) formation via glycolysis. PEP is then carboxylated to oxaloacetate which enters the tricarboxylic (TCA) cycle. As we have shown earlier (Shishkina,Trotsenko, 1982) all obligate methanotrophs are deficient in several enzymes responsible for PEP- pyruvate interconversions. It is particularly important to note the lack of pyruvate kinase that prevents PEP conversion to pyruvate and gives less energy (one mole of ATP instead of two). To compensate for such enzymatic and energy deficiencies, *M.methanica* appeared to develop the energy-saving PPi-PFK and ED-cleavage, the latter being the only possible pathway for pyruvate formation followed by its oxidation to acetyl-CoA entering the TCA-cycle. Therefore, both FBP- and ED-pathways are necessary for biosynthetic operation of the incomplete TCA-cycle in Type I methanotrophs, particularly *M.methanica*. We hypothesize that PPi-PFK as a branch-point enzyme plays the key role in regulation of the fructose-6P distribution between FBP- and ED-pathways in this organism (Fig.1).

Since the PPi-PFK of *M.methanica* is not an allosteric enzyme, its activity in vivo appears to depend only on the levels of own substrates (Beschastny et al., 1993). On the basis of rapid equilibrium random kinetic mechanism of the microbial PPi-PFKs it is suggested that this enzyme is responsible for in vivo maintenance of the ratio of its substrates and products (Bertagnolli, Cook, 1984). To understand all of these regulatory details we have to learn the factors which can influence the intracellular concentrations of fructose 6-phosphate, fructose 1,6-bisphosphate, PPi and Pi. Thus, the regulatory pattern of the RuMP cycle in *M.methanica* still remains unclear and needs further studies.

In living cells PPi is formed in many anabolic reactions and is rapidly hydrolyzed by inorganic pyrophosphatase (PPi-ase) thereby driving the biosynthesis. However, in obligate methanotrophs, including *M.methanica*, PPi-ase has a very low activity (Trotsenko, Shishkina, 1990; Shishkina, Trotsenko, 1991) thus implying the operation of PPi-PFK instead of PPi-ase. To check this suggestion we have studied the properties of PPi-ase from *M.methanica* (A.P. Beschastny, unpublished).

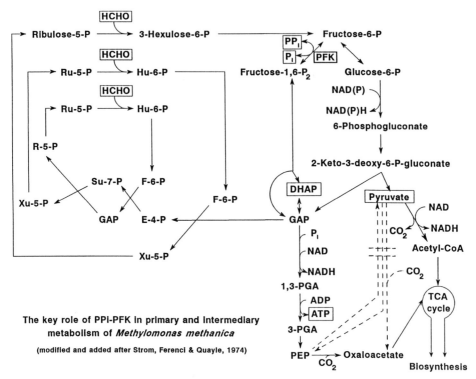

The key role of PPI-PFK in primary and intermediary
metabolism of *Methylomonas methanica*

(modified and added after Strom, Ferenci & Quayle, 1974)

As found, the PPi-ase was a very labile and localized only in the soluble fraction (no activity has been detected in the particulate fraction). This indicates the absence of the membrane-associated proton-translocating PPi-ase which is used as a PPi generator in some microorganisms. Thus, taking into account a very low activity of the cytoplasmic PPi-ase we suggest that biosynthetic reactions are the major source for intracellular PPi in *M.methanica*. Interestingly, PPi-ase activity did not depend on the growth conditions and was not induced even in the presence of exogenous PPi as a phosphate source (Khmelenina et al., 1994). In contrast, the alkaline phosphatase was expressed in phosphate-starved cells and excreted into culture medium.

Very recently, we were able to purify and preliminary characterize PPi-ase from *M.methanica*. Molecular mass of the native enzyme is about 100 kDa. It consists of 3 subunits and has pH optimum 8-9. The substrate saturation curve of the enzyme at pH 8.0 in the absence of free Mg^{2+} showed a sigmoidal relationship (apparent K_m for MgPPi=0.62 mM). The PPi-ase lability, its oligomeric structure and sigmoidal curve of the substrate saturation imply an allosteric nature of this enzyme. We propose that having a low substrate affinity the allosteric PPi-ase of *M.methanica* is unable to hydrolyze total anabolic PPi and only regulates its intracellular level. In this case the role of classical PPi-ase appears to be fulfilled by the PPi-PFK.

Some Properties of the PP$_i$-PFKs from Methylotrophic and Non-methylotrophic Microorganisms

Parameters	Methylomonas methanica (Beschastny et al, 1993)	Amycolatopsis methanolica (Alves et al, 1994)	Propionibacterium shermanii (O'Brien et al, 1975)	Entamoeba histolytica (Reeves et al, 1976)	Acholeplasma laidlawii (Pollack et al, 1986)
Molecular mass (kDa)	92	170	95	83	74
Subunits	2	4	2	2	nt
pH optimum	8.0	7.5	7.4	7.0	7.4
Temperature optimum	40°	35-46°	nt	nt	nt
V_{max} (U mg^{-1} protein) formation of:					
Fructose-1,6-P$_2$	840	107	258	45.7	38.9
Fructose-6-P	870	nt	232	nt	nt
Apparent K_m (mM)					
PP$_i$	0.051	0.2	0.069	0.014	0.11
Fructose-6-P	0.38	0.4	0.1	0.038	0.65
P$_i$	1.7	0.84	0.6	0.8	nt
Fructose-1,6-P$_2$	0.101	0.025	0.051	0.018	nt
Mg^{2+}					
forward reaction	0.012	0.04	0.008	nt	nt
reverse reaction	0.47	0.77	0.083	0.5	nt
Effectors	No	No	No	No	No

nt - not tested

CONCLUSIONS AND FOLLOW-UP

Hence, our recent findings are consistent with the idea that PPi serves as both phosphoryl and energy donor in obligate methanotrophs. Nevertheless, some general questions arise:

1. Why is PPi accumulated at such a high level and in which subcellular constituents (compartment) is the PPi localized?

2. Which enzymes are responsible for PPi generation or utilization, where they are located and how they are regulated?

3. How are the carbon, nitrogen and phosphate fluxes interlinked and regulated in various types of methanotrophs?

Also, one more general and intriguing question arises: what about facultatively methylotrophic bacteria, like *A. methanolica*, having no complex intracytoplasmic membranes but possessing the activity (albeit low) of PPi-PFK? Evidently, to answer these questions further comprehensive studies and new approaches are required for better knowledge of the peculiarities of phosphate metabolism in taxonomically different methylotrophic bacteria.

PPi has been proposed to be an evolutionary precursor of ATP in living cells (Mertens,1993). If so, the methylotrophs having the RuMP-cycle coupled with PPi-PFK can be placed in a unique position with respect to their biochemical evolution. Finally, comparative investigations of phosphate metabolism both in obligate and facultative methylotrophs are necessary to learn more about distribution and possible function of PPi as well as the corresponding enzymes. These studies will hopefully lead to a correct understanding of the operation of phosphate-controlled metabolic systems in various methylotrophs.

ACKNOWLEDGMENTS
The authors acknowledge support from ISF (NJ 2000 and NJ 2300), INTAS (93-2531) and RFFI (95-04-12514a).

REFERENCES
Alves, A.M.C.R. et al. (1994) J. Bacteriol. 176, 6827-6835.
Bertagnolli B.L., Cook P.F. (1984) Biochemistry. 23, 4101-4108.
Beschastny, A.P. et al. (1993) Biokhimiya. 57, 835-840.
Khmelenina V.N. et al. (1994) Mikrobiologiya. 63, 95-98.
Mertens E. (1991) FEBS Letters. 285, 1-5.
Mertens E. (1993) Parasitology Today. 9, 122-126.
O'Brien W.E. et al. (1975) J. Biol. Chem. 250, 8690-8695.
Pollack J.D., M.V. Williams. (1986) J. Bacteriol. 165, 53-60.
Quayle J.R., (1980) FEBS Letters. 117 (Suppl.), K16-K27.
Reeves R.E. et al. (1976) J. Biol. Chem. 251. 2958-2962.
Shishkina V.N.,Trotsenko Y.A. (1982) FEMS Microbiol.Letters 13, 237-242.
Shishkina V.N., Trotsenko Y.A. (1991) Mikrobiologiya. 59, 357-361.
Ström T. et al. (1974) Biochem. J. 144, 465-476.
Trotsenko Y.A., Shishkina V.N. (1990) FEMS Microbiol. Reviews. 87, 267-272.

METABOLIC REGULATION IN THE ACTINOMYCETE *AMYCOLATOPSIS METHANOLICA*, A FACULTATIVE METHYLOTROPH EMPLOYING THE RUMP CYCLE FOR FORMALDEHYDE ASSIMILATION

Lubbert Dijkhuizen
Department of Microbiology, Groningen Biomolecular Sciences and
Biotechnology Institute (GBB), University of Groningen, Kerklaan 30
9751 NN Haren, The Netherlands

INTRODUCTION

Facultative methylotrophs can be found abundantly among organisms employing the Calvin cycle, the serine pathway, or the XuMP cycle for the assimilation of C_1-compounds. It is only in recent years, however, that we have succeeded in the isolation of a number of versatile RuMP cycle bacteria. These facultative RuMP cycle methylotrophs are found almost exclusively among Gram-positive bacteria. Representatives are various bacilli, coryneform bacteria, and actinomycete species (Dijkhuizen et al., 1992; Dijkhuizen, 1993). Most of these organisms grow on methylated amines and only few use methanol as sole carbon- and energy source for growth. Currently we are engaged in a detailed physiological, biochemical and genetic analysis of pathways of primary metabolism in the actinomycete *Amycolatopsis methanolica*. This Gram-positive bacterium is a very versatile methanol-utilizing organism, employing the fructose-bisphosphate aldolase cleavage variant of the RuMP cycle of formaldehyde fixation (Hazeu et al., 1983; de Boer et al., 1990a). In the following sections our studies on metabolic regulation in *A. methanolica* are reviewed with emphasis on glucose and methanol metabolism, the biosynthesis of aromatic amino acids, and the development of gene cloning systems for this organism.

RESULTS

The methylotrophic actinomycete Amycolatopsis methanolica

The nocardioform actinomycete *Amycolatopsis methanolica,* isolated from soil in New-Guinea, was initially labelled *Streptomyces* sp. 239 (Kato et al., 1974), then *Nocardia* sp. 239 (Hazeu et al., 1983). Developments in the taxonomy of actinomycetes allowed its proper identification (de Boer et al., 1990c). Chemosystematic studies showed that the organism has a wall chemotype IV (meso-diaminopimelic acid, arabinose and galactose present). Unlike representatives of the genus *Nocardia*, cell walls of the organism are devoid of mycolic acids. Further chemotaxonomic and morphological data,

9

M. E. Lidstrom and F. R. Tabita (eds.), Microbial Growth on C₁ Compounds, 9–15.
© 1996 *Kluwer Academic Publishers. Printed in the Netherlands.*

and a comparison of 16S rRNA sequences, identified the organism as a member of the genus *Amycolatopsis*. On the basis of a variety of biochemical and microbiological tests it was concluded that the organism forms the nucleus of a new species, *Amycolatopsis methanolica* (de Boer et al., 1990c). At the outset of our studies of this organism we were faced with a general lack of methods for its physiological and genetic manipulation (Dijkhuizen et al., 1993). Most of these problems have been solved meanwhile and effective procedures are now available for (a) growth at relatively high biomass densities in batch and chemostat cultures, allowing purification of enzymes present at relatively low levels, (b) the isolation of various mutants, including auxotrophs, (c) transformation of whole cells, i.s.o. protoplasts as is generally the case with actinomycetes, yielding 10^6 transformants /μg DNA, (d) genetic manipulations using derivatives of the indigenous, integrative and conjugative, plasmid pMEA300 as vector systems, (e) gene cloning using a gene library of total *A. methanolica* DNA in a pMEA300-derived shuttle vector in *Escherichia coli*. We therefore have adopted *A. methanolica* as a model organism for the study of metabolic regulation in a methylotrophic acinomycete. Little is known at the moment about primary metabolism in actinomycetes. Over the years most attention has been devoted to the analysis of the genetics of pathways for secondary metabolite synthesis, and the screening and testing for new applications of the enormous variety of secondary metabolites (e.g. antibiotics) produced by many actinomycetes. Secondary metabolites, however, are derived from intermediates of central metabolic pathways, including those of glucose utilization and aromatic amino acid biosynthesis (intermediates or the amino acids themselves). Knowledge about primary metabolism therefore is considered to be important, especially for the further improvement of processes for the fermentative production of primary and secondary metabolites (Dijkhuizen, Harder, 1992; Dijkhuizen, 1993).

Methanol metabolism
Methanol oxidation in Gram-positive methylotrophic bacteria involves enzymes clearly different from those in Gram-negative bacteria (a periplasmic PQQ-dependent methanol dehydrogenase; EC 1.1.99.8) and in yeasts (a peroxisomal alcohol oxidase; EC 1.1.3.13). All thermotolerant, methanol-utilizing strains of *Bacillus methanolicus* studied were found to possess a cytoplasmic NAD-dependent methanol dehydrogenase (MDH; EC 1.1.1.1) (Arfman et al., 1989; Dijkhuizen, Arfman, 1990), which is strongly stimulated by a specific (activator) protein (Arfman et al., 1991). No NAD-dependent MDH activity could be detected in *A. methanolica*. Instead, methanol oxidation in this organism resulted in concomitant reduction of N,N'-dimethyl-4-nitrosoaniline (NDMA). The corresponding cytoplasmic enzyme has been designated methanol:NDMA oxidoreductase (MNO) (Bystrykh et al., 1993a,b,c). NDMA is known to re-oxidize pyridine nucleotides which are tightly bound to the active centers of dehydrogenases (Dunn, Bernhard, 1971; Kovář et al., 1984).
Analysis of the quaternary protein stuctures of the purified *B. methanolicus* MDH (subunit M_r 43,000) (Vonck et al., 1991) and the *A. methanolica* MNO enzyme (subunit M_r 50,000) (Bystrykh et al., 1993b) by electron microscopy and image processing revealed similar decameric structures with five-fold symmetry. The three proteins are also similar with respect to their metal composition (1-2 Zn^{2+}- and Mg^{2+}-ions per subunit) and the presence of a bound NAD(P)(H) cofactor in each subunit. The amino acid sequences

of these enzymes, deduced from the cloned genes (de Vries et al., 1992; H.J. Hektor, unpublished), show that these proteins share a high degree of identity and belong to the Family III alcohol dehydrogenases. The classical dinucleotide binding fold for NAD(P)(H) is not present in these proteins.

In addition to the methanol:NDMA oxidoreductase activity of MNO, also dye (DCPIP and MTT)-linked methanol dehydrogenase activities can be detected reproducibly in crude extracts of *A. methanolica* (van Ophem et al., 1991; Bystrykh et al., 1995). These dye-linked methanol dehydrogenases appear to represent the overall activities of multienzyme systems. The biochemistry of methanol oxidation in Gram-positive bacteria is complex and both MDH of *B. methanolicus* (Arfman et al., 1991) and MNO of *A. methanolica* (Bystrykh et al., 1995) *in vivo* require additional proteins, most likely participating in the transfer of reducing equivalents from NAD(P)H cofactors to NAD coenzyme and/or to the electron transport chain. Meanwhile the gene encoding the activator protein of *B. methanolicus* has been cloned and characterized. No clear similarities were observed, however, with any other protein sequence available in databases (H.J. Kloosterman, unpublished).

Glucose metabolism

Studies of glucose metabolism in *A. methanolica* revealed the presence of the normal set of glycolytic pathway and pentose phosphate cycle enzymes, with a few exceptions. During growth on glucose, glycolysis involved a PPi-dependent phosphofructokinase (PFK) which was completely insensitive to allosteric control (Alves et al., 1994). The amino acid sequence deduced from the cloned gene, nevertheless, revealed a strong similarity with ATP-dependent PFK enzymes from various other sources (*Bacillus stearothermophilus*) (A. Alves, unpublished). Screening of other actinomycetes revealed the presence of similar PPi-dependent PFK enzymes in other members of the family *Pseudonocardiacea* but not in other actinomycetes, e.g. *Streptomyces coelicolor* A3(2), which thus may reflect an evolutionary signature. Protein purification studies revealed a second remarkable feature, namely the presence of a 2,3-bisphosphoglycerate activated 3-phosphoglycerate mutase in *A. methanolica*, which is normally present in eukaryotes only. Glucose metabolism is regulated at the level of the PPi-dependent PFK enzyme (synthesis), at the phosphoglycerate mutase (activity) and pyruvate kinase (activity) steps. The latter step involves an allosteric enzyme regulated via feedback inhibition by ATP and Pi, and activated by AMP (Alves et al., 1994). Further studies are required to identify the possible presence of other regulatory control points (e.g. glucose uptake). No evidence was obtained for the involvement of a PEP-dependent sugar phosphotransferase system in glucose uptake, but the system involved remains to be identified.

Mixed substrate experiments in batch cultures with glucose plus methanol resulted in simultaneous utilization of these substrates. The presence of glucose repressed synthesis of the RuMP cycle enzymes HPS and HPI, and methanol was only utilized as an energy source. Similar results were found following addition of formaldehyde (fed-batch system) to a culture growing on glucose. The synthesis of enzymes involved in methanol dissimilation and assimilation in *A. methanolica* thus appears to be regulated differently. Methanol and/or formaldehyde induce the synthesis of these enzymes, but under carbon-excess conditions their inducing effect on HPS and HPI synthesis is overruled completely by glucose. Repression of HPS and HPI was of minor significance following addition of methanol to glucose-, acetate- and ethanol-limited chemostat cultures (de Boer et al., 1990a).

Biosynthesis of aromatic amino acids

Using brief ultrasonication treatments to obtain single cells of the pseudomycelium-forming actinomycete *A. methanolica*, and simple protocols for the identification of metabolic lesions, we succeeded in the isolation and characterization of a large number of auxotrophic mutants, covering virtually every step in aromatic amino acid biosynthesis (Euverink et al., 1995b). Characterization of these mutants revealed that synthesis of L-phenylalanine and L-tyrosine proceeds via single pathways, involving phenylpyruvate and L-arogenate as intermediates, respectively. Dehydroquinate (DHQ) dehydratase mutants invariably were also blocked in DHQ synthase, suggesting common control elements or gene clustering (Euverink et al., 1992). No mutants were obtained in 3-deoxy-D-arabinoheptulosonate 7-phosphate (DAHP) synthase and prephenate aminotransferase, suggesting the presence of isoenzymes, as has been reported for various other organisms. This was subsequently confirmed in biochemical studies (G.J.W. Euverink, unpublished).

L-Phenylalanine aminotransferase (aroAT) catalyzes the last step in L-phenylalanine biosynthesis. Enzyme purification studies showed the presence of a minor aroATI activity, coeluting with branchd chain aminotransferase, and a major aroATII activity using both L-phenylalanine and L-tyrosine as substrates (Abou-Zeid et al., 1995). A leaky L-phenylalanine auxotroph, strain GH141, was subsequently identified as deficient in the aroATII enzyme. In this strain the minor aroATI activity is responsible for L-phenylalanine biosynthesis; its low specific activity explains the leaky phenotype. Interestingly, strain GH141 also had lost the ability to grow on L-tyrosine as carbon source. Apparently the aroATII protein of *A. methanolica* is functioning in both L-phenylalanine biosynthesis and in L-tyrosine catabolism (Abou-Zeid et al., 1995; Euverink et al., 1995b).

Three species of prephenate aminotransferase (PpaATI-III), the first enzyme of L-tyrosine biosynthesis, were chromatographically resolved. PpaATI and PpaATII coeluted with aroATI (branched chain aminotransferase) and with aspartate aminotransferase, respectively. PpaATIII appeared highly specific for prephenate and thus appears to be the main in vivo PpaAT activity (Abou-Zeid et al., 1995). The presence of these three isoenzymes with PpaAT activity explains the failure to isolate L-tyrosine auxotrophic mutants in this step (see above).

The product of the PpaAT activity with prephenate, arogenate, is subsequently converted into L-tyrosine by an NAD(P)-dependent L-arogenate dehydrogenase. Numerous mutants blocked in this step have been isolated, confirming the presence of a single pathway from prephenate towards L-tyrosine in *A. methanolica* (Abou-Zeid et al., 1995; Euverink et al., 1995b).

The single DAHP synthase enzyme species that can be detected in wild type *A. methanolica* is sensitive to cumulative feedback inhibition by all three aromatic amino acids. Partially purified enzyme showed apparent K_i values of 3, 160 and 180 µM for L-tryptophan, L-phenylalanine and L-tyrosine, respectively. The aromatic amino acids displayed competitive inhibition with respect to E4P. L-Tryptophan and E4P showed uncompetitive and competitive inhibition towards PEP, with apparent K_i values of 11 and 530 µM, respectively (de Boer et al., 1989). Chorismate mutase functions in L-phenylalanine and L-tyrosine biosynthesis. The activity of the single chorismate mutase detectable in extracts of the wild type organism was inhibited by both L-phenylalanine and L-tyrosine (apparent K_i values of 60 and 35 µM, respectively) (de Boer et al., 1989). Prephenate dehydratase, an enzyme specifically involved in L-phenylalanine biosynthesis, was purified to homogeneity and characterized as a 150 kDa homotetrameric protein with a subunit size of 34 kDa. Kinetic studies showed that this enzyme is allosterically inhibited by L-phenylalanine and activated by L-tyrosine (Euverink et al., 1995a). L-Phenylalanine caused an increase in the $S_{0.5}$ for prephenate and a decrease in the V_{max}. L-Tyrosine caused a decrease in the $S_{0.5}$ for prephenate and an increase in the V_{max} (Euverink et al., 1995a). Anthranilate synthase, the first enzyme in the L-tryptophan specific branch, was strongly inhibited by L-tryptophan (apparent K_i value of 5 µM). Addition of the aromatic amino acids, either separately or in combinations, did not result in significant repression of the synthesis of these enzymes (de Boer et al., 1989).

A. methanolica is very sensitive to inhibition of growth by various amino acid analogs. o-Fluoro- and p-fluorophenylalanine also inhibited the activities of chorismate mutase and prephenate dehydratase in vitro (de Boer et al., 1990b). Efficient methods for the isolation of analog-resistant mutants of A. methanolica have been developed subsequently. Many analog-resistant mutant strains had become unable to grow on L-phenylalanine as carbon source and most likely had lost phenylalanine(analog) transport systems. Several mutants were found to possess either a chorismate mutase or a prephenate dehydratase enzyme which had become completely insensitive to L-phenylalanine(analog) inhibition. Some prephenate dehydratase mutants were still activated by tyrosine, while others had become insensitive to both phenylalanine and tyrosine (Euverink et al., 1995a).

Gene cloning systems for A. methanolica

A. methanolica was found to possess a conjugative plasmid (pMEA300) which is able to integrate into the chromosome at a specific site (Vrijbloed et al., 1994). Recently we have completed the 13,285 nucleotide sequence analysis of pMEA300 (deposited in the GenBank database under the accession number L36679), revealing a total of twenty open reading frames with relatively little untranslated intervening sequences. The overall G+C content of pMEA300 is 69.3%. This high value is characteristic for actinomycetes. Depending on the growth conditions, pMEA300 is maintained either integrated in the genome or occurs as an autonomously replicating plasmid, co-existing with the integrated form (Vrijbloed et al., 1994). This made the isolation of a plasmid-deficient strain rather difficult. Once the pMEA300-deficient A. methanolica strain WV1 strain had been isolated we were able to carry out a further functional analysis of pMEA300-encoded functions, by re-introducing various deletion derivatives in the original host. Initially a derivative plasmid, carrying the thiostrepton resistance gene (pMEA301), was used for the development of a whole cell transformation procedure (10^6 transformants /µg DNA; Vrijbloed et al., 1995a) with strain WV1. A further analysis of pMEA300-encoded functions resulted in identification of plasmid fragments encoding integration (Vrijbloed et al., 1994), replication (Vrijbloed et al., 1995b) and conjugation functions (Vrijbloed et al., 1995c). Suitable pMEA300 derivatives have been used as a basis for the construction

14

of *E. coli* - *A. methanolica* shuttle vectors, based on the pHSS6 colE1 replicon. Cloning of genes involved in glucose (A. Alves, unpublished) and methanol (H.J. Hektor, unpublished) utilization, and in aromatic amino acid biosynthesis (by complementation of the available *A. methanolica* mutants), using a genebank of chromosomal DNA of *A. methanolica* in *E. coli* (Vrijbloed et al., 1995d), is well under way now.

CONCLUSIONS
-The methanol-utilizing actinomycete *Amycolatopsis methanolica* is a very versatile RuMP cycle methylotroph that can be manipulated both physiologically and genetically.
-Methanol oxidation in Gram-positive bacteria involves a new and uniform type of decameric protein belonging to family III alcohol dehdyrogenases, possessing a tightly bound NAD(P)(H) cofactor, and associated with additional proteins in vivo.
-The glycolytic pathway in *A. methanolica* involves a pyrophosphate-dependent phosphofructokinase (which is not regulated at the activity level), and a 2,3-bisphosphoglycerate activated 3-phosphoglycerate mutase (normally present in eukaryotes only).
-The organization and regulation of the pathways for aromatic amino acid biosynthesis in *A. methanolica* have been studied in detail, allowing rational construction of strains overproducing primary or secondary metabolites in further work.
-Based on the indigenous plasmid pMEA300 and a plasmid-deficient strain (WV1), efficient protocols for transformation of whole cells of *A. methanolica*, and for cloning genes involved in glucose or methanol utilization, or in aromatic amino acid biosynthesis, have been developed.

ACKNOWLEDGEMENTS
These studies were supported by grant no GBI81.1510 from the Netherlands Technology Foundation (STW) which is subsidized by the Netherlands Organization for the Advancement of Pure Research (NWO), and by grant no BD808-IF90 from JNICT (Portugal). The studies on MTT-dependent methanol dehydrogenase were carried out in collaboration with J.A. Duine (TU Delft).

REFERENCES
Abou-Zeid A et al. (1995) Appl. Environm. Microbiol. 61, 1298-1302.
Alves A et al. (1994) J. Bacteriol. 176, 6827-6835.
Arfman N et al. (1989) Arch. Microbiol. 152, 280-288.
Arfman N et al. (1991) J. Biol. Chem. 266, 3955-3960.
Boer L de et al. (1989) Arch. Microbiol. 151, 319-325.
Boer L de et al. (1990a) Arch. Microbiol. 153, 337-343.
Boer L de et al. (1990b) Appl. Microbiol. Biotechnol. 33, 183-189.
Boer L de et al. (1990c) Int. J. Syst. Bacteriol. 40, 194-204.
Bystrykh LV et al. (1993a) J. Gen. Microbiol., 139, 1979-1985.
Bystrykh LV et al. (1993b) J. Bacteriol. 175, 1814-1822.
Bystrykh LV et al. (1993c) In Murrell C and Kelly DP, eds, Microbial Growth on C_1 Compounds, pp 245-251, Intercept Ltd, Andover, UK.
Bystrykh LV et al. (1995) Submitted.
Dijkhuizen L, Harder W (1992) In Balows CA et al., eds, The Prokaryotes (2nd ed.), pp

197-206, Springer-Verlag, New York, USA.

Dijkhuizen L (1993) In Rehm HJ et al., eds, Biotechnology (2nd ed.), Vol. 1, Sahm H, ed., pp 265-284, VCH, Weinheim, Germany.

Dijkhuizen L, Arfman N (1990) FEMS Microbiol. Rev. 87, 215-220.

Dijkhuizen L et al. (1993) In Murrell JC, Dalton H, eds, Methane and Methanol Utilizers, pp 149-181, Plenum Press, New York, USA.

Dijkhuizen L et al. (1993) In Murrell JC, Kelly DP, eds, Microbial Growth on C_1 Compounds, pp 329-336, Intercept Ltd, Andover, UK

Dunn MF, Bernhard SA (1971) Biochemistry 10, 4569-4575.

Euverink GJW et al. (1992) J. Gen. Microbiol. 138, 2449-2457.

Euverink GJW et al. (1995a) Biochem. J. 308, 313-320.

Euverink GJW et al. (1995b) FEMS Microbiol. Lett., in press.

Hazeu W et al. (1983) Arch. Microbiol. 135, 205-210.

Kato N et al. (1974) J. Ferment. Technol. 52, 917-920.

Kovář J et al. (1984) Eur. J. Biochem. 139, 585-591.

Ophem PW van et al. (1991) FEMS Microbiol. Lett. 80, 57-64.

Vonck J et al. (1991) J. Biol. Chem. 266, 3949-3954.

Vries GE de et al. (1992) J. Bacteriol. 174, 5346-5353.

Vrijbloed JW et al. (1994) J. Bacteriol. 176, 7087-7090.

Vrijbloed JW et al. (1995a) Plasmid 34, 96-104.

Vrijbloed JW et al. (1995b) Molec. Microbiol., in press.

Vrijbloed JW et al. (1995c) J. Bacteriol., in press.

Vrijbloed JW et al. (1995d) J. Bacteriol., in press.

Metabolism of Formaldehyde in *M. extorquens* AM1.
Molecular genetic analysis and mutant characterization

Ludmila Chistoserdova
W.M. Keck Laboratories, Environmental Engineering Science, Caltech,
Pasadena, CA, USA

INTRODUCTION

M. extorquens AM1 is one of the best studied methylotrophs, a representative of a large group of pink-pigmented facultative methylotrophic bacteria utilizing the serine cycle for formaldehyde assimilation. Reactions of the serine cycle were essentially solved using this organism in J.R. Quayle's laboratory in the 1960's and are as follows (Fig. 1). Formaldehyde in the form of methylene tetrahydrofolate (THF) is condensed with glycine by serine hydroxymethyltransferase (SHMT) to produce serine. Serine is transaminated with glyoxylate to produce hydroxypyruvate in the serine glyoxylate aminotransferase (SGAT) reaction. Hydroxypyruvate is reduced to D-glycerate by hydroxypyruvate reductase (HPR), the latter is phosphorylated by glycerate kinase (GK) to produce phosphoglycerate. Phosphoglycerate is converted into PEP by enolase, PEP is carboxylated to oxaloacetate by PEP carboxylase (PEPC), the latter is converted to malate, which is transformed to malyl-CoA by malate thiokinase (MTK) and cleaved to acetyl-CoA and glyoxylate by malyl-CoA lyase (MCL). Glyoxylate is then used to regenerate glycine. In serine methylotrophs possessing isocitrate lyase (ICL), the second molecule of glycine is regenerated from acetyl-CoA in the glyoxylate cycle. In the ICL-minus serine methylotrophs like *M. extorquens* AM1, the glyoxylate cycle is not operational, and the pathway leading to regeneration of glyoxylate from acetyl-CoA remains unknown.

The pathways in which the formaldehyde that is not directed into the serine cycle is oxidized to formate also remain questionable. It has been assumed that dye-linked or NAD-linked dehydrogenases of formaldehyde are involved, although an alternative pathway involving THF derivatives of formaldehyde and formate was proposed but never proven (Marison, Attwood, 1982).

Described in this chapter are results of our approach to the refinement of our understanding of formaldehyde metabolism based on molecular characterization of specific regions of the *M. extorquens* AM1 chromosome involved in C_1 and C_2 assimilation and formaldehyde oxidation.

M. E. Lidstrom and F. R. Tabita (eds.), Microbial Growth on C$_1$ Compounds, 16–24.
© 1996 *Kluwer Academic Publishers. Printed in the Netherlands.*

.

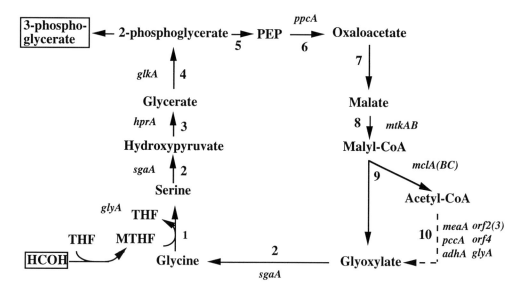

Figure 1. Reactions of the serine cycle (Anthony, 1982). 1) serine hydroxymethyltransferase, 2) serine glyoxylate aminotransferase, 3) hydroxypyruvate reductase, 4) glycerate kinase, 5) enolase, 6) PEP carboxylase, 7) malate dehydrogenase, 8) malate thiokinase, 9) malyl-CoA lyase, 10) unknown pathway for acetyl-CoA oxidation.

RESULTS

The strategy of this work was as follows. Genes specifically involved in formaldehyde metabolism have been cloned either by complementation of available chemically induced mutants (*mclA*, Fulton et al, 1984; *glkA, meaA*, Stone, Goodwin, 1989) or by using specific probes based on known amino acid sequence (*hprA*, Chistoserdova, Lidstrom, 1992) or sequences conserved in known proteins (*glyA*, Chistoserdova, Lidstrom, 1994b). Specific genes have been identified by sequencing. Sequences of adjacent DNA regions have also been determined in order to search for more genes which might be involved in C$_1$ metabolism. Putative open reading frames have been identified, and deduced amino acid (AA) sequences compared to the protein bank databases. In cases in which considerable homology was found, putative functions have been assigned to the genes. DNA fragments of appropriate size have been cloned for generation of in vitro mutated genes, and the kanamycin (Km) resistance gene cassette has been inserted into chosen sites in every open reading frame. The mutated genes cloned into a suicide vector have been recombined with the chromosome in wild type *M. extorquens* AM1 to produce specific mutants that were selected on succinate by their resistance to Km. Double-crossover recombinants not carrying any vector sequences have been identified by DNA-DNA hybridization analysis, and phenotypes of these mutants analyzed. In some cases double crossovers were not possible to obtain

18

1 kb

sgaA hprA mtdA orf4   mtkA B   ppcA   mclA (BC)   abcABCD

Figure 2. Map of the serine cycle gene cluster. Closed rectangles - serine cycle genes; checked rectangles - formaldehyde oxidation genes; darkly shaded rectangles - genes involved in C_1 metabolism at yet unknown level; lightly shaded rectangles - genes involved in methanol oxidation; open rectangles - genes whose involvement in C_1 metabolism is not proven. Arrows show direction of transcription.

and only single-crossover transconjugants have been selected, which contained the entire suicide vector inserted into the chromosome. Such recombinants contained two copies of the mutated gene, one interrupted by the Km gene cassette, and one complete copy, which in some cases has been separated from its own promoter (Chistoserdova, Lidstrom, 1994a). Mutants of this type that had lost their ability to grow on C_1 compounds were analyzed. Single-crossover mutants that retained their ability to grow on C_1 compounds have not been analyzed further. In cases when a function was proposed for a gene, the absence of this function was confirmed in the corresponding mutant, and restoration of the function has been demonstrated in mutants complemented by DNA fragments carrying the respective genes.

At this time, four unlinked regions involved in formaldehyde metabolism have been identified. Two of these remain poorly characterized and contain, respectively *glyA* and *glkA*. Two other regions have been extensively characterized. One of them is a 33 kb region containing genes encoding enzymes of the serine cycle, *mox* genes encoding methanol oxidation functions, genes participating in the oxidation of formaldehyde, C_1 genes with yet unknown function and a number of genes whose association with C_1 metabolism remains unproven (Fig. 2). The second region of 14 kb contains three genes apparently involved in the unknown pathway of regeneration of glyoxylate from acetyl-CoA and five other genes whose involvement in C_1 metabolism we were unable to prove (Fig. 3). Characteristics of specific genes and respective mutants are given below.

Serine cycle genes. glyA encodes the first enzyme of the serine cycle, SHMT. So far it has not been linked to any other serine cycle genes, and its surrounding DNA remains unstudied (Chistoserdova, Lidstrom, 1994b). The AA sequence of SHMT is highly homologous to these of heterotrophic organisms, as well as to SHMT from *Hyphomicrobium methylovorum* (Miyata et al, 1993). *glyA* seems not to be required for multicarbon metabolism, since an insertion mutant in *glyA* lacking SHMT still grows on succinate, although at somewhat reduced rates. The *glyA* mutant of *M. extorquens* AM1 not only
has a C_1-negative phenotype, it is also C_2-negative, implying a second function of this enzyme, probably in the unknown pathway of acetyl-CoA oxidation. The mutant can be complemented for growth by the addition of glyoxylate but not glycolate on C_2 but

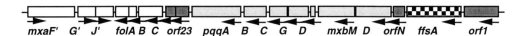

mxaF' G' J' folA B C orf23 pqqA B C G D mxbM D orfN ffsA orf1

not C_1 compounds.

sgaA encoding SGAT is the first gene in the large C_1 gene cluster shown in Fig. 2. The AA sequence deduced from *sgaA* reveals some similarity with a number of enzymes belonging to the serine:pyruvate aminotransferase family. These enzymes do not carry sequence similarities with tyrosine or aspartate aminotransferases, but apparently share with them common secondary structure, and the Lys residue responsible for the cofactor (pyridoxal phosphate) binding is conserved in all of them (Chistoserdova, Lidstrom, 1994a). *sgaA* is preceded by its own promoter. Insertion mutants in *sgaA* are only impared in their growth on C_1 compounds, confirming that SGAT is specifically involved in the serine cycle.

hprA encodes HPR. The AA sequence of HPR reveals some low similarity with NAD-linked dehydrogenases (Chistoserdova, Lidstrom, 1994a). The similarity is surprizingly low even with the counterpart enzyme from *H. methylovorum* (29% identity, Yoshida et al, 1994). Insertion mutation in *hprA* resulted in a C_1-negative phenotype (Chistoserdova, Lidstrom, 1992), contradicting a previous suggestion that it might be involved in C_2 metabolism, based on data obtained with a chemically induced mutant, 20BL (Dunstan et al, 1972). The gene is preceded by a promoter.

mtkAB encode the two subunits of MTK. The presence of this enzyme in *M. extorquens* AM1 had remained questionable due to the difficulty of measuring this reaction in the organism (Anthony 1982). Identification of *mtkAB* by the homology of their products with the two subunits of a similar enzyme, succinate thiokinase, and analysis of insertion mutants in which *mtkA* and *B* had been disrupted have proven the presence of MTK in *M. extorquens* AM1 and its specific involvement in the serine cycle (Chistoserdova, Lidstrom, 1994c).

ppcA encods an acetyl-CoA-independent PEPC. Transposon mutants in this gene have been obtained earlier and shown to have a C_1-negative phenotype (Arps et al, 1993). The sequence of the 3'-terminus of *pccA* ia available, and the deduced AA sequence reveals high homology with a number of heterotrophic PPCses (L. Chistoserdova, unpublished).

mclA encodes a subunit of MCL whose AA sequence is similar to the lyase component (β subunit) of citrate lyase (L. Chistoserdova, unpublished). Two open reading frames downstream of *mclA* putatively designated *mclBC* do not carry similarity with genes in the citrate lyase operon of *Klebsiella pneumonia* encoding the α and γ subunits of citrate lyase (Bott, Dimroth, 1994), but transposon mutants obtained in this region show a C_1-negative and multicarbon-positive phenotype (Arps et al, 1993). *mclA* is preceeded by a promoter.

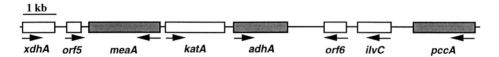

Figure 3. Map of the region containing genes for acetyl-CoA oxidation. Shaded rectangles - genes involved in acetyl-CoA oxidation; open rectangles - genes whose involvement in acetyl-CoA oxidation is not proven. Arrows show direction of transcription.

glkA codes for glycerate kinase. The gene is located on a DNA fragment not linked to other fragments carrying C_1 genes (Stone, Goodwin, 1989). The AA sequence derived from the gene reveals high homology with pyruvate kinases from various sources (L. Chistoserdova, unpublished).

Formaldehyde oxidation genes. Two genes encoding enzymes of the THF pathway have been found in the large cluster containing a number of C_1 genes (Fig. 2). Mutations in these genes have led to a phenotype that implies participation of the products of these genes in the dissimilatory pathway for formaldehyde oxidation. These are described below.

mtdA encodes methylene tetrahydrofolate dehydrogenase (MTHFDH). The homology of the AA sequence derived from the gene with available sequences of eucaryotic MTHFDses is very low (not exceeding 15% identity, Chistoserdova, Lidstrom, 1994a), although multiple alignments allowed the identification of 7 AA residues invariant in all of the sequences. The functions of these residues remain unknown. Mutants in the gene have lost their ability to grow on both methanol and methylamine, indicating that the THF pathway must be directly involved in the oxidation of formaldehyde. Only single-crossover insertion mutants were possible to obtain in *mtdA*, resulting in low uninducible levels of MTHFDH, which implied its necessity not only for C_1, but also for multicarbon metabolism. The function of MTHFDH in generating C_1 units for various cell biosyntheses is well known (MacKenzie 1984). *mtdA* is apparently cotranscribed with *hprA* (Chistoserdova, Lidstrom, 1994a)

ffsA encodes another enzyme of the THF pathway, formyl tetrahydrofolate synthase (FTHFS). The AA sequence derived from the gene is highly homologous in its N-terminal part to the sequence of formylmethanofuran tetrahydromethanopterin formyltransferase, an enzyme carrying out a similar function of formyl transfer in methanogenesis (A. Springer, unpublished). Insertion mutants in *ffsA* reveal the same phenotype as mutants in *mtdA*. They will not grow on C_1 compounds because of the lack of induction of FTHFS (L. Chistoserdova, unpublished).

Genes involved in regeneration of glyoxylate from acetyl-CoA. Three genes have been identified clustered in a 14 kb fragment of the *M. extorquens* AM1 chromosome that are involved in the unknown pathway of glyoxylate regeneration (Fig. 3). Insertion mutants in these genes all have the characteristic C_1 and C_2-negative phenotype and they are complemented for growth on these substrates by the addition of glyoxylate or glycolate. Characteristics of these genes are given below.

pccA encodes propionyl-CoA carboxylase (PCC). The AA sequence deduced from the gene is highly homologous to known sequences of eucaryotic PCCses, and a mutation in *pccA* leads to disappearance of PCC activity (L. Chistoserdova, unpublished).

meaA codes for an enzyme with yet unknown function. The amino acid sequence derived from the gene reveales high homology with subunits of methylmalonyl-CoA mutase (MCM), but insertion mutants in *meaA* reveal MCM levels similar to wild type levels, indicating that the gene might code for an enzyme with similar function but different specificity (L. Smith, personal communication; L. Chistoserdova, unpublished).

adhA encodes a NAD-linked dehydrogenase of unknown function. The AA sequence derived from *adhA* reveals homology with short-chain alcohol dehydrogenases, and the typical fingerprint for a pyridine nucleotide cofactor-binding site is found (L. Chistoserdova, unpublished). A number of short chain alcohols and aldehydes have been tested as substrates for specific activity staining in gels with crude extracts from wild type *M. extorquens* AM1 and insertion mutants, but no missing bands have been observed in the mutants.

C_1 genes with unknown functions. In addition to genes encoding known methylotrophic functions, a few genes have been identified involved in formaldehyde metabolism at an unknown level. These are listed below.

orfN is located between *ffsA* and *mxbD* and codes for a small polypeptide whose AA sequence shows no similarity with known protein sequences (A. Springer, unpublished). Only single-crossover mutants can be obtained in the gene, and those show a C_1-negative and C_2-positive phenotype (L. Chistoserdova, unpublished). The role of *orfN* remains unknown.

orf1 is located immediately upstream of *ffsA* and codes for a polypeptide of 55 kDa whose AA sequence is non-homologous to known protein sequences. Only single-crossover insertion mutants were obtained for this gene, and these show a C_1-negative and C_2-positive phenotype. Mutants in *orf1* are not only methanol-minus but methanol-sensitive (L. Chistoserdova, unpublished). Their growth on succinate is inhibited by the addition of methanol, but they will grow normally on succinate in the presence of methylamine or formate. Such a phenotype might imply a regulatory role for the gene.

orf2 and *orf3* have been identified by sequencing and they overlap by 53 nucleotides. Mutants in *orf3* are not available at this time, while mutants in *orf2* have a C_1- and C_2-negative phenotype and are complemented for growth on these substrates by the addition of glycolate or glyoxylate, making these two genes candidates for participation in the pathway for glyoxylate regeneration.

orf4 is located immediately downstream of *mtdA*. It encodes a polypeptide of 22 kDa, which might be an integral membrane protein. The AA sequence derived from the gene did not show considerable homology with known proteins, and the function of this polypeptide is not known. Insertion mutants in the gene have lost their ability to grow on C_1 and C_2 compounds, indicating the possible involvement of this gene in the unknown pathway of acetyl-CoA oxidation. The mutants are restored for growth on C_2 but not C_1 compounds when supplemental concentrations of glyoxylate are added. Glycolate cannot substitute for glyoxylate (Chistoserdova, Lidstrom, 1994c).

Other genes in C_1 clusters. A brief description of genes found in C_1 clusters whose association with C_1 metabolism remains unproven is given below.

abcABCD. Four genes apparently encoding components of an ABC-type transporter system have been identified downstream of the *mcl* genes, by the similarity of their products with other ABC-type transporters. Insertion mutants in the ATP-binding component of the transporter (*abcD*) were only obtained as single crossover recombinants, and those were C_1-positive. Thus the function of the transporter system remains unknown.

mxaF'G'J'. Downstream of *abcABCD* are located three genes homologous to *xoxF, cycB* and *xoxJ* from *Paracoccus denitrificans* (Ras, 1994) and to a lesser extent to *mxaFGJ* from a number of methylotrophs as well as genes encoding other PQQ-dependent dehydrogenases (L. Chistoserdova, unpublished). These are preceded by a promoter upstream of *mxaF'*. Insertion mutants in *mxaF'* have a C_1-positive phenotype. The search for the missing dye-linked dehydrogenase activity in these mutants has not been successful, thus the expression itself of *mxaF'* remains questionable. It is interesting to point out that no homolog was found for *xoxI* of *P. denitrificans* apparently encoding the small subunit of the unknown dehydrogenase.

folABC. Separated by only 47 base pairs from *mxaF'G'J'* and divergently transcribed are three genes whose products are homologous to the three enzymes of folate metabolism: 6-hydroxymethyl-7,8-dihydropterin pyrophosphokinase, dihydroneopterin aldolase and dihydroxypteroate synthase, respectively (L. Chistoserdova, unpublished). A mutation in *folC* resulted in isolation of only single-crossover recombinants, and those were C_1-positive, indicating that *fol* genes are not specific for C_1 metabolism. One could speculate that the presence of *fol* genes in the C_1-gene cluster might not be completely accidental since these genes are ultimately involved in the biosynthesis of THF.

pqq and **mxb** genes are described elsewhere (Morris et al, 1994; Springer et al, in press).

The genes listed below are found in a region shown in Fig. 3, containing genes for conversion of acetyl-CoA to glyoxylate. Mutations in all of them except for *katA* resulted only in single-crossover recombinants, and these all showed a C_1 and C_2-positive phenotype (L. Chistoserdova, unpublished).

xdhA encodes a NAD-linked dehydrogenase of unknown specificity. The highest homology for the product of *xdhA* was found for 3-oxoacyl-[acyl-carrier protein] reductase, glucose 1-dehydrogenase and acetoin reductase.

orf5 and ***orf6*** encode, respectively, polypeptides of 15 and 28 kDa not homologous to any known protein sequences.

ilvC encodes a polypeptide highly homologous to ketol acid reductoisomerases from various sources.

katA encodes a catalase, and double-crossover mutants in this gene have a C_1, C_2-positive phenotype and still reveal high levels of catalase activity, since *katA* is not the only gene encoding catalase (L. Chistoserdova, unpublished).

CONCLUSIONS

Serine cycle. The study presented here based on direct molecular characterization of chromosomal regions of *M. extorquens* AM1 involved in formaldehyde metabolism opens new insights on metabolic pathways operating in this organism. So far 8 genes encoding enzymes of the serine cycle have been characterized. The characteristics of insertion serine cycle mutants allowed us to prove or to correct the phenotypes of previously isolated chemically induced mutants, as well as to characterize lesions in genes for which chemical mutants were not available. We have proven that HPR is only involved in the serine cycle and does not participate in C_2 metabolism. *glyA*, encoding SHMT, was shown to be present in a single copy and not to be essential for multicarbon metabolism. Instead, its involvement in C_2 metabolism, probably in the unknown pathway of glyoxylate regeneration, was implied. The presence of MTK in *M. extorquens* AM1 was finally proven by protein sequence comparison and mutant characterization. The genes of the serine cycle described so far have been found in three unlinked regions on the *M. extorquens* AM1 chromosome, 6 of these genes are closely linked and located within the large cluster of C_1 genes.

Formaldehyde oxidation. High levels of the enzymes of the THF pathway involving THF derivatives of formaldehyde and formate have been found to be induced in *M. extorquens* AM1 on C_1 substrates (Large, Quayle, 1963), and the possible involvement of the THF pathway not only in formate assimilation but also in formaldehyde oxidation has been proposed (Marison, Attwood, 1982). Analysis of phenotypes of the insertion mutants in *mtdA* and *ffsA* has clearly indicated that the THF pathway must be directly involved in formaldehyde oxidation. This suggests that methylene THF, not formaldehyde is the branch-point intermediate in the C_1 metabolism of *M. extorquens* AM1.

Acetyl-CoA oxidation. A chromosomal region of 14 kb has been characterized containing three genes required for the pathway leading from acetyl-CoA to glyoxylate. One gene has been identified by L. Smith and P. Goodwin (personal communication) and independently in our lab as a coenzyme B_{12}-dependent mutase with an unknown specificity. Two more genes in the region code for an alcohol dehydrogenase of unknown specificity and PCC. In addition, three more genes have been identified as being involved in this pathway, two genes encoding unknown functions (*orf2* and *orf4*), and *glyA* which apparently plays a double role in C_1 assimilation, one the well defined role in the serine cycle and a second possible role in the acetyl-CoA oxidation. These data are not sufficient to allow the identification of a new acetyl-CoA oxidation pathway, but they make it clear that the pathways proposed so far are probably not correct.

REFERENCES

Anthony C. (1982) The biochemistry of methylotrophs, London: Academic Press.
Arps PJ et al (1993) J. Bacteriol. 175, 3776-3783.
Bott M, Dimroth P (1994) Molecular Microbiol. 14, 347-356.
Chistoserdova LV, Lidstrom ME (1992) J. Bacteriol. 174, 71-77.
Chistoserdova LV, Lidstrom ME (1994a) J. Bacteriol. 176, 1957-1968.
Chistoserdova LV, Lidstrom ME (1994b) J. Bacteriol. 176, 6759-6763.
Chistoserdova LV, Lidstrom ME (1994c) J. Bacteriol. 176, 7398-7404.
Dunstan PM et al (1972) Biochem. J. 128, 107-115.
Fulton GL et al (1984) J. Bacteriol. 160, 718-723.
Large PJ, Quayle JR (1963) Biochem. J. 87, 386-396.
MacKenzie RE (1984) In RL Blackey and SJ Benkovic, eds, Folates and Pterins, pp 255-306, John Wiley & Sons, NY.
Marison IW, Attwood MM (1982) J. Gen. Microbiol. 128, 1441-1446.
Miyata A et al (1993) Eur. J. Biochem. 212, 745-750.
Morris CJ et al (1994) J. Bacteriol. 176, 1746-1755.
Ras J (1995) C_1-metabolism of Paracoccus denitrificans: Genetic and physiological studies. PhD Thesis, Vrije University, Amsterdam.
Smith L, Goodwin PM (1992) Abstr. of the 7th Symposium on The Growth On C_1 Compounds, Warwick, England.
Springer A et al, Molecular Microbiol., In press
Stone S, Goodwin PM (1989) J. Gen. Microbiol. 135, 227-235.
Yoshida T et al (1994) Eur. J. Biochem. 223, 727-732.

Structure and function of the serine pathway enzymes in *Hyphomicrobium*

Y. Izumi*, T. Yoshida**, T. Hagishita**, Y. Tanaka**, T. Mitsunaga**,
T. Ohshiro*, T. Tanabe***, A. Miyata***, C. Yokoyama***, J. D.
Goldberg**** and P. Brick****
* Department of Biotechnology, Tottori University, Tottori 680, Japan
** Department of Food and Nutrition, Kinki University, Nara 631, Japan
*** Department of Pharmacology, National Cardiovascular Center Institute,
Suita, Osaka 565, Japan
****Blakett Laboratory, Imperial College, London SW7 2BZ, UK

INTRODUCTION

A certain group of methylotrophic bacteria possess the serine pathway for the assimilation of C_1 compounds, and this pathway operates to synthesize 2-phosphoglycerate as a precursor of cell constituents, comprising the following two part. In the first part, 2-phosphoglycerate is synthesized from glyoxylate and formaldehyde, and malyl-CoA is also formed by subsequent enzyme reactions. In the second part of the pathway, malyl-CoA is cleaved to glyoxylate and acetyl-CoA, and another molecule of glyoxylate is generated from acetyl-CoA. Two variants of the serine pathway have been reported (C. Anthony, 1982), in which the route for the generation of glyoxylate from acetyl-CoA is varied. In the *icl* $^+$-serine pathway, isocitrate lyase is involved in the route of the oxidation of acetyl-CoA to glyoxylate. Another variant, *icl*⁻-serine pathway, is known in methylotrophic bacteria such as *Methylobacterium* and *Hyphomicrobium* strains (C. Anthony, 1982), which have no isocitrate lyase activities, and the route for the generation of glyoxylate has not been elucidated.

The enzymes in the serine pathway are highly induced when bacteria are grown on C_1 compounds, most of which are not required during the growth on multicarbon compounds. Therefore, it is considered that the synthesis of enzymes in the serine pathway are co-regulated at the genetic level. However, little information is available concerning the regulatory mechanism at the level of enzyme activity. Moreover, there have been few reports on the systematic characterization of the serine pathway enzymes using one certain methylotrophic strain. In the present study, the characterization and structural analysis of several serine pathway enzymes in *H. methylovorum* GM2, a serine-producing methylotroph (H. Yamada et al., 1986), has been developed on the basis of the above situations concerning the serine pathway.

RESULTS

Characterization of the Serine Pathway Enzymes in Hyphomicrobium

1. Hydroxypyruvate reductase (HPR)   HPR of *H. methylovorum* GM2 was purified to complete homogeneity and crystallized (Fig. 1) (Y. Izumi et al., 1990a). The enzyme

25

M. E. Lidstrom and F. R. Tabita (eds.), Microbial Growth on C₁ Compounds, 25–32.
© 1996 *Kluwer Academic Publishers. Printed in the Netherlands.*

was found to be a dimer composed of identical subunits (38 kDa), the molecular mass of the enzyme being about 70 kDa. The enzyme was stable against heating at 25°C for 10 min at pH values between 5 and 9. Maximum activity was observed at pH 6.8 and around 45°C. The enzyme catalyzed the irreversible reduction of hydroxypyruvate to D-glycerate with the oxidation of only NADH. Other than hydroxypyruvate, only glyoxylate served as a substrate. The K_m values were found to be 0.175 mM for hydroxypyruvate and 10.8 mM for glyoxylate. The reduction of hydroxypyruvate is also catalyzed by glycerate dehydrogenase, glyoxylate reductase, tartronic semialdehyde reductase and lactate dehydrogenase. However, the HPR of *H. methylovorum* GM2 is clearly distinguished from these enzymes in its substrate specificity, optimal pH and irreversible reaction.

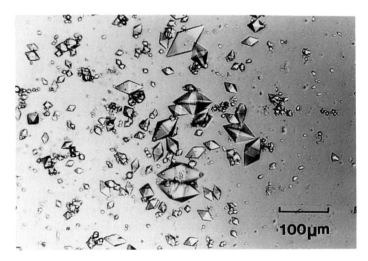

100μm

Fig. 1. Photomicrograph of crystalline hydroxypyruvate reductase

2. Serine–glyoxylate aminotransferase (SGAT)   An accurate assay method for SGAT was established by using the purified HPR of *H. methylovorum* GM2 (Y. Izumi et al., 1990b). Hydroxypyruvate, the reaction product of SGAT, was specifically determined in the presence of Tris, the residual glyoxylate in SGAT reaction being trapped with Tris by forming a stable Shiff base. Using this method, the activity of SGAT was detectable (0.21 units/mg protein) even in the cell-free extract of *H. methylovorum* GM2.

SGAT in the serine pathway was purified to complete homogeneity from *H. methylovorum* GM2 (Y. Izumi et al., 1990c). This is the first microbial SGAT to be purified. The enzyme has a molecular mass of about 140 kDa and consists of four subunits of identical mass, i.e. 40 kDa. The holoenzyme exhibited absorption maxima at 282 nm and 408 nm, and a shoulder at about 315–345 nm in potassium phosphate buffer (pH 7.0); it contained 4 mol pyridoxal 5'-phosphate/mol enzyme. The K_m values for glyoxylate and L-serine were 0.23 and 4.98 mM, respectively, and the enzyme showed high specificity for these substrates (Table 1). The transamination between L-serine and glyoxylate seems to be nearly irreversible and the equilibrium of the reaction lies toward the product, glycine and hydroxypyruvate. This results suggested that SGAT promotes L-serine metabolism with HPR in the serine pathway.

Table 1. Substrate specificity of SGAT from *H. methylovorum* GM2

Amino acceptor	Relative activity (%)	Amino donor	Relative activity (%)
Glyoxylate	100	L-Serine	100
Pyruvate	0	D-Ser, L-Thr, L-Met	
2-Oxoglutarate	0	L-Glu, L-Gln, L-His	
Oxaloacetate	0	L-Arg, L-Phe, L-Ala	trace
Oxamate	0	L-Asp, L-Trp	
Oxomalonate	13.5	*O*-Methyl-DL-Ser	
2-Oxo-*n*-butyrate	0		
2-Oxo-3-methylbutanoate	0	L-Val, L-Leu, L-Ile	
		L-Orn, L-Lys	
		2-Methyl-D-Ser	
		L-Ser-*O*-sulfate	0
		O-Acetyl-L-Ser	
		DL-Ser hydroxamate	
		O-Phospho-DL-Ser	

3. Glycerate kinase (GK)   The GK of *H. methylovorum* GM2 was purified to complete homogeneity and characterized, the first time for an enzyme from a methylotroph (T. Yoshida et al., 1992). The enzyme exists as a monomer with a molecular mass about 41–52 kDa. The enzyme was stable against heating at 35°C for 30 min at pH values over 6–10. Maximum activity was observed at pH 8.0 and around 50°C. The K_m values for D-glycerate and ATP were 0.13 mM, respectively. The enzyme showed high specificity for D-glycerate, and was highly activated by potassium and ammonium ions. The reaction product of the enzyme was identified as 2-phosphoglycerate.

As described above, HPR, SGAT and GK exhibited high substrate specificities and nearly irreversibly catalyzed the respective reactions. This suggests the exclusive involvement of these enzymes in the assimilation of C_1 compounds in *H. methylovorum* GM2. However, no potential effectors were found for these enzymes in the serine pathway, indicating that the regulation at the level of enzyme activity does not occur. It seems necessary to investigate whether these three enzymes of *H. methylovorum* GM2 are co-regulated at the level of enzymes synthesis.

4. Phosphoenolpyruvate carboxylase (PEPC)   Acetyl-CoA-independent PEPC was purified from *H. methylovorum* GM2, and compared with those from other organisms (T. Yoshida et al., 1995a). The activity of the present enzyme was not affected by acetyl-CoA, a well-known activator for enzymes from various heterotrophs, NADH and ADP, effectors for PEPC from methylamine-grown *Pseudomonas* MA (Table 2). This suggested that the *H. methylovorum* enzyme should be classified into the second group according to Utter and Kolenbrander (M.F. Utter & M. Kolenbrander, 1972) on the basis of activators and inhibitors: those which are inhibited by various di- or tricarboxylic acids (TCA cycle intermediates) or L-aspartate but not to respond to acetyl-CoA or other activator. Therefore, the regulatory mechanism of PEPC is considered to be different between *H. methylovorum* GM2 and *Pseudomonas* MA.

Table 2. Effects of various metabolites on the PEPC activity

Compound	Relative activity (%)	Compound	Relative activity (%)
None	100	Glycerol	100
Acetyl-CoA	91	DL-Glyceraldehyde	102
L-Malate	103	NADH	97
L-Aspartate	45	NAD+	100
Citrate	98	NADPH	96
Succinate	100	NADP+	101
Fumarate	98	ATP	80
Pyruvate	95	ADP	103
Glucose-6-phosphate	96	AMP	98
Fructose-1,6-bisphosphate	100	GTP	91
Glyoxylate	90	GDP	100
L-Serine	98	CTP	95
Glycine	102	CDP	96
Hydroxypyruvate	100	UTP	92
D-Glycerate	67		

5. Isocitrate lyase (ICL)   There has been a controversy concerning the occurrence of ICL in the genus *Hyphomicrobium*. Bellion and Spain reported that the *Hyphomicrobium* strains WC and B522 contained a high level of the enzyme activity when ICL was assayed by the 2,4-dinitrophenylhydrazone method (E. Bellion & J.C. Spain, 1976), whereas Attwood and Harder concluded that *Hyphomicrobium* assimilates C_1 compounds via icl^--serine pathway, based on the results when the ICL was assayed by the 1,5-diphenylformazan carboxylic acid method (M.M. Attwood & W. Harder, 1977).

Table 3. Activities of ICL and HPR in cell-free extracts of *Hyphomicrobium* and *Methylobacterium* strains
Enzyme activities were measured immediately after the disruption of cells in the presence of 1 mM dithiothreitol. In parenthesis, isocitrate lyase activities after incubation at 30°C for 60 min with an additional 1 mM dithiothreitol in 50 mM potassium phosphate buffer, pH 7.0, are indicated.

Strain	ICL (units/mg)		HPR (units/mg)
Hyphomicrobium methylovorum GM2	0.182	(0.206)	3.49
Hyphomicrobium methylovorum KM146	0.115	(0.136)	2.81
Hyphomicrobium sp. X	0.120	(0.149)	1.59
Hyphomicrobium sp. G	0.096	(0.106)	2.84
Hyphomicrobium sp. NCIB10099	0.212	(0.252)	3.60
Hyphomicrobium sp. 53-49	0.160	(0.170)	4.30
Hyphomicrobium sp. JTS-811	0.183	(0.204)	3.85
Methylobacterium extorquens AM1	0	(0)	1.71
Methylobacterium organophilum XX	0	(0)	2.12

In the course of our studies on the enzymes in the serine pathway of *H. methylovorum* GM2, ICL activity has been detected at a high level by the 1,5-diphenylformazan carboxylic acid method. Therefore, we re-evaluated ICL activity in various *Hyphomicrobium* strains (T. Yoshida et al., 1995b). ICL activities in a range of 0.096–0.212 units mg^{-1} were detected in cell-free extracts of all the tested *Hyphomicrobium* strains grown on methanol as a sole carbon source (Table 3), although the activities were rapidly lost during the storage at 4°C. When the cell-free extracts were incubated with dithiothreitol after the storage, the recovery of the activities was observed, indicating the involvement of a labile sulfhydryl group in the activities. This confirmed the distribution of unstable ICL activities in the genus *Hyphomicrobium*, and the operation of the icl^+-serine pathway was suggested for the assimilation of C_1 compounds, contrary to the previous observations.

Primary Structures of the Serine Pathway Enzymes in Hyphomicrobium

1. Serine hydroxymethyltransferase (SHMT)   The gene encoding SHMT and its flanking regions were isolated from *H. methylovorum* GM2 using a DNA fragment encoding *Escherichia coli* SHMT as a probe (A. Miyata et al., 1993). Nucleotide sequencing of the recombinant plasmids revealed the serine hydroxymethyltransferase gene codes for the 434-amino-acid protein with a calculated molecular mass of 46,068 Da. The amino acid sequence of the enzyme showed 65.9, 54.7, 55.6, 55.5, 51.5 and 43.8% similarity to the sequences of the enzymes from *B. japonicum*, *E. coli*, *S. typhimurium*, *B. stearothermophilus*, *C. jejuni*, rabbit liver and rabbit mitochondoria, respectively, and the unique amino acid sequence around active site Lys, -T/S-T/S-T-T-H-K-T/S-, is completely conserved in *H. methylovorum* GM2 SHMT. The SHMT of *H. methylovorum* GM2 was related to the enzyme of *Bradyrhizobium japonicum* based on the phylogenetic tree according to the alignment of primary structures.

2. Hydroxypyrvate reductase (HPR)   The gene encoding HPR was isolated from *H. methylovorum* GM2 (T. Yoshida et al., 1994). Nucleotide sequencing of the recombinant plasmids revealed the HPR gene codes for the 322-amino-acid protein with a calculated molecular mass of 35,726 Da. The results of the similarity search for primary structure demonstrated that the *H. methylovorum* GM2 hydroxypyruvate reductase exhibited 31.9, 28.2, 32.1, 35.2. 30.9, 30.2, 26.2 and 32.8% similarity to the HPR of cucumber, formate dehydrogenase of *Pseudomonas* sp. 101, D-3-phosphoglycerate dehydrogenase of *Escherichia coli*, D-lactate dehydrogenase of *Lactobacillus plantarum*, D-lactate dehydrogenase of *L. bulgaricus*, D-2-hydroxyisocaproate dehydrogenase of *L. casei*, erythronate-4-phosphate dehydrogenase of *E. coli* and vancomycin resistant protein VanH of *Enterococcus faecium*, respectively. All these enzymes except formate dehydrogenase utilize the D-isomers as substrates except for formate dehydrogenase. These facts clearly indicates that HPR of *H. methylovorum* GM2 belongs to D-isomer specific 2-hydroxyacid dehydrogenase family.

3. Serine–glyoxylate aminotransferase (SGAT)   The gene encoding SGAT and its flanking regions were isolated from *H. methylovorum* GM2 (T. Hagishita, unpublished). Nucleotide sequencing of the recombinant plasmids revealed the SGAT gene codes for the 405-amino-acid protein with a calculated molecular mass of 43,880 Da. The results of similarity search for primary structures demonstrated that the *H. methylovorum* GM2 SGAT exhibited 56.8, 31.2, 27.5, 26.6, 33.4 and 32.1% similarity to SGAT of *Methylobacterium extorquens* AM1, aspartate aminotransferase of

Methanobacterium thermoformicicum, alanine–glyoxylate aminotransferase of human peroxisome, mitochondrial serine–pyruvate aminotransferase of rat liver, and soluble hydrogenase small subunits of *Anabaena cylindrica* and *Synechococcus* sp. All these enzymes, except for the soluble hydrogenase small subunits of cyanobacteria, are classified into the subgroup IV aminotransferase according to Metha & Christen (P.K. Metha & P. Christen, 1993) and catalyze the transamination involved in L-serine or glyoxylate as a substrate. Although phosphoserine aminotransferase is belong to this subgroup, the reasonable similarity with SGAT was not observed.

Crystallographic Analysis of Hydroxypyruvate Reductase from H. methylovorum GM2

HPR catalyzes the NADH-linked reduction of hydroxypyruvate to D-glycerate, and the enzyme is a member of a family of NAD-dependent dehydrogenases as described above. The crystal structure of the apoenzyme form of hydroxypyruvate reductase from *H. methylovorum* GM2 was determined by the method of isomorphous replacement and refined at 2.4 Å resolution using a restrained least-squares method (J.D. Goldberg et al., 1992, 1994). The crystallographic R-factor is 19.4% for all 24,553 measured reflections between 10.0 and 2.4 Å resolution. The HPR molecule is a symmetrical dimer composed of subunits of molecular mass 38,000, and shares significant structural homology with another NAD-dependent enzyme, formate dehydrogenase.

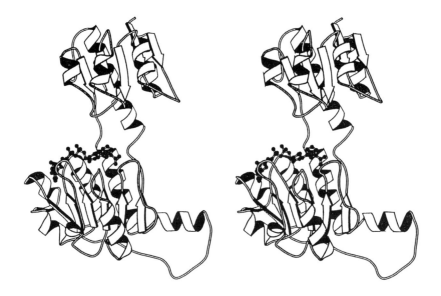

Fig. 2. A stereo diagram of the HPR holoenzyme model indicating the hypothetical juxtaposition of the NAD molecule and protein secondary structure elements.

The HPR subunit consists of two structurally similar domains that are approximately related to each other by 2-fold symmetry (Fig. 2). The domains are separated by a deep cleft that forms the putative NAD and substrate binding sites. One of the domains has been identified as the NAD-binding domain based on its close structural similarity to the NAD-binding domains of other NAD-dependent dehydrogenases (Fig. 2). The topology of the second domain is different from that found in the various catalytic domains of other dehydrogenases. A structural comparison between HPR and L-lactate dehydrogenase indicates a convergence of active site residues and geometries for these two enzymes. The reactions catalyzed are chemically equivalent but of opposing stereospecificity. A hypothesis was presented to explain how the two enzymes may exploit the same coenzyme stereospecificity and a similar spatial arrangement of catalytic residues to carry out reactions that proceed to opposite enantiomers (J.D. Goldberg et al., 1994).

CONCLUSION

The characterization and structural analysis of the following serine pathway enzymes in *H. methylovorum*, a serine-producing methylotroph. SGAT, HPR and GK exhibited high substrate specificities and irreversibly catalyzed the respective reactions, suggesting the exclusive involvement of these enzymes in methanol assimilation. Furthermore, the distribution of unstable ICL activities in *Hyphomicrobium* strains was confirmed, contrary to the previous observations, suggesting the operation of *icl* $^+$-serine pathway.

H. methylovorum GM2 SHMT was related to the enzyme of *Bradyrhizobium japonicum* based on the phylogenetic tree according to the alignment of primary structures. SGAT exhibited 26-31% similarity to serine–pyruvate aminotransferase (mammals), aspartate aminotransferase (*Methanobacterium*), and hydrogenase small subunit (cyanobacteria). With respect to HPR, the results of similarity search for the primary structure exhibited 26-36% similarity to the members of D-isomer specific 2-hydroxyacid dehydrogenase family, and a model of ternary complex of HPR was built in which putative catalytic residues are identified from crystallographic analysis of *H. methylovorum* GM2 HPR, the first time for D-isomer specific 2-hydroxyacid dehydrogenase family.

Such enzymological and structural analyses as well as the gene analysis of the serine pathway enzymes will allow us to speculate on the evolutionary history of the serine pathway: how the methylotrophs have acquired enzymes for the assimilation of C1 compounds.

REFERENCES

Anthony C (1982) In Buchana RE, ed, The Biochemistry of Methylotroph, pp 95-136, Academic Press, New York.
Attwood MM & Harder W (1977) FEMS Microbiol. Lett. 1, 25-30.
Bellion E & Spain JC (1976) Can. J. Biochem. 22, 404-408.
Goldberg JD et al (1992) J. Mol. Biol. 225, 909-911.
Goldberg JD et al (1994) J. Mol. Biol. 236, 1123-1140.
Izumi Y et al (1990a) Eur. J. Biochem. 190, 279-284.
Izumi Y et al (1990b) Eur. J. Biochem. 190, 285-290.
Izumi Y et al (1990c) Agric. Biol. Chem. 54, 1573-1574.
Metha PK & Christen P (1993) Eur. J. Biochem. 211, 373-376
Miyata A et al (1993) Eur. J. Biochem. 212, 745-750.
Utter MF & Kolenbrander HM (1972) In Boyer PD, ed, The Enzymes, 2nd Ed., Vol. 6, pp 117-136, Academic Press, New York

Yamada H et al (1986) Agric. Biol. Chem. 50, 17-21.
Yoshida T et al (1992) Eur. J. Biochem. 210, 849-854.
Yoshida T et al (1995a) Biosci. Biotech. Biochem. 59, 140-142.
Yoshida T et al (1995b) FEMS Microbiol. Lett. 126, 221-226.
Yoshida T et al (1994) Eur. J. Biochem. 223, 727-732.

Metabolism of methanesulfonic acid

Don P Kelly and J Colin Murrell[1]

Institute of Education and [1]Department of Biological Sciences,
University of Warwick, Coventry CV4 7AL, UK

Introduction

Sulfonates. These are organosulfur compounds in which sulfur at the oxidation state of +5 is covalently linked to a carbon atom , which may be part of a larger carbon based molecule: $R\text{-}CH_2\text{-}SO_3H$. The simplest sulfonate is methanesulfonic acid, where R = H, and it and the series of aliphatic sulfonates, $CH_3.(CH_2)_n.SO_3H$, in which n = 1-12, have been shown to be used as sources of sulfur for growth by various microorganisms (Thysse, Wanders 1972; Biedlingmeier, Schmidt 1982; Krauss, Schmidt 1987). More complex sulfonates include the naturally occurring cysteate, taurine, isethionate, and N-acylaminosulfonates and anthropogenic aromatic sulfonates, such as substituted naphthalenesulfonic and benzenesulfonic acids, and pH buffers such as HEPES (N-[2-hydroxyethyl]-piperazine-N'-[2-ethanesulfonate]). Some of these can be degraded by bacteria for use as sources of sulfur, and in some cases also as sources of both carbon and energy for growth (Locher et al. 1989; Seitz et al. 1993; Kertesz et al. 1994; Chien et al. 1995).

Methanesulfonic acid (MSA). This compound is of particular importance for two reasons: (i) it is a product of the oxidation in the atmosphere of dimethyl sulfide, the major biogenic sulfur gas entering the atmosphere; and (ii) MSA is a C_1-compound capable of being used as an energy and carbon substrate for the growth of some specialized methylotrophs.

Biogenic sources of methanesulfonic acid. Some phytoplankton (e.g. *Phaeocystis* and *Emiliana*), macroalgae, and the marsh grass, *Spartina,* produce an osmolyte (acting as a cryoprotectant in some 'ice algae'), dimethylsulfoniopropionate, which undergoes enzymatic hydrolysis on release into the ocean (or grazing of phytoplankton) to produce dimethyl sulfide (Keller et al. 1989; Kelly, Smith 1990; Kelly et al. 1993; Dacey et al. 1994; Kelly et al. 1994a; Otte, Morris 1994; Wolfe et al. 1994; De Souza, Yoch 1995), a proportion of which is vented to the atmosphere (40-70 Tg DMS, with a mean of around 26 Tg DMS-sulfur per annum). Depending on the latitude at which DMS reacts in the atmosphere, 25-70% of it is oxidized to MSA (Mihalopoulos et al. 1992; Kelly et al. 1993, 1994b; Thompson et al. 1995), meaning that the annual rate of MSA formation is in the mean range 19.5-54.6 Tg MSA per annum. MSA is a very stable strong acid and does not undergo further degradative reaction in the atmosphere, from which it is lost by dry deposition and precipitation in rain and snow.

M. E. Lidstrom and F. R. Tabita (eds.), Microbial Growth on C₁ Compounds, 33–40.

There is thus deposition of very large amounts of this potential methylotrophic substrate, of the order of 40 million tonnes per year, with perhaps 70% falling into the oceans and the rest depositing on land or the ice caps. Ice core measurements reveal MSA deposition to have been continuous for millennia (Saigne, Legrand 1987; Whung et al. 1994), but there is no evidence of its progressive accumulation in the biosphere, thus indicating it to be recycled as a component of the global biogeochemical sulfur cycle.

Biodegradation of methanesulfonic acid. In our laboratories, we and our colleagues have isolated the only organisms described to date that are capable of aerobic methylotrophic growth on MSA, and have established that a common mechanism for the initial oxidation of MSA is shared by aerobic bacteria from terrestrial and marine sources, namely a MSA mono-oxygenase (MSAMO). In this paper we summarize the observations available to date on the use of MSA as a methylotrophic substrate and as a sulfur source, and present hypotheses of the kinds of organisms, able to use MSA, which potentially remain to be isolated.

Results

Methylotroph strain M2. The isolation of this soil organism (Kelly et al. 1994b) demonstrated for the first time that bacteria exist which are able to degrade MSA for growth, causing its complete mineralization and thus closing this atmospheric link between the marine and terrestrial environments (Baker et al. 1991). This bacterium is strictly aerobic and grows on several one-carbon substrates, including methanol, mono-, di- and tri-methylamine, formaldehyde and formate, as well as MSA. It is also capable of slower growth on other sulfonates, including the n-alkane sulfonates, ethane-, propane, butane- and pentane-sulfonate, and on isethionate and cysteate, but not on monomethyl sulfate, taurine or benzenesulfonate. Classification of the organism, a gram negative rod, with 61 mol % G+C, was not achieved by using standard biochemical tests as it grew so poorly on many of the standard diagnostic media. Use of 16S rRNA analysis, however, has indicated that it is a novel species, and may be a unique genus, being phylogenetically closer to *Agrobacterium.* than *Methylobacterium* (A. Holmes et al., unpublished).

Methanesulfonate metabolism by strain M2. MSA was only oxidized by bacteria previously grown on MSA, and not after growth on other C_1-compounds including methanol, methylamines or formate. Growth on MSA in both batch and chemostat culture showed a maximum specific growth rate (μ_{max}) of 0.1 h^{-1} (i.e. a doubling time of 7 h), and the growth yield was about 10 g dry (mol MSA)$^{-1}$ in chemostat culture (D = 0.07 h^{-1}). Using either ^{14}C-MSA or ^{14}C-carbon dioxide, batch cultures growing on MSA were shown to obtain at least 70% of their carbon directly from MSA, and the balance (experimentally 10-18%) from carbon dioxide (Kelly et al. 1994b). Enzymatic analysis of cell-free extracts of MSA-grown bacteria showed the presence of high activities of NADPH-dependent hydroxypyruvate reductase (HPR; crude extract V_{max} > 1.0 μmol NADPH oxidized min^{-1} (mg protein)$^{-1}$) but the absence of hexulose phosphate synthase and ribulose bisphosphate carboxylase, indicating that the serine pathway of carbon assimilation *via* formaldehyde operates in strain M2. Comparable activities of HPR were found in strain M2 after growth on MSA, methanol, formaldehyde, formate and methylamine (V_{max} 0.9-1.2) and the apparent K_m for MSA was similar in all extracts (at 3.8-4.4 μM MSA), indicating the serine pathway was likely to operate during growth on all these substrates.

Suspensions of MSA-grown bacteria oxidized MSA completely to sulfate and carbon dioxide, with a stoichiometry of MSA:O_2 = 1.0:2.0 and the production of increased acidity, consistent with the equation:

$$CH_3SO_3^- + H^+ + 2\,O_2 = CO_2 + H_2O + SO_4^{2-} + 2\,H^+$$

Suspensions of MSA-grown bacteria also oxidized amino-MSA and ethane-sulfonic acid (at 86% and 36% of the rate of MSA oxidation), but had little or no ability to oxidize propane- (8%), butane- (0%), or pentane- (0%) sulfonic acids (Kelly et al. 1994b). The kinetics of oxidation of MSA indicated K_m and V_{max} values of 20 µM MSA and 229 nmol O_2 min^{-1} (mg dry wt)$^{-1}$ respectively.

Crude cell-free extracts of strain M2 previously grown on MSA exhibited MSA-dependent oxidation of NADH, with a 1:1 stoichiometry for MSA:NADH, suggesting the presence of an NADH-dependent mono-oxygenase specifically induced by MSA and catalyzing the initial step in MSA oxidation (Kelly, Baker 1990; Kelly et al. 1994b):

$$CH_3SO_3H + O_2 + NADH + H^+ = HCHO + NAD^+ + H_2SO_3 + H_2O$$

This enzyme has subsequently been proved to be the principal dissimilatory enzyme for MSA and partially purified (Higgins et al. 1995). The formation of sulfite as a product was shown by means of Ellman's reagent [5-5'-dithiobis(2-nitrobenzoic acid)], which gave a yellow color with the enzymatically-produced sulfite. Crude extracts oxidized NADH at 59 nmol min^{-1} (mg protein)$^{-1}$ with MSA as substrate, and at about 47 and 17 nmol NADH min^{-1} (mg protein)$^{-1}$ with ethane- or propane-sulfonic acids, but showed no activity with butane- or pentane-sulfonic acids. K_m and V_{max} values observed for MSA in crude extract assays were 48 µM and 66 nmol NADH oxidized min^{-1} (mg protein)$^{-1}$ respectively, which were similar to, but somewhat better in kinetic terms, than the values for ethanesulfonic acid [62 µM and 39 nmol min^{-1} (mg protein)$^{-1}$]. This suggested that both ethane- and propane-sulfonic acids were also substrates for the MSA mono-oxygenase, but for which the enzyme had apparently had less affinity and a lower rate of catalysis. Monomethyl sulfate was also a substrate for the crude extract assay (94% of MSA rate).

The MSA mono-oxygenase of strain M2 (MSAMO). The activity observed in crude cell-free extracts was inhibited by the metal chelators, bathophenanthroline, bathocuproine, neocuproine, dipyridyl and Na-EDTA, with bathophenanthroline, bathocuproine and Na-EDTA causing 88-89% inhibition of NADH oxidation at 1.0, 0.5 and 5.0 mM respectively. This suggested the activity to be that of an enzyme with associated metal ions, as is found with other oxygenases (Mason, Cammack 1992). The essentiality of an associated electron transport system was shown by the inhibition of the MSA-dependent NADH oxidation by azide, arsenate, cyanide (66-72% inhibition by 1.0 mM) and by carbon monoxide.

Purification of the MSAMO was achieved using Q-Sepharose ion exchange chromatography and elution with a linear NaCl gradient (Higgins et al. 1995). This resulted in no individual fractions from the ion exchange column showing any oxygenase activity. The activity was, however, restored when three successive fractions (Proteins A, B and C, eluting with 200, 300 and 440 mM NaCl respectively) from the column chromatography were recombined. The activity of the reconstituted enzyme [26 nmol NADH oxidized min^{-1} (mg protein)$^{-1}$] was stimulated three-fold by adding 2 µM

FAD, and five-fold if both FAD and 0.1 mM Fe^{2+} were added. These observations indicated the requirement for FAD and Fe^{2+}, as has been found with other oxygenases (Locher et al. 19991; Schlafli et al. 1994). Further purification and analysis of the three components (T Higgins, JC Murrell unpublished) by gel filtration protocols has indicated that (i) Protein A (36-38 kDa; pI 5.7) is the reductase component of the MSAMO; (ii) Protein B is a 200 kDa complex (pI 4.8), probably a tetramer of two proteins (70 and 25 kDa), and is the hydroxylase component; and that (iii) Protein C is the ferredoxin component (16 kDa; [2Fe-2S] per mol of protein), for which N-terminal sequence analysis showed some homology to the ferredoxin component of benzene dioxygenase.

The purified and reconstituted enzyme preparation (supplemented with FAD and Fe^{2+}) exhibited a more restricted range of substrate specificity than MSA-grown whole cells or crude extracts made from them (Higgins et al. 1995). The only substrates oxidized (compared to the MSA control rate of 113 nmol NADH oxidized min^{-1} (mg $protein)^{-1}$ = 100%) were ethanesulfonic acid (98%), propanesulfonic acid (36%) and monomethyl sulfate (102%). The ability of MSA-grown cells to oxidize amino-MSA and isethionate (Kelly et al. 1994b) and the oxidation of NADH by crude extracts supplemented with amino-MSA (36% of the rate with MSA), or isethionate (46%), taurine (34%) or cysteate (23%) were thus not due to direct use of these compounds as substrates, but must have required prior processing (e.g. by enzymatic deamination of amino-MSA to release free MSA) or catalysis by different enzyme systems for those substrates.

Marine methylotrophic bacteria growing on MSA. While no data exist on the concentrations of MSA in sea water, or indeed on whether there are marine sources of MSA other than its input from the atmospheric oxidation of dimethyl sulfide, the ocean must be a sink for MSA as most of the rain carrying MSA from the atmosphere falls on the oceans. Recently two strains of MSA-degrading methylotrophs were isolated from British coastal waters (Thompson et al. 1995), using MSA in combination with methanol or methylamine as enrichment substrates. Both were gram negative rods growing optimally on MSA as sole source of carbon and energy with NaCl at concentrations around 0.4 M, and characteristically producing a gelatinous material in which the cells were embedded, as is not uncommon for freshly isolated marine bacteria. Both also grew on a wide range of sugars and organic acids as well as C_1 compounds, and could use the sulfonates taurine or isethionate, but not cysteate or ethanesulfonate, for growth. Growth on MSA was faster (μ = 0.10-0.13 h^{-1}) than on methanol or methylamine (μ = 0.05), and growth yields up to 5.7 g dry wt (mol $MSA)^{-1}$ were observed in pH-controlled batch cultures. These strains were very similar to each other, both exhibiting mol % G+C values of 57%, and MSA-grown suspensions oxidized MSA, formaldehyde and formate with substrate: oxygen stoichiometries that showed oxidation was complete to carbon dioxide (and sulfate in the case of MSA). Cultures growing on MSA produced sulfite (detected with Ellman's reagent), which was presumed to oxidize spontaneously in the culture medium.

Cell-free extracts of both contained activities for NADH-dependent MSAMO [11.4 and 12.5 nmol NADH oxidized min^{-1} (mg $protein)^{-1}$], hydroxypyruvate reductase, and serine-glyoxylate aminotransferase, but lacked ribulose bisphosphate carboxylase and hexulose phosphate synthase, indicating that their dissimilatory and assimilatory C_1 metabolism was similar in principle to that the soil methylotroph strain M2, from which they are, however, morphologically distinct. The apparent crude cell-free extract K_m values for MSA were rather high (at 640 and 100 μM, compared to 48 μM for strain

M2), which needs to be considered when assessing the concentrations of MSA likely to be available in solution in the marine habitats from which the bacteria came: the possibility of *in situ* production of significant localized surface concentrations of MSA from dissolved DMS needs evaluation. SDS-PAGE of extracts of these new strains showed bands specific for organisms grown on MSA, and having M_r values (kDa) of 40-45, 35 and 25, which resemble the pattern seen with strain M2 (T.P. Higgins, J.C. Murrell, unpublished).

Anaerobic use of methanesulfonate. To date no bacteria have been described that can use MSA without also using oxygen. We have not, so far, had success in isolating denitrifying methylotrophs able to use MSA as carbon and nitrogen source under anaerobic conditions. There are, however, reports of the fermentative growth (on glucose or sucrose) of *Klebsiella* or *Clostridium* with sulfonates as the sole source of sulfur for biosynthesis (Chien et al. 1995). These strains were isolated from anaerobic enrichment cultures provided with cysteate, taurine or isethionate as sulfur sources. The *Klebsiella* (probably *K. oxytoca*) grew aerobically on glucose with any one of taurine, isethionate, cysteate, HEPES or MSA (or sulfate) as sulfur source, but could only grow fermentatively with cysteate (or sulfate). The *Clostridium* (*C. pasteurianum*) would, however, grow fermentatively with taurine, isethionate or *p* -toluenesulfonate as sole sulfur sources. It could not use MSA as an anaerobic sulfur source. The ability thus exists among fermentative heterotrophs to degrade some sulfonates as sources of biosynthetic sulfur, presumably reducing the sulfonate group (to sulfide and thence to cysteine) either after cleavage to release free sulfite from them, or even prior to cleavage of the C-S bond. Whatever the mechanism, it must involve a hydrolytic split of the C-S bond and not oxygenation as has been described so far for all known sulfonate-cleaving organisms.

Alternative biochemistries for MSA metabolism and the hunt for novel MSA-using methylotrophs. It is now well-established that diverse heterotrophic, phototrophic and methylotrophic microorganisms can use sulfonates, including MSA, as sources of sulfur and in some cases carbon and energy, but that all of these are aerobes and are likely to use a mono-oxygenase to split the C-S bond of the sulfonate, as was first observed with the degradation of C_4-C_8 n-alkane-1-sulfonates by *Pseudomonas* strains (Thysse, Wanders 1974). Two lines of enquiry are indicated: (i) further marine and terrestrial aerobes need to be isolated in order to determine whether the mono-oxygenase mechanism for aerobic methylotrophic MSA degradation is the only one extant in modern bacteria; (ii) further anaerobic studies are needed both to seek denitrifying MSA-methylotrophs and bacteria using MSA as an anaerobic sulfur source, and to establish the nature of the mechanisms by which sulfonates can be used by heterotrophs as sulfur sources.

Hydrolysis of MSA to produce methanol does not occur in any of the methylotrophs isolated so far, indeed methanol is involved in no way as an intermediate in MSA breakdown and mutants of strain M2 lacking methanol dehydrogenase have been found to be unimpaired in ability to use MSA (Higgins et al. 1995). MSA hydrolysis as a means of splitting the C-S bond is, however, a hypothesis worthy of experimental test:

$$CH_3SO_3H + H_2O = CH_3OH + H_2SO_3$$

MSA is an analogue of methanephosphonate, which has been shown to be degraded by *Pseudomonas testosteroni* to produce methane and phosphate by hydrolytic

38

cleavage of the C-P bond (Daughton et al. 1979). The analogous reaction for MSA would be:

$$CH_3SO_3H + H_2O = CH_4 + H_2SO_4$$

This has not yet been demonstrated, but the feasibility of the reaction, perhaps among anaerobes is worth investigating.

A reaction catalysed by *Pseudomonas* MS during growth on trimethylsulfonium chloride is the cleavage of the methyl groups from the sulfonium compounds by means of a tetrahydrofolate transfer, producing methyl-tetrahydrofolate and sulfite. Again, the analogous process for MSA has not been found but might be possibility among anaerobes or aerobes.

Finally, it is not inconceivable that MSA could act as a substrate anaerobically for either or both of methanogens and sulfate-reducing bacteria, being used as a respiratory hydrogen acceptor:

$$CH_3SO_3H + 4 H_2 = CH_4 + H_2S + 3 H_2O$$

The products of anaerobic degradation of [14]C- and ^{35}S-labeled MSA by anaerobic sediment slurries might reveal whether labeled methane and sulfide appear, which would be a first step in establishing whether the C-S bond of MSA is broken, by whatever means or whatever bacteria, to give rise to substrates usable by methanogens and/or sulfate-reducers. The primary organisms responsible for that cleavage (which might not be the methanogens or sulfate-reducers) would then be known to exist and could be sought by appropriate enrichment culture methods.

Conclusions

Thus far MSA has only been shown to be completely mineralized, and used as sole carbon plus energy substrate, by a very small number of specialized methylotrophs, all of which use a MSA mono-oxygenase for the initial cleavage of the C-S bond of MSA. They are all gram negative, rod-shaped, serine pathway methylotrophs with DNA containing 57-61 mol % G+C. Other aerobic MSA-users undoubtedly exist but have not been further characterized (Baker et al. 1991). The pathways for MSA dissimilation and assimilation in the well-studied strains are summarized in Fig. 1.

We believe that it is likely that many microorganisms, both aerobic and anaerobic, are likely to be able to use MSA, either as a growth substrate or just as a sulfur source. Extrapolation from observations of the scale of modern oceanic DMS production and of MSA formation in the atmosphere, coupled with the evidence from ice core analyses demonstrating MSA production to been continuous over thousands of years, leads to the conclusion that many trillions of tonnes of it have been deposited on the earth over geological time. In that DMS-producing phytoplankton have existed on a multi-million year time scale, the period during which DMS, and consequently MSA, has been part of the atmospheric sulfur cycle is likely to be immensely long, allowing much time for evolution and dissemination of MSA-consuming abilities among diverse organisms.

An interesting recent hypothesis was stimulated by correlation of the known superabundance of coccolithophorids in the shelf seas of the Late Cretaceous period (70-90 million years before the present) with the increasing rate of extinctions in that period: increased environmental acidification as a consequence of acid rain (including MSA) from DMS may have been a contributory factor in these extinctions (Robinson 1995). In more modern times, the contribution of biogenic sulfur acids to acid rain has also been considered as augmenting anthropogenic sources of sulfur dioxide which contribute to environmental degradation (see Kelly, Smith 1990).

Clearly, with such a long history of steady injection of MSA into the biosphere, the possibilities for the development of diverse organisms exploiting it for carbon, energy and sulfur are considerable, and potentially exotic biochemistries that may have evolved to effect its degradation remain to be explored.

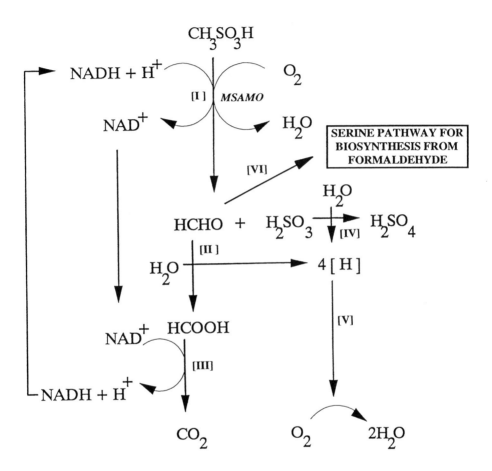

Figure 1. Pathways for the degradation and assimilation of methanesulfonic acid by the methylotroph strains M2, TR3, and PSCH4 (based on Kelly et al. 1994; Thompson et al. 1995). [I] MSA mono-oxygenase (NADH-dependent); [II] formaldehyde dehydrogenase; [III] formate dehydrogenase (NAD-dependent); [IV] sulfite oxidation, (probably non-biological); [V] electron transport to oxygen from formaldehyde oxidation and sulfite; [VI] assimilation of C1 from formaldehyde, involving serine-glyoxylate aminotransferase and hydroxypyruvate reductase.

References

Baker SC et al (1991) Nature (London) 350, 627-628.
Biedlingmeier S, Schmidt, A (1982) Biochim. Biophys. Acta 86, 95-104.
Chien CC et al (1995) FEMS Microbiol. Lett. 129, 189-194.
Dacey JWH et al (1994) Mar. Ecol. Prog. Ser. 112, 67-74.
Daughton CG et al (1979) FEMS Microbiol. Lett. 5, 91-93.
De Souza MP, Yoch DC (1995) Appl. Environ. Microbiol. 61, 21-26.
Higgins TP et al (1995) Microbiology (UK), submitted.
Krauss F, Schmidt A (1987) J. Gen. Microbiol. 133, 1209-1219.
Keller MD et al (1989) In Saltzman ES and Cooper WJ, eds, Biogenic Sulfur in the Environment, pp 167-182, American Chemical Society, Washington, DC, USA.
Kelly DP, Baker SC (1990) FEMS Microbiol. Rev. 87, 241-246.
Kelly DP, Smith NA (1990) Adv. Microb. Ecol. 11, 345-385.
Kelly DP et al (1993) In Murrell JC and Kelly DP, eds, Microbial Growth on C$_1$ Compounds, pp 47-63, Intercept, Andover, England.
Kelly DP et al (1994a) Biochem. Soc. Trans. 22, 1011-1015.
Kelly DP et al (1994b) Microbiology (UK) 140, 1419-1426.
Kertesz MA et al (1994) FEMS Microbiol. Rev. 15, 195-215.
Locher HH et al (1989) Appl. Environ. Microbiol. 55, 492-494.
Locher HH et al (1991) J. Bacteriol. 173, 3741-3748.
Mason JR, Cammack R (1992) Annu. Rev. Microbiol. 46, 277-305.
Mihalopoulos N et al (1992) J. Atmos. Chem. 14, 459-477.
Otte ML, Morris JT (1994) Aquatic Bot. 48, 239-259.
Robinson ND (1995) J Geol Soc 152, 4-6.
Saigne C, Legrand M (1987) Nature (London) 330, 240-242.
Schlafli HR et al (1994) J. Bacteriol. 176, 6644-6652.
Seitz A et al (1993) Arch. Microbiol. 159, 440-444.
Thompson AS et al (1995) Appl. Environ. Microbiol. 61, 2388-2393.
Thysse GJE, Wanders TH (1972) Antonie van Leeuwenhoek 38, 53-63.
Thysse GJE, Wanders TH (1974) Antonie van Leeuwenhoek 40, 25-37.
Whung PY et al (1994) J. Geophys. Res. 99, 1147-1156.
Wolfe GV et al (1994) Mar. Ecol. Prog. Ser. 111, 111-119.

Dimethylsulfide as an electron donor in *Rhodobacter sulfidophilus*

Alastair G. McEwan, Tze-Hsien Toh, Peter S. Solomon, Anthony Shaw and
*Stephen P. Hanlon
Department of Microbiology, The University of Queensland, Brisbane 4072,
Australia
*Department of Biochemistry, University of Leicester, Leicester LE1 7RH,
U.K.

1. SUMMARY

Marine bacteria have been described which grow photoautotrophically using dimethylsulfide (DMS) as an electron donor. In this process DMS is oxidised to dimethylsulfoxide (DMSO) which accumulates in the growth medium. The use of DMS as an electron donor has been investigated further in the purple non-sulfur phototroph *Rhodobacter sulfidophilus*. An assay for DMS:acceptor oxidoreductase activity has been developed and the enzyme has been shown to be localised to the periplasm. DMS:acceptor oxidoreductase is a distinct enzyme from the periplasmic DMSO reductase which is commonly found in purple non-sulfur phototrophs. DMS:acceptor oxidoreductase has been purified and is composed of a single polypeptide (M_r = 94,000). The enzyme contains a molybdenum pterin cofactor and a haem b with an α-absorption maximum at 562nm. The level of DMSO in the surface water of the ocean has been shown to be around 10nM suggesting that it constitutes a significant pool of organic carbon. This paper ends with a consideration of how DMSO might be degraded.

2. INTRODUCTION

Dimethylsulfide (DMS) is now recognised as the major volatile sulfur compound released from the marine environment to the atmosphere (Andreae, 1990). This molecule is currently the subject of much attention because of its central position in the biogeochemical cycling of organic sulfur and its postulated role in climate regulation (Charlson et al, 1987). Dimethylsulfoniopropionate (DMSP) is the main source of DMS, and the former is used as an osmolyte and compatible solute in certain marine algae (Reed, 1983). DMSP is degraded via two distinct routes (Taylor, 1992). One of these generates DMS and this has been estimated to account for about 30% of the soluble DMSP which is turned over (Kiene, 1993). Microbial cycling of DMS is of major significance since it has been calculated that only about 10% of the DMS produced in the ocean is released into the atmosphere (Kiene, 1993). It follows that there must be pathways for the degradation of DMS and the outline of these is known (Taylor, 1992). A few chemolithotrophic bacteria have been isolated which can use DMS as a carbon and energy source under aerobic conditions. Examples of such bacteria are *Thiobacillus thioparus* (Smith and Kelly, 1988)

M. E. Lidstrom and F. R. Tabita (eds.), Microbial Growth on C₁ Compounds, 41–48.
© 1996 *Kluwer Academic Publishers. Printed in the Netherlands.*

and certain *Hyphomicrobium* strains (de Bont et al, 1981; Suylen et al, 1986). A second possible route of DMS removal is its oxidation to dimethylsulfoxide (DMSO) (Kiene, 1993). Since the E_o' of the DMSO/DMS couple is +160mV (Wood, 1981), then DMS is a potential electron donor in bacterial respiration and photosynthesis. In this paper we discuss the use of DMS as an electron donor in bacteria and focus on one particular organism, *Rhodobacter sulfidophilus*.

3. ISOLATION OF BACTERIA WHICH OXIDISE DMS TO DMSO

Zeyer et al (1987) grew enrichment cultures of purple phototrophic bacteria from a coastal salt pond and a salt marsh. These cultures oxidised DMS to DMSO under anaerobic conditions in the light and a pure culture from the enrichments was tentatively identified as a *Thiocystis* sp. (Zeyer et al, 1987). Independently, Visscher and van Gemerden (1991) isolated a strain of *Thiocapsa roseopersicina* from a marine microbial mat which was able to grow photoautotrophically using DMS as electron donor. These results established that marine purple sulfur bacteria were able to utilise DMS as an electron donor in photosynthesis. In experiments begun in 1992 we isolated bacteria from an intertidal zone on the North Norfolk coast (England). Growth curves for one of these isolates, strain SH1, demonstrated that this bacterium grew photoautotrophically under anaerobic conditions and chemoautotrophically with oxygen as electron acceptor with DMS as reductant. DMSO accumulation was followed during cell growth by NMR spectroscopy (Hanlon et al, 1994) and about 70% conversion of DMS to DMSO was observed.

Strain SH1 was characterised by a variety of biochemical tests. The absorption spectrum of its photopigments indicated that it contained bacteriochlorophyll *a* and that it was probably from the genus *Rhodobacter* (Hanlon et al, 1994). Examination of periplasmic cytochromes showed that the most abundant component was ascorbate-reducible and had an α-absorption maximum at 556nm, as in *Rb. sulfidophilus*. The isolate was identified as a strain of *Rb. sulfidophilus* by a phylogenetic analysis of a partial 16S rRNA gene sequence of strain SH1 (Hanlon et al, 1994). In further work it was shown that the type species *Rb. sulfidophilus* DSM 1374 was also able to use DMS as an electron donor during autotrophic growth (Hanlon and McEwan, unpublished observation). Recently, it has been proposed that *Rb. sulfidophilus* and other marine *Rhodobacter* be transfered to a new genus *Rhodovulum*, with *Rhodovulum sulfidophilum* as the type species (Hiriashi and Ueda, 1994). In this review the established nomenclature will be retained. *Rb. sulfidophilus* is recognised as a bacterium that can utilise and tolerate relatively high levels of sulfide compared to most other purple non-sulfur bacteria (Hansen and Veldkamp, 1973). However, we have also established that strain SH1 is also a methylotroph since it can utilise methanol, methylamine, dimethylamine and trimethylamine as carbon sources under phototrophic conditions (Hanlon et al, 1994). Since serine pathway enzymes were not detected in strain SH1 it is assumed that carbon assimilation proceeds via the Calvin cycle.

DMS is a reductant which may be used by a wide variety of marine bacteria during phototrophic growth. However, chemolithotrophic growth of *Rb. sulfidophilus* is the first example of the direct use of DMS as an electron donor during autotrophic growth.

4. CHARACTERISATION OF THE DMS-OXIDATION SYSTEM OF *Rb. sulfidophilus*

4.1 Identification of a periplasmic DMS:acceptor oxidoreductase   A spectrophotometric assay was developed for DMS-oxidising activity in cell-free extracts of *Rb. sulfidophilus* (Hanlon et al, 1994). This assay measured the reduction of dichlorophenolindophenol (DCPIP) by DMS. The reaction was dependent on the presence of the redox dye phenazine ethosulfate (PES) and it is assumed that DMS oxidation is catalysed by a DMS:acceptor oxidoreductase which can reduce PES. So far, no other electron acceptors which can mediate DMS-dependent reduction of DCPIP have been identified.   Using this assay, DMS:acceptor oxidoreductase activity was shown to be located in the periplasm of *Rb sulfidophilus* (Hanlon et al, 1994). This activity was distinct from a periplasmic DMSO reductase which was also present in *Rb. sulfidophilus* (Hanlon et al, 1994). The periplasmic DMSO reductase of *Rb. sulfidophilus* cross-reacted with anti-DMSO reductase (*Rhodobacter capsulatus*) antibodies and had a subunit M_r = 82,000 (Hanlon, 1994). This indicates that it is very similar to the well-characterised DMSO reductases of *Rb. capsulatus* and *Rhodobacter sphaeroides* (McEwan et al, 1991; Bastian et al, 1991).

4.2 Purification of DMS:acceptor oxidoreductase   DMS:acceptor oxidoreductase was purified from a periplasmic fraction of *Rb. sulfidophilus* which had been grown photoheterotrophically in the presence of DMSO. Enzyme activity was precipitated in a 30%-60% ammonium sulfate fractionation. Purification was completed by chromatography on DEAE-Sepharose, hydroxyapatite and gel filtration (Sephacryl S300). Analyis of purified DMS:acceptor oxidoreductase by SDS PAGE showed that the enzyme was composed of a single polypeptide, M_r = 94,000. Gel filtration of the native form of the enzyme indicated that it was probably a dimer.

4.3 DMS:acceptor oxidoreductase contains b-type heme and molybdenum-pterin cofactor   The oxidation of DMS to DMSO in *Rb. sulfidophilus* occurred under anaerobic conditions. This immediately indicated that the reaction could not be catalysed by a cytochrome P450-type monooxygenase which has previously been shown to oxidise thioethers to sulfoxides (Light, et al, 1982). It seemed most likely that DMS:acceptor oxidoreductase would be a molybdenum oxotransferase. This class of enzymes catalyses the transfer of an oxo group to or from a substrate in a reaction which involves a two electron oxidation/reduction and, most importantly, water is the source of oxygen in the oxo group transfer reaction (Rajagopalan, 1991).

$$DMS + H_2O \rightarrow DMSO + 2H^+ + 2e$$

Fluorescence excitation and emission spectra of material released from DMS:acceptor oxidoreductase by boiling at pH 2.5 were collected. An excitation maximum at 370nm and emission maximum at 470nm was observed and this is characteristic of the Form B pterin derivative of the Mo-pterin cofactor (Johnson and Rajagopalan, 1982). The presence of molybdenum in DMS:acceptor oxidoreductase was confirmed by atomic absorption spectroscopy (Hanlon, 1994). All Mo-pterin cofactors possess a common core structure in which an organic component, a 2-amino, 4-hydroxy (hydro) pteridine, is linked via a dithiolene side chain to an Mo ion (Rajagopalan, 1991). The form of the cofactor found in eukaryotes ends in a phosphate group and it is known as molybdopterin (MPT). However, in almost all enzymes examined from bacteria and archaea the MPT is extended via a pyrophosphate linkage to a nucleoside monophosphate (Rajagopalan, 1991). The most common form of this modified Mo-pterin cofactor in bacteria appears to be molybdopterin guanine dinucleotide (MGD) (Fig. 1). MGD was first identified in the DMSO reductase of *Rb. sphaeroides* (Johnson et al, 1991) and, although further analysis is required to support this view, it seems probable that this form of the cofactor will be present in DMS:acceptor oxidoreductase.

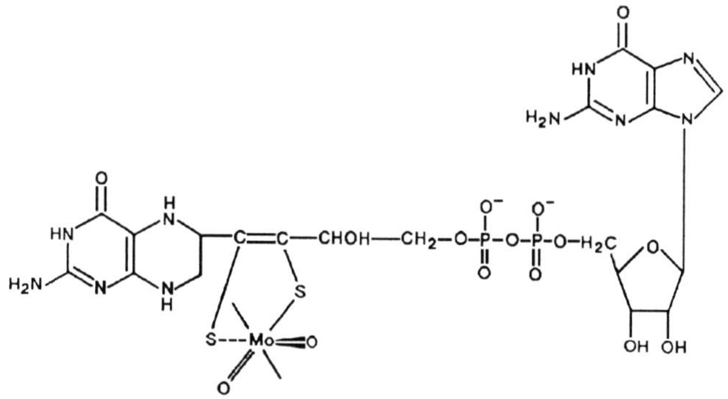

Figure 1. Proposed structure of molybdenum pterin guanine dinucleotide (Johnson et al, 1991)

The intensity of the visible absorption spectrum of the Mo-pterin cofactor is very weak, relative to other cofactors such a haem and flavin. As a result, the only intact enzyme in which the absorption spectrum of a Mo-pterin cofactor has been observed is *Rhodobacter* DMSO reductase, an enzyme which lacks additional prosthetic groups (McEwan et al, 1991; Bastian et al, 1991). The absorption spectrum of resting DMS:acceptor oxidoreductase is shown in Fig. 2. The spectrum is dominated by maxima at 562nm, 533nm and 428nm which are characteristic of heme α, β and γ absorption bands.

Further reduction of the enzyme with dithionite showed that it was 90% reduced in its resting state. Pyridine hemochrome analysis showed that the heme in DMS:acceptor oxidoreductase was a b-type. This is a rare example of a periplasmic enzyme with a non-covalently bound heme group. In further experiments it was demonstrated that DMS was able to reduce the heme group of DMS:acceptor oxidoreductase.

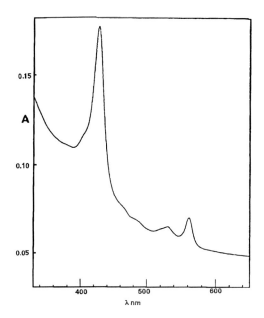

Figure 2. Visible absorption spectrum of the resting form of DMS: acceptor oxidoreductase (protein concentration 0.35mg ml^{-1}).

4.4 Organisation of the DMS oxidising pathway in Rb. sulfidophilus
Although the cyclic electron transfer in *Rb. sulfidophilus* has not been investigated in detail it is probably very similar to the well-characterised systems of *Rb. capsulatus* and *Rb. sphaeroides* (reviewed by McEwan, 1994). There are two potential sites at which electrons from periplasmic DMS oxidation could enter the cyclic photosynthetic electron transfer system of *Rb. sulfidophilus*. Reduction of ubiquinone seems the least likely in view of the standard redox potentials of UQH$_2$/UQ (E$_o$' = +90mV) compared to the DMS/DMSO redox couple (E$_o$' = +130mV) and so it is suggested that a periplasmic c-type cytochrome accepts electrons from DMS:acceptor oxidoreductase (Fig. 3). During photoautotrophic growth c-type cytochromes are involved in electron transfer between the cytochrome bc$_1$ complex and the photochemical reaction center (RC). In *Rb. capsulatus* and *Rb. sphaeroides* cytochrome c$_2$ is known to have such a role but the c-type cytochrome(s) involved in electron transfer to the RC in *Rb. sulfidophilus* have not been characterised. During chemoautotrophic growth the only site of energy conservation would be the cytochrome oxidase and reduction of NAD$^+$ by reduced cytochrome c catalysed by an energy-linked reverse electron flow via the cytochrome bc$_1$ complex and NADH dehydrogenase.

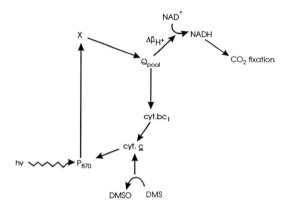

Figure 3. Organisation of electron transfer in *Rb. sulfidophilus* during phototautotrophic growth with DMS as electron donor. Light-driven cyclic electron transfer, involving the RC (P_{870}/X), ubiquinone pool (Q-pool), cytochrome bc_1 complex and periplasmic cytochrome c, acts as a generator of a proton electrochemical gradient ($\Delta \mu_{H+}$). Electrons from DMS enter cyclic electron transfer at the level of cytochrome c. Reverse electron transfer from ubiquinol to NAD^+ consumes $\Delta \mu_{H+}$ and is catalysed by NADH dehydrogenase.

5. DMSO IN THE MARINE ENVIRONMENT

5.1 Determination of DMSO in seawater DMSO is now established as a major methylated sulfur molecule in the marine environment and recent experiments by Hatton et al (1994) and Kiene and Gerard (1994) have demonstrated that DMSO is present at concentrations of about 10-20nM. The method used by Hatton et al (1994) was an enzyme-linked assay for DMSO in which DMSO reductase from *Rb. capsulatus* was used to convert DMSO to DMS then measured DMS by gas chromatography. This method allows DMSP, DMS and DMSO to be measured in a single sample. In the analysis of Hatton et al (1994) the depth profile for DMSO rose steeply from about 30m towards the surface water. One explanation for this may be that light is required for most transformations of DMS to DMSO. It has been demonstrated that abiotic photochemical oxidation of DMS to DMSO can occur in seawater (Brimblecombe and Shooter, 1986) but biological oxidation of DMS to DMSO by phototrophic bacteria may also take place. A recent assessment of the phylogenetic diversity of marine bacterial assemblages has shown that the α proteobacterial group (which includes *Rb. sulfidophilus*) are far more abundant in coastal and open sea environments than previous estimates had indicated (DeLong et al, 1993). Dark aerobic routes of oxidation of DMS to DMSO might include its oxidation during co-metabolism (Zhang et al, 1991) and the gratuitous oxidation of DMS to DMSO by ammonium monooxygenase in nitrifying bacteria (Juliette et al, 1993) although the importance of both is unknown. Clearly, a great deal of work needs to be done in order to understand how DMSO is formed in marine systems.

5.2 What is the fate of DMSO in seawater ? There are several possible routes of DMSO removal. It is well-established that under anaerobic conditions DMSO can be reduced to DMS by respiratory electron transfer (Zinder and Brock, 1978). The significance of this process in the marine environment is unknown although the presence of γ proteobactcria in aggregate-attached assemblages suggests that the process might be possible in anaerobic microenvironments associated with these aggregates (DeLong et al, 1993). Among free-living bacteria, DMSO can be reduced under aerobic conditions back to DMS then metabolised as a carbon and energy source as described in *Hyphomicrobium* (DeBont et al, 1981) or *Thiobacillus* (Smith and Kelly, 1988). If this route of DMSO cycling is a major one then it would suggest that in some environments bacterial DMSO assimilation contributes to the microbial loop of the ocean's food web in a significant way. Finally, DMSO could be oxidised to dimethylsulfone ($DMSO_2$). This reaction has been suggested to occur via disproportionation of DMSO (Harvey and Lang, 1986):

$$DMSO + DMSO \rightarrow DMS + DMSO_2$$

A biological route may also be possible since the redox potential of the $DMSO/DMSO_2$ couple ($E_0' = -240$ mV) makes it potentially a strong electron donor (Wood, 1981). By analogy with the use of DMS as an electron donor it should be possible to isolate photoautotrophs which oxidise DMSO to $DMSO_2$. Since $DMSO_2$ is a relatively stable molecule its removal would be dependent upon bacterial metabolism. Whatever the route of DMSO degradation, the preliminary data of Kiene and Gerard (1994) suggest that it is turned over in seawater over a period of only a few days.

ACKNOWLEDGEMENTS
AGM thanks John Fuerst for many useful discussions. This work was supported between 1990 and 1994 by SERC (U.K.) & NERC (U.K.) and since 1995 by the Australian Research Council.

REFERENCES
Andreae, M.O. (1990) Mar. Chem. **30**, 1-29.
Bastian, N.R., Kay, C.J., Barber, M.J. and Rajagopalan, K.V. (1991) J. Biol. Chem. **226**, 45-51
Brimblecombe, P. and Shooter, D. (1983) Mar. Chem. **19**, 343-353.
Charlson, R.J., Lovelock, J.E., Andreae, M.O. and Warren, S.G. (1987) Nature **326**, 655-661.
De Bont, J.A.M., van Dijken, J.P. and Harder, W. (1981) J. Gen. Microbiol. **127**, 315-323.
DeLong, E.F., Franks, D.G. and Alldredge, A.L. (1993) Limnol. Oceanogr. **38**, 924-934.
Hanlon, S.P., Holt, R.A., Moore, G.R. and McEwan, A.G. (1994) Microbiology **140**, 1953-1958.
Hansen, T.A. and Veldkamp, H. (1973) Arch. Microbiol. **92**, 45-58
Harvey, G.R. and Lang, R.F. (1986) Geophys. Res. Lett. **13**, 49-51.
Hatton, A.D., Malin, G.R., McEwan, A.G. and Liss, P.S. (1994) Analyt. Chem. **66**, 4093-4096.
Hiraishi, A. and Ueda, Y. (1994) Int. J. Syst. Bacteriol. **44**, 15-23.
Johnson, J.L. and Rajagopalan, K.V. (1982) Proc. Nat. Acad. Sci. (USA) **79**, 6856-6860
Johnson, J.L., Bastian, N.R. and Rajagopalan, K.V. (1991) Proc. Nat. Acad. Sci. (USA)

87, 3190-3194.

Julliette, L.Y, Hyman, M.R. and Arp, D.J. (1994) Appl. Env. Microbiol. **59,** 3718-3727.

Kiene, R.P. (1993) In: Microbial Growth on C$_1$ Compounds (J.C. Murrell and D.P. Kelly eds.) pp 15-33.

Kiene, R.P. and Gerard, G. (1994) Mar. Chem. **47,** 1-12.

Light, D.R., Waxman, D.J. and Walsh, C. (1982) Biochemistry **21,** 2490-2498.

McEwan, A.G. (1994) . Antonie van Leeuwenhoek **66,** 151-164

McEwan, A.G., Ferguson, S.J. and Jackson, J.B. (1991) Biochem. J. **274,** 305-307.

Rajagopalan, K.V. (1991) Adv. Enzymol. Rel. Areas Mol. Biol. **64,** 215-290

Reed, R.H. (1983) Marine Biol. Lett. **34,** 173-181

Smith, N.A. and Kelly, D.P. (1988) J. Gen. Microbiol. **134,** 1407-1417.

Suylen, G.M.H., Stefess, G.G. and Kuenen, J.G. (1986) Arch. Microbiol. **146,** 192-198.

Taylor, B.F. (1992) In: Biogeochemistry of global change: radiatively active trace gases (R. S. Oremland, ed) pp 745-781, Chapman and Hall, New York.

Visscher, P.P. and van Gemerden, H. (1991) FEMS Microbiol. Lett. **81,** 247-250.

Wood, P.M. (1981) FEBS Lett. **164,** 223-226

Zeyer, J., Eicher, P., Wakeham, S.G. and Schwarzenbach, R.P. (1987) Appl. Env. Microbiol. **53,** 2026-2032.

Zhang, L., Kuniyoshi, I., Hirai, M. and Shoda, M. (1991) Biotechnol. Lett. **13,** 223-228.

CYANOBACTERIAL MUTANTS DEFECTIVE IN HCO$_3^-$ UPTAKE

Michal Ronen-Tarazi, Judy Lieman-Hurwitz, David J. Bonfil
Vera Shinder and Aaron Kaplan

Department of Botany, The Hebrew University of
Jerusalem, 91904 Jerusalem, Israel
Fax: 972 2 6584425, E-Mail: aaronka@vms.huji.ac.il

Introduction

Photosynthetic microorganisms are capable of adapting to a wide range of CO$_2$ concentrations. When cyanobacterial cells are transferred from high to low concentrations of CO$_2$, they undergo an adaptation process. This includes modulation of the expression of certain genes, some of which are involved in the operation of the inorganic carbon (Ci)-concentrating mechanism (CCM, see Miller et al. 1990; Kaplan et al. 1991; Raven 1991; Badger, Price 1992; Ogawa 1993 and Kaplan et al. 1994 for recent reviews and literature citations). The active accumulation of Ci to levels as high as 50-100 mM Ci, consequent on the activity of the CCM, enables the cells to perform efficient photosynthesis in spite of the relatively low affinity of their ribulose 1,5-bisphosphate carboxylase (rubisco) for CO$_2$. Moreover, the elevated concentration of CO$_2$ in close proximity to rubisco, within the carboxysomes, activates the enzyme, reduces competition by O$_2$ and inhibits photorespiration (Schwarz et al. 1995).

It is now widely accepted that the CCM consists of two major components, the energy-dependent mechanisms whereby Ci accumulates within the cytoplasm; and the carboxysomes where the elevated CO$_2$ concentration produced by carbonic anhydrase (CA) activity, confined to these bodies, enables proper activation of rubisco and efficient carboxylation (Schwarz et al. 1995; Ronen-Tarazi et al. 1995). The saturation kinetics observed when accumulation of Ci internally was measured as a function of external CO$_2$ concentration suggested that CO$_2$ uptake is not merely the result of passive diffusion (see Fridlyand et al. 1995). Uptake of bicarbonate also displays saturation with rising external HCO$_3^-$ concentration but the kinetics parameters are different than those observed in the case of CO$_2$ uptake. Bicarbonate uptake and CO$_2$ uptake differ not only in their kinetic parameters but also in certain other characteristics. The presence of sodium (in the mM range) is required for the uptake of HCO$_3^-$ but a very low concentration of sodium (in the μM range) is large enough to saturate the response of CO$_2$ uptake to sodium (Miller et al. 1990). Bicarbonate uptake is associated with hyperpolarization of the cytoplasmic membrane, not observed for CO$_2$ uptake (Kaplan et al. 1991). Thus, there are numerous physiological indications that uptake of the two Ci species is mediated by two distinct mechanisms.

M. E. Lidstrom and F. R. Tabita (eds.), Microbial Growth on C$_1$ Compounds, 49–55.

It is important to note that regardless of which Ci species is supplied, bicarbonate is the species which accumulates internally, and the Ci species are not at chemical equilibrium in the cytoplasm. Expression of a human CA gene in *Synechococcus* sp. PCC 7942 resulted in a high-CO_2-requiring mutant due to excessive leak of CO_2 from the cells (Badger, Price 1991), confirming that CO_2 and HCO_3^- are not at chemical equilibrium within the cytoplasm. These findings led to the suggestion that uptake of CO_2 involves a vectorial CA-like moiety which converts CO_2 to HCO_3^- in transit. The system would have to depend on a metabolic energy supply since the release of bicarbonate in the cytoplasm would occur against its chemical potential gradient. The discovery of a plasma membrane-associated CA activity in *Synechocystis* sp. PCC 6803 (Bedu et al. 1993) lends support to this suggestion. The presence of CA-like activity would also serve the important function of scavenging CO_2 molecules in the cytoplasm and thus minimizing their leak to the medium. Moreover, this postulate obviates the earlier suggestion that uptake of CO_2 in cyanobacteria is mediated by a membrane-located CO_2 transport mechanism. Diffusion of CO_2 across the cytoplasmic membrane, and subsequent energy-dependent conversion and release of bicarbonate, might account for the reported saturation kinetics and apparent uphill transport of CO_2 (Fridlyand et al. 1995).

Ogawa (1993) has recently isolated a mutant of *Synechocystis* PCC 6803, SC, impaired in CO_2 uptake but not in that of bicarbonate. These data provided supporting evidence that uptake of the two Ci species is mediated by two distinct systems. In this study we isolated new high-CO_2-requiring mutants of *Synechococcus* sp. PCC 7942 with the aid of an "inactivation library". Analysis of three of them indicated that they are unable to raise the HCO_3^- uptake capability upon exposure to low CO_2.

Results

The mutants isolated and partly characterized in this study were obtained with the aid of an inactivation library by a modification of the technique of Dolganov and Grossman (1993, see Fig. 1). The library was constructed by a complete digestion of the genomic DNA with *Taq*I and subsequent ligation of the small (100-400 bp) fragments within the *Acc*I site of a modified Bluescript SK plasmid. In the latter, we inactivated the gene conferring ampicillin resistance by the insertion of a cartridge encoding kanamycin resistance (Kan^r) within the *Sca*I site. Since the genomic fragments ligated in the vector were short and in most cases did not cover the entire length of a relevant gene, single crossover events between the genomic DNA and the homologous fragment in the library inactivated different genes. *Synechococcus* sp. PCC 7942 cells were transfected with the library and grown in the presence of kanamycin. The Kan^r cells were exposed to the desired selection conditions (e.g. low CO_2) for 5 hours of adaptation followed by ampicillin (400 μg/ml) treatment for 12 hours. Cells capable of adapting to low CO_2 were eliminated by this treatment. Mutants which were unable to divide under the selection conditions, survived and were isolated following transfer of the culture to high CO_2 concentration in the absence of ampicillin. The ampicillin-enrichment procedure was also applied for the isolation of high-CO_2-requiring mutants obtained following chemical mutagenesis (Marcus et al. 1986). Basically, this procedure may be applied for the isolation of various mutants and identification of the mutated genes, if the host is capable of inserting heterologous DNA by a single event of cross-over recombination and the selection by the ampicillin treatment, i.e. cessation of growth of the mutant under different selection conditions, is applicable.

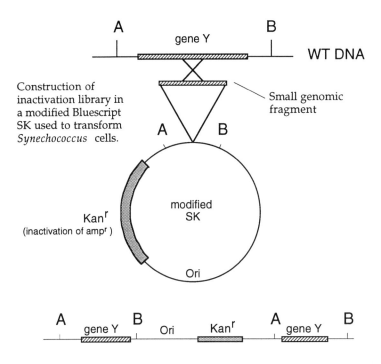

Figure 1. Schematic presentation of the procedure used to isolate mutants of *Synechococcus* sp. PCC 7942 where genes (designated Y) involved in the ability to grow under low CO_2 concentration) were inactivated. The figure also shows the organization of the genomic DNA in the mutants (see text for detailed explanation).

Several genes involved in the ability of *Synechococcus* sp. PCC 7942 to grow in the presence of low CO_2 were identified and mapped in the genomic region of *rbcLS*, encoding the large and small subunits of rubisco (see Kaplan et al. 1994). The high-CO_2-requiring mutants isolated by the technique described here were transfected with DNA bearing the genomic region of *rbcLS*. Those which regained ability to grow under low CO_2 and became sensitive to kanamycin were discarded since they were mutated in the genomic region already characterized (Kaplan et al. 1994; Ronen-Tarazi et al. 1995). The major advantage of the method used here is that the inactivated genes are "tagged" with the vector and Kan[r] sequences. Therefore, it is rather easy to clone them. Figure 1 shows the expected (and verified in Southern analyses) genomic region of the mutants. The (A) and (B) represent restriction sites located on either side of the *Acc*I site where the genomic fragments were inserted, in the vector. The DNA of the mutants were digested by either (A) or (B), typically *Eco*RI or *Apa*I, respectively, and self-ligated. Kan[r] colonies of *E. coli* were isolated following transfection of competent DH5α cells with the newly obtained plasmid. These colonies carried part of the inactivated genes plus the vector sequences bearing the origin of replication and the Kan[r] cartridge. This procedure was used to clone the two ends of the relevant gene in each of the mutants. Obviously, the other fragments obtained following the digestion with (A) or (B) were lost since they did not carry the vector sequences, including the origin of replication and the Kan[r] cartridge, and therefore did not confer growth of *E. coli* in the presence of kanamycin.

Digestion of the plasmid isolated from the bacteria with (A) *and* (B) resulted in the isolation of the genomic fragments flanking the inactivated gene. These fragments were used for northern analyses demonstrating that the relevant genes are transcribed in the wild type. The same fragments were also used for the identification of the relevant clones in the genomic EMBL3 library of *Synechococcus* sp. PCC 7942. It is important to note that in some of the cases we could not clone *Sal*I fragment bearing the relevant inserts from EMBL3. Cloning of these genes from the genomic library was only possible following their inactivation or digestion with an enzyme which cleaved the gene. This finding might explain the failure of the earlier attempts to isolate genes involved in HCO_3^- transport (Yu et al. 1994). Sequence analyses of the genes inactivated in the mutants is in progress. However, the reader should be aware of possible artifacts in the identification of the inactivated gene. In one case, mutant IL-7, the inactivating fragment in the library was composed of two different, unrelated, *Taq*I fragments ligated during the insertion into the *Acc*I site. Consequently, the fragment obtained following the digestion with (A) and (B), as above, contained sequences which hybridized to a clone in the EMBL3 library which was not homologous to the gene inactivated in the mutant.

Several high-CO_2-requiring mutants, unable to grow in the presence of air level of CO_2 (including IL-3, IL-7 and IL-11, presented here) were isolated. Mutants IL-3 and IL-11 were able to grow like the wild-type in the presence of 5% CO_2. The growth curves (not shown) were similar to those observed for other high-CO_2-requiring mutants of *Synechococcus* sp. PCC 7942 (Marcus et al. 1986; Schwarz et al. 1988). Mutant IL-7 grew considerably slower than the wild type even under 5 % CO_2 (not shown). Analyses of the rate of photosynthesis as a function of Ci indicated very low affinity for ambient Ci in the mutants. The apparent photosynthetic $K_{0.5}$ (Ci) exhibited by the mutants was 10-20 mM, similar to those observed in several other high-CO_2-requiring mutants of *Synechococcus* sp. PCC 7942 of the type bearing aberrant carboxysomes (Marcus et al. 1986; Schwarz et al. 1988; Friedberg et al. 1989). When grown in the presence of high CO_2, some of the mutants exhibited normal ability to take up CO_2. However, when the mutants were exposed to low CO_2 conditions, they were unable to adapt to the new conditions by raising the uptake of bicarbonate, as does the wild type (Table 1). These data suggested that the mutants were either impaired in the process of adaptation to low CO_2 or in the ability to transfer HCO_3^- and accumulate Ci within the cells.

| | CO₂ Uptake | | HCO₃⁻ Uptake | |
	High CO₂	Low CO₂	High CO₂	Low CO₂
WT	31.6	53.9	30.9	182
IL-3	16.8	38.9	30.6	43.1
IL-7	18.5	30.5	28.0	41.1
IL-11	20.3	37.4	27.3	48.8

Table 1
Uptake of CO_2 and of HCO_3^- in *Synechococcus* sp. PCC 7942 and the mutants thereof. The cells were grown under high CO_2, and than some of them were transferred to low CO_2 for 12 hours. The cells were aerated with air in the light and then supplied with either ^{14}C-CO_2 or ^{14}C-HCO_3^- (provided as a tracer with minimal effect on Ci concentration in the medium) for 5 sec. Data are presented as µmole Ci within the cells$*$mg^{-1} Chl$*$h^{-1}.

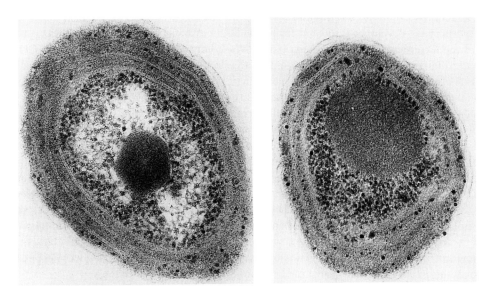

Figure 2
Electronmicrographs of *Synechococcus* sp. PCC 7942 (left) and mutant IL-3 (right). Note the larger carboxysome in the mutant.

High-CO_2-requiring mutants impaired in adaptation are expected to exhibit photosynthetic characteristics similar to those of high-CO_2-grown wild-type. This was the case in mutant D4 where *purK*, essential for purine biosynthesis in the presence of low CO_2, was inactivated (Schwarz et al. 1992). The mutants isolated in the present study, on the other hand, exhibited 50- to 100-fold lower apparent photosynthetic affinity than the wild type suggesting that they are impaired in HCO_3^- uptake or its accumulation.

The electronmicrographs (Fig. 2) indicated that some of the mutants (mutant IL-3) contained very large carboxysomes while others lacked visible carboxysomes (mutant IL-11, not shown). The carboxysomes in mutant IL-7 were normal even though the cells were considerably longer than the wild type (not shown). The low apparent photosynthetic affinity for Ci, exhibited by the earlier isolated carboxysome-defective mutants has been attributed to the low state of activation of rubisco when the cells were exposed to low concentration of CO_2 (Schwarz et al. 1995). It is important to note, however, that Ci uptake, where reported, was normal in the carboxysome-defective mutants characterized before, unlike the case in mutants IL-3 IL-7 and IL-11.

Conclusions

High-CO_2-requiring mutants are being used to study the adaptive response of cyanobacteria to changing ambient CO_2 concentration and to identify and characterized genes involved in the operation of the CCM and CO_2-dependent genes. In this study we developed a method for the isolation of novel CO_2-dependent mutants and for the identification of the relevant genes. We briefly presented some of the high-CO_2-requiring mutants obtained which are defective in the ability to accumulate Ci internally when the cells are supplied with bicarbonate. It is not known whether the smaller accumulation of Ci in the mutants was due to a defective uptake mechanism. Carbonic anhydrase activity, released from the aberrant carboxysomes could also result in the observed phenotype. Finally, though the apparent photosynthetic affinity for Ci was low, we can not rule out the possibility that some of these mutants are defective in the process of adaptation to low CO_2. This is the case in the high-CO_2-requiring mutant, O221, which does not contain carboxysomes. Substitution of a G with an A in *ccmN*, in mutant O221, resulted in inability to induce transcription from *cmpA* under low CO_2. Therefore, O221 does not accumulate the 42 kDa polypeptide following exposure to low CO_2 (see Kaplan et al. 1994). The data presented here provided strong evidence that uptake of CO_2 and of HCO_3^- is mediated by means of two distinct mechanisms. Northern analyses demonstrated that the relevant genes inactivated in the mutants are transcribed. Sequence analyses, in progress, will enable further characterization of these genes.

References

Badger MR and Price GD (1992) Physiol Plant 84, 606-615.
Bedu S et al (1992) In Murata N, ed, Research in Photosynthesis pp 819-822, Kluwer Academic Publishers Dordrecht, The Netherlands.
Dolganov N and Grossman AR (1993) J Bacteriol 175, 7644-7651.
Fridlyand L et al (1995) BioSystems (in press).

Friedberg D et al (1989) J Bacteriol 171, 6069-6076.

Kaplan A et al (1991) Plant Physiol 97, 851-855.

Kaplan A et al (1994) In Bryant DA, ed, The Molecular Biology of the Cyanobacteria, pp 469-485, Kluwer Academic Publishers Dordrecht, The Netherlands.

Marcus Y et al (1986) Plant Physiol 82, 610-612.

Miller AG et al (1990) Can J Bot 68, 1291-1302.

Ogawa T (1993) In Yamamoto H and Smith S, eds, Photosynthetic Responses to the Environment, pp 113-125, American Society of Plant Physiologist Series, Rockville, MD.

Price GD et al (1993) J Bacteriol 175, 2871-2879.

Raven JA (1991) Can J Bot 69, 908-924.

Ronen-Tarazi M et al. (1995) Plant Physiol (in press).

Schwarz R et al (1988) Plant Physiol 88, 284-288.

Schwarz R et al (1992) Plant Physiol 100, 1987-1993.

Schwarz R et al (1995) Plant Physiol 108, 183-190.

Yu J-W et al (1994) Plant Physiol 104, 605-611.

This research was supported by grants from the USA-Israel Binational Science Foundation (BSF), Jerusalem, Israel and the New Energy and Industrial Technology Development Organization (NEDO), Japan.

CARBOXYSOMES: THE GENES OF *THIOBACILLUS NEAPOLITANUS*

Jessup M. Shively, Stanley C. Lorbach, Songmo Jin, Stefanie H. Baker
Department of Biological Sciences, Clemson University, Clemson,
SC 29634- 1903 USA

INTRODUCTION

Polyhedral inclusion bodies have been observed in all of the cyanobacteria thus far examined and in many, but not all, chemoautotrophic bacteria (Codd 1988, Shively 1974, Shively et al. 1988, Shively,English 1991). The bodies were first isolated from *Thiobacillus neapolitanus* and shown to consist of a monolayer shell surrounding multiple molecules of the enzyme ribulose bisphosphate carboxylase/oxygenase, RuBisCO (Shively et al. 1973a,b). The inclusions were subsequently named carboxysomes. Presumably, all of the polyhedral inclusion bodies of similar appearance are carboxysomes. To date, evidence has been gathered which identifies the bodies of *Nitrosomonas* sp., several *Nitrobacter* sp., *Anabaena cylindrica, Chlorogleopsis fritschii, Anacystis nidulans, Synechococcus* sp., *Synechocystis* sp., *Prochloron*, and *Prochlorothrix* as carboxysomes (Shively,English 1991).

In addition to *T. neapolitanus*, carboxysomes have been isolated from *Chlorogleopsis fritschii, Nitrobacter* species, and *Synechococcus* PCC7942 (Biederman,Westphal 1979, Lanaras,Codd 1981, Price et al. 1992, Shively et al. 1977). The carboxysomes of *T. neapolitanus* have been purified to apparent homogeniety in two separate laboratories (Cannon,Shively 1983, Holthuijzen et al. 1986). Based on these data the carboxysomes of *T. neapolitanus* are reportedly composed of eight major peptides with M_r x 1000, as determined by SDS-PAGE, of 120, 85, 56, 54, 36, 15, 13, and 10. The 56,000 and 10,000 peptides are the large (CbbL) and small (CbbS) subunits of RuBisCO, respectively (Snead,Shively 1978). The peptides with M_r x 1000 of 120, 85, 54, 15, and 10 are glycopeptides and appear to be components of the shell. The function of the 36, 000 peptide is unknown.

Initially most researchers believed that the carboxysome represented an enzyme storage body (Codd 1988, Shively,English 1991). However, over the years a wealth of evidence has been gathered which supports the theory that the RuBisCO of the carboxysome is active in CO_2 fixation (Beudeker et al, 1980, Beudeker et al. 1981, Codd 1988, Shively,English 1991, Turpin et al. 1984). The most compelling evidence has been recently garnered by genetic analysis using *Synechococcus* and *T. neapolitanus* (English et al. 1994 1995, Friedberg et al. 1989, Lieman-Herwitz et al. 1991, Marco et al. 1994, Price,Badger 1989 1991, Price et al. 1992 1993). Several putative carboxysome genes have been located in *Synechococcus* upstream of the gene (*cbbL*) for the RuBisCO large subunit (Friedberg et al. 1989, Marco et al. 1994, Price, Badger 1989 1991, Price et al. 1992 1993). Mutation of these "CO_2 concentrating mechanism" genes, labeled *ccmK*,

M. E. Lidstrom and F. R. Tabita (eds.), Microbial Growth on C₁ Compounds, 56–63.

ccmL, ccmM, ccmN, and *ccmO,* results in either a complete loss of carboxysomes or the formation of aberrant carboxysome structures. Furthermore, and most importantly, the mutants require CO_2 supplementation for growth. Unfortunately, to date, not one of these genes has been shown to code for a particular *Synechococcus* carboxysome peptide. Recently, we purified one of the major Carboxy$\underline{SO}$me Shell peptides, M_r 10,000, from *T. neapolitanus* and accomplished N-terminal sequencing. The gene for this peptide was located and isolated using the technique of reverse genetics (English et al. 1994)). This gene, labeled *csoS1A,* codes for a 98 amino acid, 9.9 kDa, highly hydrophobic peptide. Sequence analysis revealed an ORF immediately upstream and one immediately downstream of *csoS1A.* These ORFs, labeled *csoS1C* and *csoS1B,* respectively, are repeats of *csoS1A* (22). The deduced amino acid sequences of CsoS1A and CsoS1C differ in only the amino acid at position #3. The deduced sequence of CsoS1B possesses 12 additional amino acids at the C-terminus in comparison to CsoS1A and CsoS1C; the rest of the peptide is 90% identical to CsoS1A and CsoS1C. This was the first correlation of a carboxysome peptide with its gene. Insertional inactivation of *csoS1A* with a Km[r] gene cartridge resulted in a mutant which possesses a drastically reduced number of carboxysomes and which requires CO_2 supplementation for growth (English et al. 1995). Interestingly, the deduced amino acid sequence of CsoS1A shows significant homology to the peptides coded for by the *ccmK* and *ccmO* of *Synechococcus* further substantiating the possibility that these genes might code for carboxysome peptides (English et al. 1994). Thus, it is now widely accepted that the carboxysome is a primitive CO_2 fixing organelle.

Herein, we report on the identification, organization, and location of carboxysome genes in *T. neapolitanus* ATCC 23641. In addition, the *csoS1* genes are expressed in *Escherichia coli* and the presence of *csoS1*-like genes is demonstrated in other autotrophic prokaryotes.

RESULTS

The location of csoS1A relative to cbbLS. Screening a λEMBL3 library of *T. neapolitanus* ATCC 23641 genomic DNA with the *cbbL* from *Anacystis nidulans* resulted in the isolation of 10 positive clones (Shively et al. 1986). Screening of these clones with *csoS1A* yielded three positive clones (English et al., 1994). One of these clones, designated λTnI, has been restriction mapped and sequenced (English et al., 1994). The genes, *csoS1C. csoS1A,* and *csoS1B* were determined, in this order, to be located approximately 5 kb downstream of *cbbLS.* Based on nucleotide sequence the molecular weights of CbbL and CbbS are 52.6 and 12.9 kDa, respectively. Four other ORFs were noted between *cbbS* and *csoS1C.* All of the genes, *cbbLS,* the four ORFs, *csoS1C, csoS1A, and csoS1B* are oriented in the same direction and are preceded by putative ribosome binding sites. A putative promoter and a potential termination hairpin loop/AT-rich region is present immediately upstream of *cbbL* and immediately downstream of *csoS1B,* respectively.

The identification of genes for additional carboxysome peptides. Carboxysomes were purified as previously reported and subjected to SDS-PAGE (Cannon,Shively 1983. Eight major peptide bands were revealed (Fig. 1). Minor bands which vary in size and quantity from preparation to preparation are thought to be contaminants, i.e. not carboxysome components. Although the relative position of the peptide bands on the gel is essentially the same as previously demonstrated, the M_r are somewhat different. We are now using a improved electrophoresis system, ultra-pure reagents, and Broad range molecular weight markers (BioRad). It is essential that one keep in mind that many of the peptides are glycosylated; peptide migration on SDS-PAGE gels can be markedly altered

58

by glycosylation. Two additional shell peptides, M_r 15,000 and 80,000, were isolated and N-terminal sequenced (English,Shively 1994). The N-terminal sequences of these two peptides perfectly matched the amino acid sequences deduced from the nucleotide sequences of *csoS1B* and the ORF residing just below cbbS, respectively. Since the peptide from this ORF has been reported to be part of the shell, the gene was subsequently

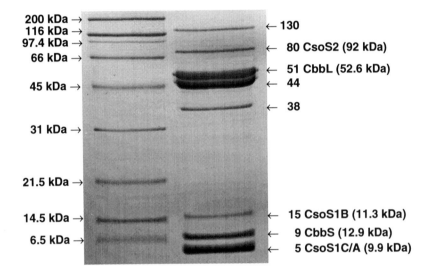

Fig. 1. Computer generated scan 15% SDS-PAGE gel of purified carboxysome of *T. neapolitnaus*. Left panel:BioRad broad range molecular weight markers. Right panel: carboxysomes. Molecular weights of standards are listed at left. M_r of carboxysome peptides are listed immediately to the right. Peptides for genes and their deduced molecular weight (parantheses) are listed to the far right.

labeled *csoS2*. The molecular weights of *csoS1B* and *csoS2* deduced from the nucleotide sequences are 11.3 and 92 kDa, respectively.

The other, currently unidentified ORFs, are labeled ORF1, ORF2A and ORF2B. ORF 1 is 1500 bp in length and codes for a peptide with a deduced molecular weight of 51 kDa. ORF 2A and 2B are gene repeats with 252 and 246 bp, respectively.

The organization of carboxysome genes in T. neapolitanus. The organization of the carboxysome genes of *T. neapolitanus* relative to the putative carboxysome genes in *Synechococcus* is shown in Fig. 2. Although ORF1, 2A, and 2B have not been identified it seems likely that they are carboxysome genes.

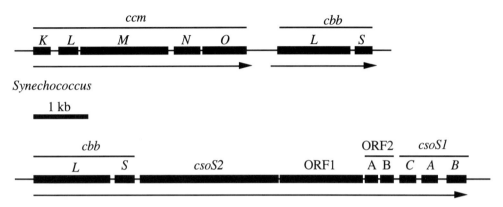

Fig. 2. Organization of carboxysome genes in *T. neapolitanus* and *Synechococcus* PCC7942. Gene sizes are only approximate. The *Synechococcus* organization is re drawn using published information plus data from this laboratory. Arrows indicate direction of transcription. The arrows also indicate possible operons.

The expression of csoS1C, csoS1A, and csoS1B. The c*soS1A, csoS1B,* and *csoS1C* genes carried on a single restriction fragment ligated into pT7-6 were expressed in *E. coli* K38. The expression system utilizes a temperature sensitive inducible promoter for the T7 polymerase gene which directs transcription from the T7 promoter upstream from the gene of interest (Tabor 1990). The incorporation of ^{35}S-Methionine during the induction process allows the preferential tagging of the cloned protein product. Autoradiography of the SDS-PAGE gel gives clean and definitive identification of the cloned gene product. The upper band in Fig. 3 is the product of *csoS1B* and the lower band, the products of both *csoS1A* and *csoS1C*; they are essentially identical in size. It must be noted that the migration of the peptides from *T. neapolitanus* are not identical to those expressed in *E. coli;* the peptides are glycosylated in *T. neapolitnaus.* Thus, the M_r of the peptides in *E. coli* more accurately represents their molecular weight.

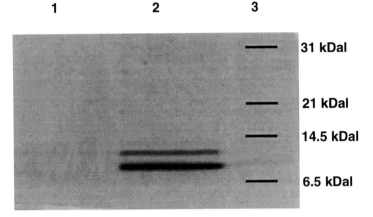

Fig. 3. Computer scan of autoradiogram of SDS-PAGE gel of the peptides expressed from the *csoS1A, csoS1B,* and *csoS1C* genes in *E. coli*. Left lane: control. Middle lane: expression; top band, CsoS1B; bottom band, CsoS1A and CsoS1C. Right lane: molecular weight markers. See text for details.

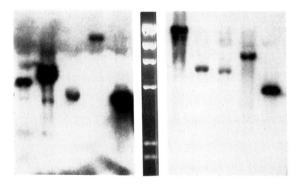

Fig. 4. Autoradiograms of Southern blots of *Eco*RI digested genomic DNAs from various thiobacilli and nitrifying bacteria probed with *csoS1A*. Middle lane: EtBr stained agarose gel of *Hind*III digested λDNA. Left panel: thiobacilli from left to right, *T. intermedius, T. thiooxidans, T. denitrificans, T. ferrooxidans,* and the control, *T. neapolitanus*. Right panel: nitrifying bacteria from left to right, *N. halophila, Nb. vulgaris, Nb. hamburgensis, Nb. winogradskyi,* and the control *T. neapolitanus*. The lower part of the gels were removed since no positive signals were visible.

Presence of csoS1A-like genes in other autotrophs (Fig. 4). Positive hybridization was obtained when Southern blots of genomic DNAs of *Thiobacillus denitrificans* ATCC 25259, *Thiobacillus ferrooxidans* ATCC 23270, *Thiobacillus intermedius* K12, and *Thiobacillus thiooxidans* ATCC 8085 were probed with *csoS1A* (Shively et al. 1986). Results with *Thiobacillus versutus* ATCC 25364 were negative (data not shown). Positive signals were obtained with the nitrifying bacteria, *Nitrosomonas halophila*, *Nitrobacter hamburgensis, Nitrobacter vulgaris*, and *Nitrobacter winogradskyi* (Fig. 4) Several cyanobacteria, *Cyanothece, Microcystis, Nostoc, Spirulina, Synechococcus*, and *Synechocystis* gave positive hybridization, but results were negative with *Anabaena* and *Fremyella* (data not shown). The strongest signal was obtained with *Synechococcus*. As expected, restriction fragments containing *ccmK* and *ccmO* hybridized with *csoS1A*.

DISCUSSION

If one includes both CsoS1A and CsoS1C, the carboxysomes of *T. neapolitanus* are composed of nine major peptides. Six of these peptides are hypothesized to be components of the shell and two are the CbbL and CbbS of RuBisCO. The function of the other peptide (M_r 38,000-See Fig. 3) is unknown. The genes for six (counting both *csoS1A* and *csoS1C*) have now been isolated and located on the *T. neapolitanus* chromosome. Preliminary evidence suggests that the peptide with an M_r of 44,000 is coded for by ORF1. Also, the 130,000 peptide might be produced by a "read through" of the stop codon for *csoS2* ending at a stop codon in ORF1. If these facts regarding peptides and genes prove to be the case only the functions of the 38,000 peptide, and ORF 2A and 2B remain to be clarified. Experiments dealing with the three peptides and the three ORFs are in progress.

All of the genes are oriented in the same direction and each of the genes (or ORF) is preceded by a ribosome binding site. A putative promoter is present just above *cbbL* and a termination sequence is present just below *csoS1B*. Thus, a carboxysome operon is hypothesized. An identical carboxysome cluster has been identified in and isolated from *T. intermedius* (Stoner,Shively 1993, Shively, unpublished). The carboxysome cluster of *Synechococcus* is somewhat different from the one in *T. neapolitanus* and *T. intermedius* (Friedberg et al. 1989, Marco et al. 1994, Price, Badger 1989 1991, Price et al. 1992 1993). See Fig. 3. In particular, in *Synechococcus,* the cluster is located above *cbbLS*, not below, the organization of the small molecular weight shell genes appears to be different, and the *cbbLS* aren't part of the operon. Although *T. denitrificans* does not possess carboxysomes it does have a *csoS1A*-like gene (see below). Preliminary evidence suggests that this gene is located close to the RuBisCO genes, but the organization will be different (English et al. 1992). The organization must also be different in *Nb. vulgaris*; other Calvin cycle genes are located immediately downstream of cbbLS in this organism (Strecker et al. 1994). The organization in *Synechocystis* may also be different (Ogawa et al. 1994).

The three CsoS1 peptides were produced in *E. coli* from a gene cluster of *csoS1C*, *csoS1A*, and the *csoS1B*. Obviously, the ribosome binding sites of *T. neapolitnaus* are functioning in *E. coli*. The migration of the peptides in SDS-PAGE gels closely mirror their deduced molecular weights supporting the observation that the peptides are post-translationally modified in *T. neapolitanus*.

Since the vast majority of data suggests that the carboxysome is a well conserved structure, we decided to determine if *csoS1A*-like genes could be detected in other carboxysome formers. The gene, *csoS1A* hybridized with the genomic DNAs of four species of *Thiobacillus* including our initially presumed negative control, *T. denitrificans*. Some time ago we accomplished extensive electron microscopy of both aerobically and anaerobically grown *T. denitrificans*; no carboxysomes were ever observed (Shively et al. 1970). Therefore, we can conclude that *T. denitrificans* either lacks some genetic components necessary to produce carboxysomes or is unable to produce carboxysomes under the growth conditions we employed. Experiments to explain these results are in progress. The results of probing the DNA of a second non-carboxysome former, *T. versutus* were negative as expected. *N. halophila* and three *Nitrobacter* species all yielded positive results. Although we previously demonstrated by sequence analysis that CsoS1A has significant homology with CcmK and CcmO of *Synechococcus* (English et al. 1994), it seemed appropriate to undertake detection of the two *Synechococcus* genes using *csoS1A* as heterologous probe. Both *ccmK* and *ccmO* were detected by hybridization. In addition to two *Synechococcus* sp., six other cyanobacterial genomic DNAs yielded positive hybridization signals. The DNAs of three *Anabaena* cultures and *Fremyella* gave negative results. The large genome size of the organisms in these two genera probably was the major factor contributing to this result, i.e. a considerably larger amount of DNA would need to be blotted in order to provide adequate copies of the gene to be detected.

Obviously, much remains to be elucidated regarding carboxysome occurrence, composition, structure, assembly, and function.

CONCLUSIONS

The carboxysomes of *T. neapolitanus* appear to be composed of nine major peptides. The genes for most of these peptides are organized into what appears to be a "carboxysome operon." The ribosome binding sites for the *csoS1* genes are functional in *E. coli*. Our data are consistent with the theory that *csoS1A* is a well conserved gene and suggest, along with previously reported data, that carboxysomes, from a variety of autotrophs, are structurally and functionally essentially the same.

ACKNOWLEDGMENTS

We are indebted to the following people who supplied genomic/cloned DNA and cultures: Prof. E. Bock, Hamburg Univ. Germany; Dr. D. Bryant, Pennsylvania State Univ.; Dr. G. Cannon, Univ. of Southern Mississippi; Dr. G. Codd, Univ. of Dundee, Scotland; Dr. S. Haegelin, Texas A&M Univ.; Dr. P Koops, Hamburg Univ., Germany; and Dr. L. Sherman, Purdue Univ. We thank Sally Brock, Dept. of Biol. Sci., Clemson Univ., for assistance with figure preparation. This material is based upon work supported by the Cooperative State Research Service, U.S. Department of Agriculture, under Agreement No. 92-37306-7663.

REFERENCES

Beudeker RF et al. (1980) Arch. Microbiol. 124,185-189.
Beudeker RF et al. (1981) Arch. Microbiol. 129,361-367.
Biedermann M, Westphal K (1979) Arch. Microbiol. 121,187-191.
Cannon GC, Shively JM (1983) Arch. Microbiol. 134,52-59.
Codd GA (1988) Adv. in Microbial Physiol. 29,115-164.
English RS et al. (1992) FEMS Microbiol. Lett. 64,111-120.
English RS et al. (1994) Molecular Microbiol. 12,647-654.
English RS et al. (1995) Appl. Environ. Microbiol. 61, In press.
Friedberg D et al. (1989) J. Bacteriol. 171,6069-6076.
Holthuijzen YA et al. (1986) Arch. Microbiol. 144,398-404.
Lanaras T, Codd GA (1981) Planta 153, 279-285.

Lieman-Herwitz J et al. (1991) Can. J. Bot. 69,945-950.
Marco E et al. (1994) Appl. Environ. Microbiol. 60,1018-1020.
Ogawa T. et al. (1994) J. Bacteriol. 176,2374-2378
Price GD, Badger MR (1989) Plant Physiol. 91,514-525.
Price GD, Badger MR (1991) Can. J. Bot. 69,967-973.
Price GD et al. (1992) Plant Physiol. 100,784-793.
Price GD et al. (1993) 175,2871-2879.
Shively JM et al. (1970) J. Bacteriol. 96:2138-2143.
Shively JM et al. (1973a) Science 182,584-586.
Shively JM et al. (1973b) J. Bacteriol. 116,1405-1411.
Shively JM (1974) Ann. Rev. Microbiol. 28,167-187.
Shively JM et al. (1977) J. Bacteriol. 132,673-675.
Shively JM et al. (1986) FEMS Microbiol. Lett. 37:251-257.
Shively JM et al. (1988) Intl. Rev. Cytol. 113,35-100.
Shively JM, English RS (1991) Can. J. Bot. 69,957-962.
Snead RM, Shively JM (1978) Current Microbiol. 1,309-314.
Tabor S (1990) In Ausubel F, Brent R, Kingston R, Moore D, Seidman J, Smith J and
    Struhl K, eds, Current Protocols in Molecular Biology, Unit 16.2.1, Greene
    Publishing Associates and Wiley Interscience, New York.
Stoner MT, Shively JM (1993) FEMS Microbiol. Lett. 107,287-292
Strecker M et al. (1994) Arch. Microbiol. 120,45-50.
Turpin DH et al. (1984) J. Phycol. 20,249-253.

Analysis of genes in the pathway for the fermentation of acetate to methane by *Methanosarcina thermophila*

J. G. Ferry and J. A. Maupin-Furlow

Department of Biochemistry and Molecular Biology, Pennsylvania State University, University Park, Pennsylvania, 16802-4500, USA.

INTRODUCTION

Most acetotrophic anaerobes from the *"Bacteria"* domain cleave acetate and then oxidize the methyl and carbonyl groups to CO_2 and reduce exogenous electron acceptors. Acetate-utilizing methane-producing *"Archaea"* also cleave acetate; however, the carbonyl group is oxidized to CO_2 to provide an electron pair for reduction of the methyl group to CH_4. The phylogenetic extremes between the two domains raises questions concerning the comparative genetics and biochemistry of acetate utilization in diverse anaerobes.

Figure 1 outlines the pathway for the methanogenic fermentation of acetate in *Methanosarcina thermophila* which is similar to the pathway in other species of *Methanosarcina* and the genus *Methanothrix* (Ferry 1993). In the first step, acetate is activated to acetyl-CoA by the activities of acetate kinase and phosphotransacetylase. Cleavage of the C-C and C-S bonds (decarbonylation) of the acetyl group is catalyzed by the nickel/iron-sulfur (Ni/Fe-S) component of the CODH (CO dehydrogenase) enzyme complex. The Ni/Fe-S component also oxidizes the carbonyl group to CO_2 and reduces ferredoxin. The methyl group is transferred to the corrinoid/iron-sulfur (C/Fe-S) component of the CODH complex which transfers the methyl group to H_4SPT (tetrahydrosarcinapterin). Coenzyme M (HS-CoM) is methylated by a membrane-bound $H_4SPT:HSCoM$ methyltransferase complex. A methylreductase reduces the methyl group of CH_3-S-CoM to CH_4 with electrons derived from the sulfur atoms of CH_3-S-CoM and HS-CoB forming CoM-S-S-CoB. The heterodisulfide is reduced to the active sulfhydryl forms of the cofactors with electrons ultimately derived from ferredoxin. Electron transport from ferredoxin to the heterodisulfide reductase involves membrane-bound carriers which include cytochrome b. It is proposed that energy is conserved by phosphorylation coupled to electron transport. Acetate-grown cells contain a carbonic anhydrase CA which is hypothesized to be located outside the plasma membrane where it facilitates removal of CO_2 by conversion to HCO_3^-. A recent review (Ferry 1993) details the pathway for the methanogenic fermentation of acetate.

M. E. Lidstrom and F. R. Tabita (eds.), Microbial Growth on C₁ Compounds, 64–71.
© *1996 Kluwer Academic Publishers. Printed in the Netherlands.*

RESULTS

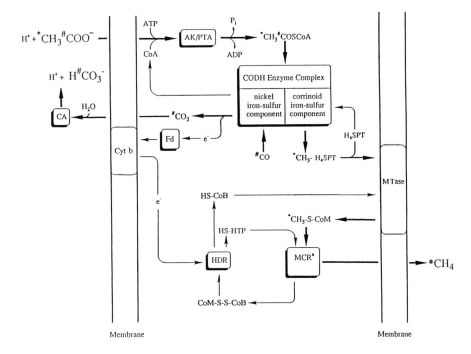

Figure 1. Proposed pathway for the conversion of acetate to CO_2 and CH_4 in *M. thermophila*. AK, acetate kinase; PTA, phosphotransacetylase; H_4SPT, tetrahydrosarcinapterin; CA, carbonic anhydrase; MCR, methylreductase; HDR, heterodisulfide (CoM-S-S-CoB) reductase; Fd, ferredoxin; CODH, carbon monoxide dehydrogenase; Cyt b, cytochrome b; MTase, methyltransferase system. The carbon atoms of acetate are marked with a (*) and a (#) symbol to distinguish between the carboxyl and methyl groups.

Acetate Activation to Acetyl-CoA. The genes (*pta* and *ack*) encoding the 35-kDa monomeric phosphotransacetylase and 90-kDa homodimeric acetate kinase are arranged in an operon with *pta* upstream of *ack* (Latimer, Ferry 1993). The activities of both enzymes are elevated 8- to 11-fold in acetate-grown *M. thermophila* compared to cells grown on methanol, monomethylamine, dimethylamine or trimethylamine (Singh-Wissmann, Ferry 1995). Northern blot analysis shows that this regulation of enzyme synthesis is at the mRNA level. The transcriptional start site is located 27 bp upstream from the translational start of the *pta* gene and 24 bp downstream from a consensus archaeal box A promoter sequence. S1 nuclease protection assays detect transcripts with four different 3' ends, each of which map to the beginning of four consecutive direct repeats downstream of *ack*. Northern analysis with an *ack*-specific probe detects both a 2.4-kb polycistronic transcript and a smaller 1.4-kb transcript which is the estimated size of monocistronic *ack* mRNA; however, *pta*-specific probes detect only the 2.4-kb mRNA. The 5′ end of the primer extension product detected with an *ack*-specific primer is in the intergenic region between the *pta* and *ack* genes but does not follow a consensus archaeal box A sequence. The results suggest processing of the 2.4-kb transcript with greater stability of the 1.4-kb *ack*-specific over the *pta*-specific species which implies an increased expression of *ack* relative to *pta*. Although not investigated, processing of this type could adjust the relative amounts of the gene products to coordinate the activities of the two enzymes. The turnover number for the monomeric phosphotransacetylase is greater than the homodimeric acetate kinase; thus, the two enzyme activities could be coordinated if the *ack* gene product were synthesized at a level greater than the *pta* gene product.

Comparative analysis of the sequences deduced from the *M. thermophila* and *Escherichia coli ack* genes show 44% amino acid identity and several conserved arginine, cysteine, and glutamic acid residues. Site-specific alteration of the *M. thermophila* acetate kinase heterologously-produced in *E. coli* identifies a glutamate residue essential for catalysis (Singh-Wissmann, Ferry unpublished), a result consistent with the proposed formation of a glutamyl phosphate intermediate during catalysis (Todhunter, Purich 1974). Comparative analysis of the sequences deduced from the *M. thermophila* and *E. coli pta* genes indicate 43% amino acid identity with 332 residues in the C-terminal half of the 72-kDa monomeric *E. coli* phosphotransacetylase (Matsuyama *et al.* 1994). Site-specific alteration of the *M. thermophila* phosphotransacetylase heterologously-produced in *E. coli* identifies a cysteine residue essential for catalysis (Rasche, Ferry unpublished). A mechanism has been proposed in which a cysteine functions to extract a proton and stabilize ⁻SCoA for nucleophilic attack on acetyl phosphate (Henkin, Ables 1976).

Cleavage of Acetyl-CoA. CODH is central to the pathway for the fermentation of acetate in the methanogenic *Archaea*. CODHs are wide spread among anaerobes from both the *Archaea* and *Bacteria* domains and function in a variety of energy-yielding pathways. Comparative studies of CODHs from these widely divergent organisms will enhance an understanding of the mechanism and molecular evolution of this universal enzyme. Much of our understanding of acetate cleavage in the methanogenic *Archaea* is based on previous knowledge of the CODH (acetyl-CoA synthase) from the homoacetogen *Clostridium thermoaceticum* which belongs to the *Bacteria* domain (Ragsdale, this volume). In the energy-yielding pathway of this anaerobe, one CO_2 is reduced to the methyl level forming CH_3-THF (methyl-tetrahydrofolate) and another CO_2 is reduced to enzyme-bound CO by the CODH. The methyl group of CH_3-THF is transferred to a corrinoid/iron-sulfur (C/Fe-S) protein which then transfers the methyl group to the CODH which catalyzes acetyl-CoA synthesis from the CODH-bound methyl, CO, and CoA moieties.

Because the enzyme complex from *M. thermophila* oxidizes CO to CO_2, it is commonly referred to as the CODH enzyme complex; however, the primary function is cleavage of acetyl-CoA during growth on acetate. The five-subunit CODH complex can be resolved into two enzyme components: the Ni/Fe-S component which contains 89,000-Da (α) and 19,000-Da (ϵ) subunits, and the C/Fe-S component containing 60,000-Da (γ) and 58,000-Da (δ) subunits. The 71,000-Da (β) subunit of the complex has not been characterized. The Ni/Fe-S component has a Ni-Fe-C center with spectroscopic properties nearly identical to the Ni-Fe-C center of the CODH from *C. thermoaceticum*. The Ni-Fe-C center in the *C. thermoaceticum* enzyme is the proposed site for synthesis of the acetyl moiety of acetyl-CoA; thus, the Ni/Fe-S component from *M. thermophila* is the proposed site for cleavage of the acetyl moiety of acetyl-CoA. The Ni/Fe-S component contains three EPR-recognizable species of Fe-S clusters with properties similar to the Fe-S clusters in the *C. thermoaceticum* CODH. Two of the Fe-S centers in the *M. thermophila* Ni/Fe-S component have EPR spectra typical of bacterial-like 4Fe-4S centers; however, the third center is atypical. The function of these 4Fe-4S centers is unknown but are likely to be involved in the transfer of electrons to ferredoxin or the C/Fe-S component. The C/Fe-S component of the CODH enzyme complex from *M. thermophila* contains a corrinoid cofactor (factor III) in the "base-off" configuration. This configuration facilitates the cobalt atom to the Co^{1+} state which is necessary to accept the methyl group from the Ni/Fe-S component. The C/Fe-S component contains a 4Fe-4S center which is probably involved in electron transfer from the Ni/Fe-S component to the cobalt atom. The EPR spectroscopic properties are similar to the 4Fe-4S center present in the *C. thermoaceticum* C/Fe-S protein. In summary, the biochemical properties of the enzyme components from the *M. thermophila* enzyme complex have a striking resemblance to properties of the *C. thermoaceticum* acetyl-CoA synthase and C/Fe-S protein. The CODH from the methanogenic archaeon *Methanothrix soehngenii* is an $\alpha_2\beta_2$ oligomer composed of subunits with molecular masses of 79,400 and 19,400 Da, and contains Ni and Fe. A recent review (Ferry, 1995) further details the biochemistry of CODHs from acetogenic and methanogenic anaerobes.

The *cdh* genes encoding the five CO dehydrogenase subunits have been cloned and sequenced (Fig. 2) (Maupin-Furlow, Ferry unpublished).

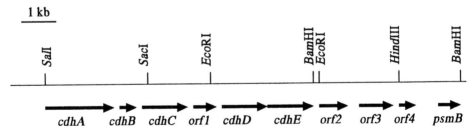

Figure 2. Physical map of *M. thermophila* DNA containing genes of the *cdh* operon (*cdhABCDE, orf1*) and the downstream *psmB* gene.

Northern blot and primer extension analyses indicate that the *cdh* genes are cotranscribed during growth on acetate. The putative CdhA (α) and CdhB (ϵ)

sequences have high identity (40 and 47%) to the sequences deduced from the genes encoding the α and β subunits of the CO dehydrogenase from *M. soehngenii* (Eggen *et al.* 1991). Two 4Fe-4S ferredoxin-like motifs [$CX_2CX_2VC(P/K)$] are conserved in the archaeal CODH α subunits. The CdhA sequence from *M. thermophila* has lower identity (22 and 26%) to the sequences deduced from the genes encoding the Ni-containing CODH from *Rhodospirillum rubrum* (Kerby *et al.* 1992) and the β subunit of the CODH from *C. thermoaceticum* (Morton *et al.* 1991). Several regions are conserved among these proteins including a four-cysteine cluster ($CX_2CX_4CX_9CG$) and a HX_2HX_2H motif in the N-terminus, vicinal cysteines [G(L/V/I)CCT(A/G)X(D/E)(L/V)], and other cysteine clusters [VDXQC(V/I), V(V/L)(G/A)TGC and GSCVXN]. Thus, the Ni and Fe-S centers of the Ni/Fe-S enzyme from *M. thermophila* are apparently located in the α subunit based on the conserved cysteine and histidine regions, as well as the absence of cysteine and presence of only one conserved histidine residue in the ε subunit. The *cdhC* gene encodes the uncharacterized β subunit of the CODH complex. The N-terminal 397 amino acids of CdhC are highly identical to the C-terminus of the large (α) subunit of CODH from *C. thermoaceticum* with both proteins containing a conserved four-cysteine motif ($CX_2CX_8CX_9CGA$). The putative CdhD and CdhE sequences have significant identity with the sequences deduced from the genes encoding the β and α subunits of the C/Fe-S enzyme from *C. thermoaceticum*. The CdhE sequence has a four-cysteine motif ($CX_2CX_4CX_{16}CP$) with the potential to bind a 4Fe-4S cluster previously identified in the C/Fe-S enzymes from *M. thermophila* and *C. thermoaceticum* by EPR spectroscopy.

A T7 RNA polymerase/promoter system has been used to independently produce CdhD and CdhE in *E. coli* (Maupin-Furlow, Ferry unpublished). The purified CdhE protein binds cobalamin in the base-off conformation, is methylated by CH_3-H_4SPT and contains an Fe-S center, results which suggest that this subunit is catalytically active. Furthermore, the results suggest that the C/Fe-S enzyme of *M. thermophila* catalyzes methyl transfer from the Ni/Fe-S enzyme to H_4SPT during growth on acetate; thus, it appears that the CODH complex and H_4SPT are the only components necessary to cleave acetyl-CoA and transfer the methyl group to the membrane-bound CH_3-H_4SPT:CoM methyltransferase. The deduced protein sequence from the open reading frame (*orf1*) located between the *cdhC* and *cdhD* genes (Fig. 2) has no significant identity to known protein sequences and the function is unknown.

Electron Transport. A novel CoM-S-S-CoB reducing system operates in *M. thermophila* during acetate conversion to CH_4 and CO_2 (Peer *et al.* 1994). In this system, a ferredoxin transfers electrons from the CODH complex to membrane-bound electron carriers (including cytochrome *b*) that are required for electron transfer to the heterodisulfide reductase. The gene (*fdxA*) encoding the ferredoxin (Clements, Ferry 1992) reveals a deduced amino acid sequence which contains eight cysteines in a spacing characteristic of 2[4Fe-4S] ferredoxins from the *"Bacteria"* domain and consistent with EPR and biochemical characterization of the *M. thermophila* ferredoxin (Clements *et al.* 1994). An open reading frame with the potential to encode a second 2[4Fe-4S] bacterial-like ferredoxin has been identified (Clements, Ferry 1992). The genes encoding both ferredoxins are present as single copies in the genome and each is transcribed on a monocistronic mRNA. The apparent transcriptional start sites for both genes are 21-28 bases downstream of sequences with high identity to the consensus box A methanogen promoter. The *fdxA* gene is transcribed in cells grown on either acetate, trimethylamine or methanol, a result which suggests a requirement for this ferredoxin in the metabolism of each substrate; however, the gene encoding the second ferredoxin is only transcribed in methanol- and trimethylamine-grown cells suggesting a specific requirement for this ferredoxin during growth on these substrates.

Carbonic Anhydrase. Growth of *M. thermophila* on acetate induces CA activity but the function of this enzyme in the conversion of acetate to methane is only speculative (Jablonski *et al.* 1990). The CA from acetate-grown *M. thermophila* is a 84-kDa enzyme with a subunit molecular mass of 40-kDa. The gene encoding this CA has been cloned and sequenced (Alber, Ferry 1994). Comparison of the deduced amino acid sequence with the N-terminal sequence of the purified protein shows that the gene encodes an additional 34 N-terminal residues with properties characteristic of signal peptides in secretory proteins; thus, location outside the cytoplasmic membrane is proposed (Fig. 1). Hydration of CO_2 to a charged species outside the membrane could facilitate removal of CO_2 from the cytoplasm and improve the thermodynamic efficiency of the pathway. The deduced amino acid sequence has no significant identity to any known CAs; thus, the *M. thermophila* enzyme is the prototype of a novel class of CAs. The deduced sequence does, however, have significant identity with several open reading frames from the *Bacteria* domain for which the function is unknown. The gene encoding CA from *M. thermophila* is hyperexpressed in *E. coli* (Alber, Ferry unpublished). The heterologously-produced enzyme has properties identical to the authentic enzyme isolated from *M. thermophila*; however, a discrepancy in molecular mass suggests that the CA is posttranslationally modified in either *E. coli* or *M. thermophila*.

Proteasome. The sequence immediately downstream of the *cdh* operon (Fig. 2) reveals several open reading frames of unknown function followed by the gene (*psmB*) encoding the β subunit of the 645-kDa proteasome purified from *M. thermophila*. The *psmB* gene is transcribed *in vivo* as a monocistronic message from a consensus archaeal promoter. The proteasome has chymotrypsin-like and peptidylglutamyl-peptide hydrolase activities and contains α (24-kDa) and β (22-kDa) subunits (Maupin-Furlow, Ferry unpublished). The gene (*psmA*) encoding the α subunit resides at a location remote from *psmB*. Processing of both subunits is suggested by comparison of N-terminal sequences with the sequences deduced from *psmA* and *psmB*. Alignment of deduced sequences for the α and β subunits reveals high similarity; however, the N-terminal sequence of the α subunit contains an additional 24-amino acids that are not present in the β subunit. The α and β subunits have high sequence identity with α- and β-type subunits of proteasomes from eukaryotic organisms and the distantly related archaeon *Thermoplasma acidophilum* (Puhler *et al.* 1994). The results suggest that proteasomes are more widespread in prokaryotes than previously proposed and are likely to play a role in protein turnover in methanogenic microbes. Southern blotting experiments suggest the presence of ubiquitin-like genomic sequences in *M. thermophila*; thus, eukaryotic-like degradation of ubiquitinated proteins by a proteasome may be operable in *M. thermophila*.

CONCLUSIONS

The pathway for the fermentation of acetate to methane in *M. thermophila* requires several enzymes that are isofunctional with enzymes present in microbes from the *Bacteria* domain. Comparison of amino acid sequences deduced from the genes, as well as biochemical properties, reveals surprisingly high identity suggesting that these enzymes were highly evolved in the last common ancestor to the *Bacteria* and *Archaea* domains or horizontal gene transfer has occurred. Either possibility has important implications for the evolution of microbes; however, until additional sequences become available, it is not possible to distinguish between these possibilities. In the mean time, comparison of the *Bacterial* and *Archaeal* sequences has revealed conserved residues and domains providing a better understanding of the active sites and helping to guide site-directed mutagenesis to investigate enzyme mechanisms.

REFERENCES

Alber BA, Ferry JG (1994) Proc. Nat. Acad. Sci. USA 91, 6909-6913.
Clements AP, Ferry JG (1992) J. Bacteriol. 174, 5244-5250.
Clements AP *et al.* (1994) J. Bacteriol. 176, 2689-2693.
Eggen RIL *et al.* (1991) J. Biol. Chem. 266, 6883-6887.
Ferry JG (1993) *In* JG Ferry, ed., Methanogenesis, pp. 304-334, Chapman and Hall, New York.
Ferry JG (1995) Ann. Rev. Microbiol. (in press).
Henkin J, Abeles RH (1976) Biochemistry 15, 3472-3479.
Jablonski PE *et al.* (1990) J. Bacteriol. 172, 1271-1275.
Kerby RL *et al.* (1992) J. Bacteriol. 174, 5284-5294.
Latimer MT, Ferry JG (1993) J. Bacteriol. 175, 6822-6829.
Lu W-P, Ragsdale SW (1991) J. Biol. Chem. 266, 3554-3564.
Lu W-P *et al.* (1990) J. Biol. Chem. 265, 3124-3133.
Matsuyama A *et al.* (1994) Biochim. Biophys. Acta 1219, 559-562.
Morton TA *et al.* (1991) J. Biol. Chem. 266, 23824-23828.
Peer CW *et al.* (1994) J. Bacteriol. 176, 6974-6979.
Puhler G *et al.* (1994) Syst. Appl. Microbiol. 16, 734-741.
Singh-Wissmann K, Ferry JG (1995) J. Bacteriol. 177, 1699-1702.
Todhunter JA, Purich DL (1974) Biochem. Biophys. Res. Comm. 60, 273-280.

ACKNOWLEDGEMENTS

This work was supported by Grant 1RO1GM44661-01A1 from the NIH and Grant DE-FG05-87ER13730 from the DOE to J.G.F. J. A. M.-F. is the recipient of NIH individual national research service award 1F32GM15877-02.

REGULATION OF THE C_1 METABOLISM OF ACETOGENS: METABOLIC BY-PASSES AND ECOLOGICAL IMPLICATIONS

Harold L. Drake, Lehrstuhl für Ökologische Mikrobiologie, BITÖK, Universität Bayreuth, 95440 Bayreuth, Germany

INTRODUCTION

The central energy-conserving process for acetogenic bacteria involves the use of CO_2 as a terminal electron accepter. The end product of CO_2 fixation is acetate. The overall process by which CO_2 is reduced to acetate by acetogens is termed the acetyl-CoA "Wood/Ljungdahl" pathway and involves the reduction of two molecules of CO_2 to either the CH_3-level via formate dehydrogenase and tetrahydrofolate-mediated reactions, or the CO-level via acetyl-CoA synthase (i.e., CO-dehydrogenase) (Wood, Ljungdahl 1991; Ragsdale 1994; Diekert, Wohlfarth 1994; Drake 1994). This 8 electron reduction of two molecules of CO_2 culminates in the synthesis of acetyl-CoA via acetyl-CoA synthase, acetyl-CoA being subsequently converted to acetate via phosphotransacetylase and acetate kinase. Energy conservation occurs by both substrate phosphorylation and ion translocation/ATPases (Ljungdahl 1994; Müller, Gottschalk 1994). For acetogens, the acetyl-CoA pathway is used for autotrophic carbon assimilation as well as energy conservation (Drake 1994), as illustrated in the following reaction:

$$CO_2 + [8 \text{ reducing equivalents}] \dashrightarrow [\textbf{acetyl-CoA}] \dashrightarrow \text{acetate} + \text{biomass}$$

Historically, acetogens were viewed somewhat narrowly and were thought of as an interesting but obscure group of bacteria. Indeed, for nearly 30 years, only one acetogen was available for study. Though not seriously evaluated, for many decades it was generally assumed that acetogens did essentially only one thing: make acetate. This somewhat narrow view has changed in recent years. Although acetate synthesis from CO_2 is the hallmark of acetogens and reason enough to study them, the diverse metabolic potentials that run tangent to or intersect with the reduction of CO_2 collectively form the basis for the competitiveness of acetogens in nature, as well as their potential usefulness in application. Acetogens appear to be ubiquitous in nature and are believed to

72

M. E. Lidstrom and F. R. Tabita (eds.), Microbial Growth on C₁ Compounds, 72–79.

play a central role in the flow of carbon in many habitats. Recent studies have revealed a large array of new metabolic potentials of acetogens, including the use of aromatic and halogenated compounds (Diekert, Wohlfarth 1994; Frazer 1994; Schink 1994; Gößner et al. 1994; Drake et al. 1994).

The regulation of acetogenesis has not been explored in much detail. The acetyl-CoA pathway is generally considered to be a constitutive process. This is only partially correct. Independent of how acetogenesis is regulated, acetogens do not always engage the acetogenic pathway in pure culture. Thus, acetate is not the only reduced end product they form. Although is has been noted in the literature for some time that acetogens vent trace amounts of alternative reduced end products (e.g., H_2, CH_4, or formate), it has only been realized in recent years that alternative routes for reductant flow may constitute major, rather than minor, processes, at least on a per acetogen basis. Indeed, many acetogens can switch from one reductive process to another. Thus, acetogenesis and CO_2 fixation may be regulated by environmental factors. The present statement evaluates recent results that address some of these alternative processes, their regulation, and ecological implications.

RESULTS

Historical considerations. Earlier studies with the acetogens *Clostridium formicoaceticum* and *Acetobacterium woodii* hinted to the potential importance of alternative routes for reductant flow when it was observed that fumarate and aromatic acrylates were used as electron sinks (Dorn et al. 1978; Tschech, Pfennig 1984). The use of alternative acceptors may influence the capacity of acetogens to compete with methanogens and sulfate reducers for H_2 (Cord-Ruwisch et al. 1988). Experimentally-induced CO_2 deprivation engages the use of alternative stategies for reductant flow, suggesting that acetogens may regulate reductant flow under certain environmental conditions (Drake 1993); the capacity to divert electron flow away from CO_2 has been confirmed with many acetogens (Drake 1994).

Effect of CO_2 and Reductant Load on Mixed Acid and Solvent Fermentation. With *Peptostreptococcus productus*, an acetogen found in sewage and mamalian gastrointestinal tracts, the switchover between acceptors is quite complex because of this acetogen's capacity to divert reductant towards 5 alternative reductant sinks: (i) CO_2 reduction via acetogenesis, (ii) aromatic acrylate dissimilation, (iii) lactate fermentation, (iv) ethanol fermentation, and (v) dissimilation of fumarate to succinate (Misoph, Drake 1995). The engagement of these competing processes is regulated by the availability and type of both the terminal acceptor and oxidizable substrate. It is thus of interest to note that recent human isolates that grow acetogenically with H_2/CO_2 appear to switch to lactate fermentation when grown at the expense of glucose (Bernalier et al. 1995). With *P. productus*, the regulation of lactate dehydrogenase appears to play a key role in reductant flow (Misoph, Drake 1995).

Nitrate-dependent Regulation of the Acetyl-CoA Pathway. *Clostridium thermoaceticum*, the most characterized acetogen to date, was long thought to be an obligate heterotroph. We now know that this acetogen is also capable of chemolithoautotrophic growth on H_2/CO_2 or CO/CO_2 (Daniel et al. 1990). Surprisingly, it was recently shown that nitrate is in fact a favored terminal electron acceptor of *C. thermoaceticum* (and also the closely related acetogen *Clostridium thermoautotrophicum*); nitrate dissimilation to nitrite and ammonium is the preferred energy-conserving process (Seifritz et al. 1993; Fröstl, Drake 1994). Ironically, these so-called "homoacetogens" are actually facultative acetogens: even trace levels of nitrate block the engagement of the acetyl-CoA pathway. Two points are noteworthy concerning regulation, i.e., nitrate-dependent metabolic blockage of the reduction of CO_2 via the acetyl-CoA pathway:

i) all acetyl-CoA pathway enzymes appear to be present and active when grown at the expense of nitrate, yet the pathway is metabolically blocked by nitrate, and

ii) autotrophic growth (CO_2 assimilation) is blocked by nitrate.

Regulation appears to occur at the level of electron flow (Figure 1):

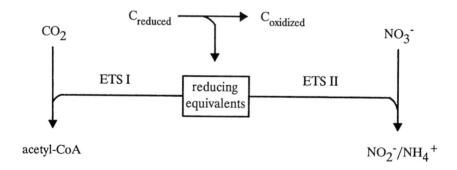

Figure 1. Bidirectional flow of reductant by *C. thermoaceticum* and *C. thermoautotrophicum*. ETS, electron transport system.

In this regard, cytochrome b is not detectable in membranes of *C. thermoaceticum* and *C. thermoautotrophicum* cultivated in the presence of nitrate. Expression of membranous cytochromes of the acetogen *Sporomusa ovata* is also regulated by growth substrate (Kamlage et al. 1993). In addition to CO_2 and nitrate, *C. thermoaceticum* is able to use thiosulfate and dimethylsulfoxide as terminal electron acceptors (Beaty, Ljungdahl 1991); how these acceptors are engaged bioenergetically has not been resolved. It is nonetheless noteworthy that growth in the presence of various inorganic sulfur compounds affects the specific activities of formate and CO dehydrogenases of *C. thermoaceticum* (White et al. 1989).

Effect of Autotrophic Conditions on Enzyme Activities. Although autotrophic growth has made the acetogens a somewhat famous bacteriological group, our understanding of the enyzmology of acetogenesis is primarily based on information that has been obtained with heterotrophically- (i.e., glucose-) cultivated cells. It is thus relevant to note that our appreciation of how carbon and energy flow might be regulated under autotrophic conditions is not well refined. Certain enzymes appear to be subject to regulation under autotrophic conditions. Hydrogenase levels of *C. thermoaceticum* are 15-fold higher under H_2-enriched autotrophic conditions (Daniel et al. 1990). Electrophoretic and spectral analyses with *C. thermoaceticum* and *Acetogenium kivui* indicate that autotrophic conditions may induce multiple hydrogenases and membranous chromophores (Kellum, Drake 1984; Braus-Stromeyer, Drake 1995). In addition, membranes of autotrophically-cultivated *A. kivui* appear to contain high levels of formate dehydrogenase (Braus-Stromeyer, Drake 1995). To date, membranous formate dehydrogenase activity has not been reported. Formate dehydrogenase catalyzes the first step in the reduction of CO_2 on the methyl branch of the acetyl-CoA pathway, and selective localization of this enzyme in membranes may play a heretofore unrecognized regulatory role in the bioenergetics and carbon flow of certain acetogens.

Acetogenic Potentials in Oxic Terrestrial Soils. Because of the diverse metabolic potentials of acetogens, the functional capacity of this bacteriological group in nature may be enormous. However, this capacity is not well resolved, perhaps because the main product we think they make (i.e., acetate) is difficult to assess in situ. This difficulty is in contrast to the production of methane or N_2 via methanogenesis and denitrification, respectively; these anaerobic processes are comparatively well examined at the ecosystem level, in part because the end products are gaseous and easily detected.

Acetogens are known to be syntrophic partners in many anaerobic habitats, including sediments and various gastrointesinal systems (Breznak 1994; Mackie,

Bryant 1994; Schink 1994; Zinder 1994). Although terrestrial soils are not generally regarded as appropriate habitats for acetogens, the first acetogen to be described, *Clostridium aceticum*, was isolated from soils (Wieringa 1936); additional acetogenic isolates have also been obtained from various soils (Andreesen et al. 1970; Wiegel et al. 1981). We have recently found that forest soils have a large potential to form acetate from endogneous matter, and to also synthesize acetate via the acetogenic fixation of CO_2 (Küsel, Drake 1994; Küsel, Drake 1995; Drake et al. 1994). Up to 25% of the organic carbon of forest soil can be converted to acetate under experimental, anaerobic conditions. Although such amounts may have little relevance in situ, they illustrate the potentials of soil microflora to convert soil organic matter to acetate.

The collective synthesis of acetate in soils is likely accomplished by a complex microbial community, each community member functioning in concert with the whole. The in situ roles and magnitude of acetogenesis and CO_2 fixation are not resolved. Nonetheless, the occurrence of this process in such habitats reinforce the general concept that acetogenic CO_2 fixation may occur in habitats not generally regarded as appropriate for acetogenic bacteria. Both the acetate-forming and acetogenic capacities of temperate forest soils has also been confirmed with temperate prairie soil (Wagner et al. 1996). Unlike forest soils, prairie soils have a higher capacity to form a closed methane cycle between methanogens and methanotrophs; however, the occurrence of a closed methane cycle in these oxic soils is unlikely under most in situ conditions and would likely not significantly contribute to carbon turnover. In this regard, under anerobic conditions, supplemental H_2 is always directed towards the reductive synthesis of acetate from CO_2 by forest and prairie soils (Küsel, Drake 1995; Wagner et al. 1996). That H_2 is not subject to methanogenic turnover by soils is corroborated by the low number of culturable methanogens in oxic soils (Peters, Conrad 1995). Under experimental conditions, the anaerobic acetate-forming and acetogenic capacities of prairie soil are stable to aerobic drying and periods of O_2 enrichment (Wagner et al. 1996), suggesting that these anaerobic activities can respond rapidly to in situ anaerobiosis. We thus postulate that oxic soils, in particular those of warmer climates, might contain heretofore unrecognized acetogenic microsites under certain in situ conditions. Conditions of water saturation and high organic soil carbon would likely stimulate both acetate-forming and acetogenic activities.

In contrast to obligately anaerobic habitats in which acetate forms a trophic link to methanogens, oxic terrestrial soils oxidize anaerobically synthesized acetate primarily via the reduction of O_2, nitrate, or alternative inorganic electron acceptors (Küsel, Drake 1995; Wagner et al. 1996) (Figure 2). Thus, the oxidative turnover of acetate in terrestrial soils is conceptulaized to form trophic links to various biological cycles via both aerobic and anerobic intercycle coupling (i.e., the metabolic fusion of the biological cycles). How extensive this coupling is would depend on in situ conditions. Transient engagement of the dissimilation of nitrate, sulfate, iron, manganese, etc., would theoretically be possible in soils that contain these compounds and elements. The diffusion/flow of acetate and the coupling oxidant (e.g., O_2 or nitrate) would clearly vary from soil to soil, and season to season. Selective stimulation of acetogens by cobalt (Florencio et al. 1994) or other factors such as co-substrate turnover (Breznak 1994; Drake 1994; Liu and Suflita 1993) might influence the in situ magnitude of acetogenesis. Soil pH might also influence the amount of acetate productivity in anaerobic microsites (Küsel, Drake 1995). That acetogenic bacteria might be active in oxic soils lends support to the concept that aerobic and anaerobic activities occur in close proximity to one another in nature, and may form tightly coupled trophic links. Recent observations on the microcompartmentalization of the termite gut corroborate the potential co-existence and co-activities of acetogens and aerobic organisms (Brune et al. 1995).

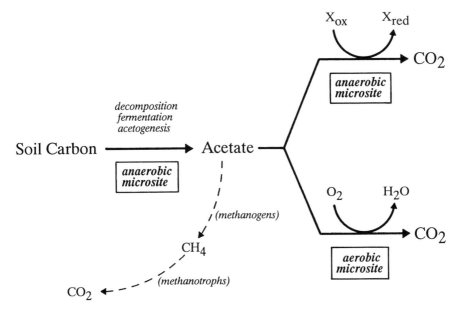

Figure 2. Model for the aerobic and anaerobic oxidation of acetate produced in anaerobic microsites in temperate forest and grassland soils. X_{ox} and X_{red} are the oxidized and reduced forms, respectively, of hypothetical terminal acceptors (e.g., $NO_3^- \dashrightarrow N_2$ or NH_4^+). The broken lines indicate that the flow of acetate towards methane is likely negligible under most in situ conditions.

CONCLUSIONS

- The metabolic versatility of acetogens may make them very competitive in nature, and their in situ activities and roles may be under estimated.

- Because relatively few acetogens have been evaluated in detail, many of the basic enzymes and electron transport systems of most acetogens remain uncharacterized. Thus, the enzymological and molecular mechanisms of regulation engaged by acetogens also remain poorly resolved.

- The occurrence and activity of acetogens in oxic terrestrial soils indicates that acetogenesis can occur in habitats not generally recognized as conducive for such obligately anaerobic processes.

Acknowledgement. These studies were supported by the Bundesministerium für Bildung, Wissenschaft, Forschung, und Technologie (0339476AO).

REFERENCES

Andreesen JR et al (1970) Arch. Microbiol. 72, 154-174.
Beaty PS, Ljungdahl LG (1991) Abstr. 91th Gen. Meet. Am. Soc. Microbiol. p 236, K-131.
Bernalier A et al (1995) Abstr. 95th Gen. Meet. Am. Soc. Microbiol. p 466, Q-377.
Brune, A et al (1995) Appl. Environ. Microbiol. 61, 2681-2687.
Braus-Stromeyer S, Drake HL (1995) Abstr. 8th Intnl. Symp. Microbial Growth on C_1 Compds. p 11, S 9.
Breznak JA (1994) In: Drake HL, ed, Acetogenesis, pp 303-330, Chapman and Hall, New York, USA.
Cord-Ruwisch R et al (1988) Arch. Microbiol. 146, 350-357.
Daniel SR et al (1990) J. Bacteriol. 172, 4464-4471.
Diekert G, Wohlfarth G (1994) Antonie van Leeuwenhoek 66, 209-221.
Dorn M et al (1978) J. Bacteriol. 133, 26-32.
Drake HL (1993) In Murrell JC, Kelly DP, eds, Microbial Growth on C_1 Compounds, pp 493-507, Intercept Limited, Andover, England.
Drake HL (1994) In Drake HL, ed, Acetogenesis, pp 3-60, Chapman and Hall, New York, USA.
Drake HL et al (1994) In Drake HL, ed, Acetogenesis, pp 273-302, Chapman and Hall, New York, USA.
Florencio L et al (1994) Appl. Envrion. Microbiol. 60, 227-234.
Frazer AC (1994) In, Drake HL, ed, Acetogenesis, pp 445-483, Chapman and Hall, New York, USA.
Fröstl JM, Drake HL (1994) Abstr. 94th Gen. Meet. Am. Soc. Microbiol. p 279, K-22 (manuscript submitted to J. Bacteriol.).
Gößner A et al (1994) Arch. Microbiol. 161, 126-131.
Kamlage B et al (1993) Arch. Microbiol. 159, 189-196.
Kellum R, Drake HL (1984) J. Bacteriol. 160, 466-469.
Küsel K, Drake HL (1994) Appl. Environ. Microbiol. 60, 1370-1373.

Küsel K, Drake HL (1995) Appl. Environ. Microbiol. (in press).

Liu S, Suflita JM (1993) Appl. Environ. Microbiol. 59, 1325-1331.

Ljungdahl LG (1994) In, Drake HL, ed, Acetogenesis, pp 63-87, Chapman and Hall, New York, USA.

Mackie RI, Bryant MP (1994) In Drake HL, ed, Acetogenesis, pp 331-364, Chapman and Hall, New York.

Misoph M, Drake HL (1995) Abstr. 95th Gen. Meet. Am. Soc. Microbiol. p 536, K-2 (manuscript in prep).

Müller V, Gottschalk G (1994) In, Drake HL, ed, Acetogenesis, pp 127-156, Chapman and Hall, New York, USA.

Peters V, Conrad R (1995) Appl. Environ. Microbiol. 61, 1673-1676.

Ragsdale SW (1994) In Drake HL, ed, Acetogenesis, pp 88-126, Chapman and Hall, New York, USA.

Schink B (1994) In Drake HL, ed, Acetogenesis, pp 88-126, Chapman and Hall, New York, USA.

Seifritz C et al (1993) J. Bacteriol. 175, 8008-8013.

Tschech A, Pfennig N (1984) Arch. Microbiol. 137, 163-167.

Wagner C et al (1996) Appl. Environ. Microbiol. (submitted).

Wiegel J et al (1981) Curr. Microbiol. 5, 255-260.

Wieringa KT (1936) Antonie van Leeuwenhoek 3, 236-273.

White H et al (1989) Eur. J. Biochem. 184, 89-96.

Wood HG, Ljungdahl LG (1991) In Shivley JM, Barton LL, eds, Variations in Autotrophic Life, pp 201-250, Academic Press, San Diego, USA.

Zinder SH (1994) In Drake HL, ed, Acetogenesis, pp 386-415, Chapman and Hall, New York, USA.

Mechanism of N-Oxidation and Electron Transfer in the Ammonia Oxidizing Autotrophs

Todd Vannelli, David Bergmann, David M. Arciero and Alan B. Hooper

University of Minnesota, St. Paul, MN 55108 USA

1 SUMMARY   In *Nitrosomonas* electrons flow from hydroxylamine oxidoreductase to ammonia monoxygenase, to NAD or to cytochrome oxidase or a nitrite-reducing system. The pathway includes the extrinsic membrane tetraheme cytochrome c-554 and, possibly, a membrane-anchored tetraheme c-cytochrome (CycB). *Cyc*554 and *cyc*B share an operon in the HAO gene cluster. CycB has high homology with nirT, napC and torC suggesting a role in anaerobic metabolism. A small fraction of active HAO is found in the membrane. In the AMO operon, the gene-derived amino acid sequence suggests, for subunits-A and -B, 6 or 2 membrane-spanning helices, respectively. AmoA could provide a proton wire or pore.
        Use of substrate analogs and specifically-deuterated substrates reveal aspects of the structure of the active site and nature of the mechanism of AMO as follows: (a) a long tubular, hydrophobic active site with an oxygen-activating center near the terminus; (b) location of the oxygen-activating site so as to provide regio- and enanteromeric selectivity in the hydroxylation of ethylbenzene; (c) an isotope-sensitive step in the hydroxylation of ethylbenzene; (d) a radical rebound reaction involving a planar intermediate.
        AMO is inactivated by phenylacetylene. The latter may be a useful specific inhibitor of AMO and sMMO but not pMMO.

2 INTRODUCTION   The known enzymes and electron transfer pathways utilized by the autotrophic nitrifying bacteria for the oxidation of ammonia to nitrite are illustrated in Fig. 1. Hydroxylamine, produced by the membrane-bound ammonia monoxygenase (AMO), is oxidized by the removal of 4 electrons by the soluble, periplasmic hydroxylamine oxidoreductase (HAO) with the tetraheme cytochrome c-554 (Cyc554) as electron acceptor (Hooper,1989; Yamanaka and Shinra, 1974, Arciero et al.,1991). Most of the latter is bound to membrane by electrostatic interactions (McTavish et al., 1995). Two electrons are obligatorily returned to AMO by unidentified carriers. Some electrons are used in reduction of pyridine nucleotide by "reverse electron flow" (Aleem, 1966). The remaining electrons are used (a) in terminal electron pathways, possibly involving the soluble monoheme cytochrome c-552, leading to a dicopper, cytochrome aa_3 oxidase (DiSpirito, et al.,1986) or a nitrite (and presumably a NO⁻ and N_2O^-) reductase (Goreau et al., 1980) or (b) a detoxifying cytochrome c peroxidase (Arciero and Hooper, 1994). Cyt P460 is a periplasmic NH_2OH-oxidizing enzyme (Bergman and Hooper, 1994a) which is also found in *M. capsulatus* (Zahn, et al., 1994).

M. E. Lidstrom and F. R. Tabita (eds.), Microbial Growth on C₁ Compounds, 80–87.
© *1996 Kluwer Academic Publishers. Printed in the Netherlands.*

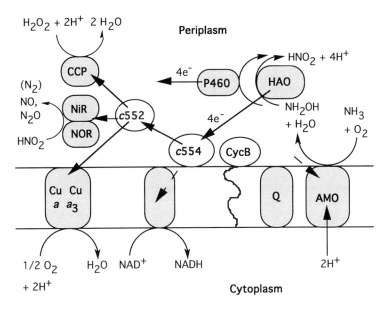

Fig. 1 N-oxidation and Electron Transport in *Nitrosomonas*.

Two proteins, amoA and amoB, which contain iron and copper and are probably subunits of AMO, have been isolated. Their genes, which share an operon, were sequenced (McTavish et al., 1993; Bergmann and Hooper, 1994b). Two copies of the genes are present. Three clusters contain separate copies of genes for HAO and Cyc554. In two of the three clusters, the latter gene is in the same operon with the gene, *cyc*B for a putative tetraheme cytochrome (Bergman and Hooper, 1994).

We report recent observations which apply to the above-mentioned electron transfer chains and the structure and catalytic mechanism of AMO.

3 THE CycB GENE PRODUCT.   CycB has high homology with the nirT, napC and torC gene products, putative tetraheme cytochromes involved in electron transport to cytochrome cd_1 nitrite reductase of *P. stutzeri* (Jungst, et al., 1991), the periplasmic nitrate reductase of *Thiosphaera pantotropha* (Berks, et al., 1994) and pentaheme cytochrome of the trimethylamine N-oxide reductase pathway of *E.coli* (Mejean, et al., 1994), respectively (Fig. 2). We have recently isolated, from salt-washed membranes of *Nitrosomonas*, a cytochrome which contains the N-terminal sequence SIGTLLTGAL. The latter is identical to part of the N-terminal gene region of cycB (Fig. 2). The serine residue at which cleavage of the putative signal sequence occurs is conserved in all members of the *nir*T gene family. This appears to be the clearest identification of the gene product of a cytochrome in the gene family. The N-terminal amino acid sequence thus contains a hydrophobic region which probably serves as a membrane anchor.

The fact that cycB shares 2 of the 3 operons containing *cyc*554 suggests a central role in dispersion of electrons originating from HAO. The homology with nirT, napC and torC suggests a role in a periplasmic anaerobic terminal electron pathway in *Nitrosomonas*. If so, its prevalence in the genome suggests that the latter mode of growth may be more important than previously thought. Recent reports describe anaerobic growth of *Nitrosomonas* on nitrite with pyruvate (Abeliovich and Vonshak, 1992) or hydrogen as electron donor (Bock, et al., 1995). On the other hand, the nitrite reductase isolated from *Nitrosomonas* is a Cu enzyme (Dispirito, et al., 1985) and cytochrome cd_1 is not found in extracts as the cells are currently grown. Further, nitrate reductase activity was not detected in *Nitrosomonas* (Wallace and Nicholas, 1968). Although the tetraheme-containing region of cytochromes in this family are similar, the C-terminal regions differ very significantly. The C-terminal end of cycB, napC and nirT lack the large domain containing the fifth heme of torC. CycB contains a 50 aa segment highly enriched in acidic amino acids which is absent in nirT, napC and torC. The cytochrome may have a novel role, for example in the electron pathway to AMO.

```
nirT    MTDKDGNKQQ KGGILALL.. ..RRPSTRYS LGGILIVGIV AGIVFWGGFN TALEATNTET FCISCHEMGD
napC    MGWIRASIRW IWGRVTWFWR VISRPSSFLS IGFLTLGGFI CGVIFWGGFN TALEITNTEK FCTSCHEMRD
torC    .......... ....MRKLWN.ALRRPSARWS VLALVAIGIV IGIALIVLPH VGIKVTSTTE FCVSCHSM.Q
cycB    .......... .......... ..MTRLQKGS IGTLLTGALL .GIVLVAVVF GGEAALSTEE FCTSCHSMTY


nirT    NVYPEYKETI HYANRTGVRA TCPDCHVPRD WTH....KMV RKVEASKELW GKIVG.TIDT AEKFEAKRLT
napC    NVYQELMPTV HFSNRSGVRA SCPDCHVPHE WTD....KIA RKMQASKEVW GKIFG.TIST REKFLEKRLE
torC    PVVYEEYKQSV HFQNASGCEL NCHDCHIPPD IPG....MVK RKLEASNDIY QTFIAHSIDT PEKFEAKRAE
cycB    PQ.AELKQST HYGA.LGVNP GCKDCHIPQG IENFHLAVAT HAIDGARELY LELVN.DYST LEKFNERRLE


nirT    LARREWARMR ASDSRECRNC HSLESMSSDM QKQRARKQHE M.AREDNLTC IACHKGIAHH ...LPEGMTE
napC    LAKHEWARLK ANDSLECRNC HAAVAMDFTK QTRRAPQIHE RYLISGEKTC IDCHKGIAHQ ...LPD MTG
torC    LAEREWARMK ENNSATCRSC HNYDAMDHAK QHPEAAR.QM KVAAKDNQSC IDCHKGIAHQ ...LPDMSSG
cycB    MAHDARMNLK KWDSITCRTC H.....KKPA PPGESAQAEH KKMETEGATC IDCHQNLVHE EVPMTDLNAS


nirT    EDED*
napC    IEPGWLEPPE LRGEEQAWLG GGAEAVHRYL ATVETR*
torC    FRKQFDDVRA SANDSGDTLY SIDIKPIYAA KGDKEASGSL LPASEVKVLK RDGDWLQIEI TGWTESAGRQ
cycB    IAQGKLVLKP EDDGDDEEAD EDEDEETEEA DDSSDSESAS SSDNSDNEDD NNDE*


torC    RVLTQFPGKR IFVASIRGDV QQQVKTLEKT TVADTNTEWS KLQATAWMKK GDMVNDIKPI WAYADSLYNG


torC    TCNQCHGAPE IAHFDANGWI GTLNGMIGFT SLDKREERTL LKYLQMNASD TAGKAHGDKK EEK*
```

Fig. 2 Sequence of Gene Products of the *nir*T Family

4 A MEMBRANE FORM OF HAO.   Turnover of ammonia by AMO equivalent to as much as 10% of the *in vivo* rate can be observed with membranes of *Nitrosomonas* (Tsang and Suzuki, 1982; Ensign et al., 1993). Hydroxylamine, hydrazine or NADH can serve as a source of electrons. It is a reasonable speculation that electron transfer from NADH is catalyzed by the enzyme which, *in vivo*, reduces NAD. We have observed this activity using nitrobenzene as substrate as catalyzed by membranes from which cytochrome c-554 has been removed by salt washing (McTavish et al., 1995). The addition of purified cytochrome c-554 was necessary for turnover, in keeping with the original observations of Tsang and Suzuki (1982). The addition of HAO was not stimulatory. Hence, enough active HAO is present that the removal of electrons from hydroxylamine was not rate limiting to AMO. We have found that the membranes contain a form of HAO which is not removed by salt washing but can be partially removed at low ionic strength and completely removed by proteolysis. Hence this fraction of HAO is significantly different in its location than the majority (~85%) of HAO in the cell. *In vitro*, electrons from this form of HAO are utilizable by AMO. We speculate that, *in vivo*, this fraction of HAO may (a) be in a late stage of processing before release from the membrane or (b) remain attached to the membrane in a functionally significant manner to donate electrons for a specific function such as NAD reduction.

5 STRUCTURE OF AmoA AND AmoB.   Suicide inhibitors such as acetylene (Hyman and Wood, 1985) or substituted propargyl amine specifically bind to a membrane protein, amoA, of *Nitrosomonas*. AmoA was isolated in an aggregate with a second protein, amoB and the gene for the latter was found to be in the same operon as *amo*A. The two proteins were hypothesized to be subunits of AMO and were sequenced (McTavish, et al., 1993; Bergmann and Hooper, 1994). The sequence of amoA is also known for a *Nitrosospira* sp and *Nitrosolobus multiformus* ATCC 25196 (Klotz and Norton, 1995a,b). Similar genes and proteins (pmoA and pmoB) have been found in *Methylococcus capsulatus*, which contains the membrane methane monoxygenase (Semrau, et al., 1995).

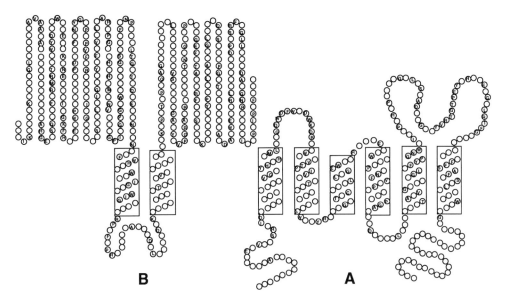

Fig. 3 Model of AMO

Analysis of conserved residues in the 4 A genes and 2 B genes and hydrophobicity plots (Jones, et al., 1994) provide a working hypothesis for the orientation of the two proteins in the membrane (Fig. 3). The A protein appears to have 6 transmembrane segments and highly conserved periplasmic loops between transmembrane segments I and II and V and VI. The B protein, which has at least two transmembrane domains, has a large periplasmic N-terminal region and a highly charged C-terminal region which appears more likely to be periplasmic. The model suggests that much of the business of the enzyme is carried out on the periplasmic side. The structure of amoA could certainly accomodate a proton wire or pore. Incorporation of protons from cytoplasm into the product hydroxylamine which is then released into the periplasm would promote an electrochemical gradient favorable for energy transduction.

Although copper and iron have been observed in the isolated but inactive aggregate of amoA and amoB (McTavish, 1992), existing evidence has not established a catalytic or electron transfer (as opposed to structural) role for these metals.

6 CATALYSIS BY AMO.   The enormously broad range of substrates acted upon by AMO suggests that, as with cytochrome P-450 (Guengrich and MacDonald, 1990) and

sMMO (Fox, et al., 1990), dioxygen is first reduced by the enzyme to a reactive species which then reacts with substrate. Despite the absence of active purified enzyme, several aspects of the structure of the active site and mechanism of catalysis have been elucidated with the use of substrates other than ammonia. Some of the organic substrates were specifically deuterated to enable the investigation of enzyme activity *in vivo* without the interference of possible factors such as differential permeability or solubilty of substrates. The weakness of the approach is that the mechanism of oxidation of the organic substrates may not be the same as the oxidation of ammonia.

6.1 THE ACTIVE SITE IS HYDROPHOBIC AND TUBULAR.   The high pH optimum for oxidation of ammonia (Suzuki, et al., 1974) and the relatively low reactivity of charged substrates or inhibitors suggests a hydrophobic active site. The large size of the active site is emphasized by its reactivity with naphthalene (Vannelli and Hooper, 1995a). The preferential oxidation of 4- to 8-carbon alkanes at the 1- or 2-postion (Hyman, et al., 1988) suggests a tubular active site with the activated oxygen near the deepest part. This model is further supported by the fact that aromatics are oxidized either on reactive substituents or the *para*-position in preference to either *ortho-* or *meta-* (Keener and Arp, 1994; Vannelli and Hooper, 1995); for example anisole is converted to a mixture of *p*-hydroxyanisole and phenol. The observation suggests that one end or the other of the substrate is allowed to approach the reactive center. These properties of the active site are also illustrated in the unusual reductive dehalogenation of Nitrapyrin (Vannelli and Hooper, 1993) where the trichloromethyl-goup is hypothesized to be positioned and react in place of oxygen with reduced metals of the oxygen-activating center (Fig. 4).

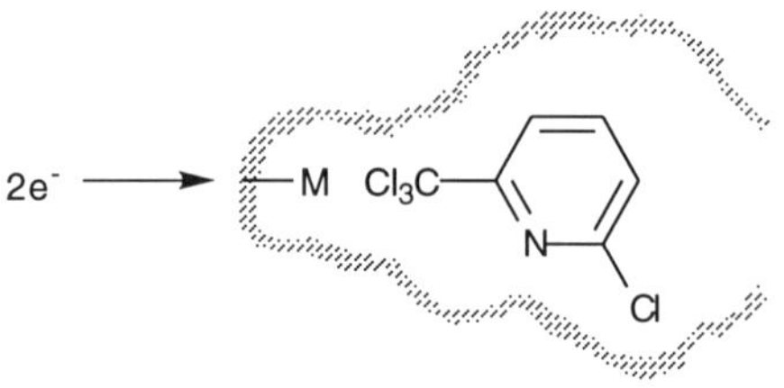

$$2e^- \longrightarrow M \quad Cl_3C-$$

Fig. 4 Reductive Dechlorination of 2-chloro-6-trichloromethylpyridine.

6.2 REGIOSELECTIVITY AND "MOLECULAR SWITCHING" IN THE OXIDATION OF ETHYLBENZENE.   As indicated in Table 1. AMO exhibits a strong regioselectivity towards the hydroxylation of C_1 vs C_2 of ethylbenzene (Vannelli and Hooper, 1995b). This further suggests the small distance between the reactive oxygen center and the deep end of the active site. This regioselectivity is dramatically altered by substitution of deuterium for hydrogen of C_1; C_2 is now being hydroxylated in preference over C_1 (Table 1). This "molecular switching" due to substitution of D for H reflects an isotopically-sensitive breakage of a C-H(D) bond during catalysis. It suggests that the orientation of the substrate by residues of the active site places C_1-D closest to the reactive oxygen but the C_2-H close enough that the latter is selected over the more stable C_1-D bond. The presence of deuterium did not increase the proportion of 4-ethylphenol produced in the reaction. Hence the affinity of ethylbenzene for the active site is apparently great enough that the switch to an attack on the non-deuterated C_2 was faster than the possible exit, inversion and re-entry of the substrate molecule. This conclusion is also illustrated by the fact that $[\alpha,\alpha,\alpha-{}^2H_3]$-*p*-xylene was hydroxylated equally well at either methyl group (data not shown).

Table 1. Hydroxylation of Ethylbenzene by AMO in *Nitrosomonas europaea.*

Substrate	Amount of Products (%)*				
	R-1-Phenyl-ethanol	S-1-Phenyl-ethanol	2-Phenyl-ethanol	4-Ethyl-phenol	Other
$phCH_2CH_3$	48	15	17	4	16
$phCD_2CH_3$	24	10	53	4	9

*Turnover was measured as in Vannelli and Hooper, 1995a, as catalyzed by cells in the presence of 10 mM ammonia.

6.3 ENANTIOMERIC SELECTIVITY IN THE OXIDATION OF ETHYLBENZENE.   Additional detail of the geometry of the active site in the vicinity of the oxygen-activating center is revealed by the enantiomeric selectivity of 1-phenylethanol production (Table 1); the proR hydrogen is abstracted 3-fold more frequently than the proS hydrogen (regardless of whether the substrate is -H_2 or -D_2). When D is present in the proR position (the chiral substrate R-phCHDCH$_3$) the preference for pro-R abstraction is diminished by a factor of ~ 2.5 (Table 2) (Vannelli and Hooper, 1995c). This isotope effect is not seen with the substrate S-phCHDCH$_3$.

Table 2. Pro-R Abstraction and Inversion of 1-Phenylethanol Products*

Substrate	Pro-R Abstraction (%)	Inversion of Products	
		Probability of R to S (%)	Probability of S to R (%)
R-phCHDCH$_3$	32	14	28
S-phCHDCH$_3$	85	14	57

*Abstraction and inversion were estimated as in White et al., 1986.

We assume that the methyl group is perpendicular to the benzene plane of ethylbenzene and that the R or S hydrogens project upwards or downwards relative to the methyl group (Fig. 5)

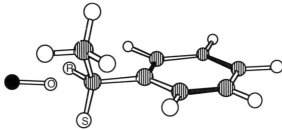

Fig.5. Possible Orientation of Ethylbenzene Relative to the active Oxygen of AMO.

As suggested by the regioselectivity, the methyl hydrogens may be positioned further from the active oxygen than either C_1-hydrogen. As suggested by the enantiomeric selectivity, the R hydrogen is positioned closer to the active oxygen than the S hydrogen. Importantly, this model requires: (a) restriction of rotation around the C-benzyl bond (perhaps due to the interaction of the methyl and benzyl moieties) and (b) restriction of rotation of the benzyl moiety within the active site. The latter suggests that the active site is a flattened tube or that the benzene is stabilized strongly by one wall of the pocket. This also supports the "deep tube" model since the benzene ring would not be stabilized if it projected from a shallow pocket.

6.4 THE OXIDATION OF ETHYLBENZENE OCCURS BY A RADICAL REBOUND MECHANISM.

The products of oxidation of proR or proS deutero ethylbenzene are shown in Table 2. In 14% of the cases of abstraction of the proR hydrogen, inversion occured around the C_1-benzyl bond so that the hydroxyl group is now the "downward" pointing position of Fig. 5. The value was the same with either H or D in the proR postion. Following proS abstraction, 28 or 57% of the molecules underwent inversion depending on whether the substrate was D or H, respectively, in the proR position. The simplest interpretation of the observed inversion is that substrate oxidation is initiated when H· (or D·) is abstracted by the activated oxygen and that the resulting substrate radical "rebounds" to attack the metal-OH center with formation of product. This mechanism is thought to occur in cytochrome P-450 (Guengerich and MacDonald, 1990) and soluble methane monoxygenase (Priestly, et al., 1992). According to this hypothesis the rate of the rebound process is slow enough to allow rotation of some part of the substrate radical (Fig. 6). The isotope effects, for which we have no explanation, show that inversion occurs more frequently when H (rather than D) remains with the substrate radical (and D rather than H is in the putative hydroxyl radical).

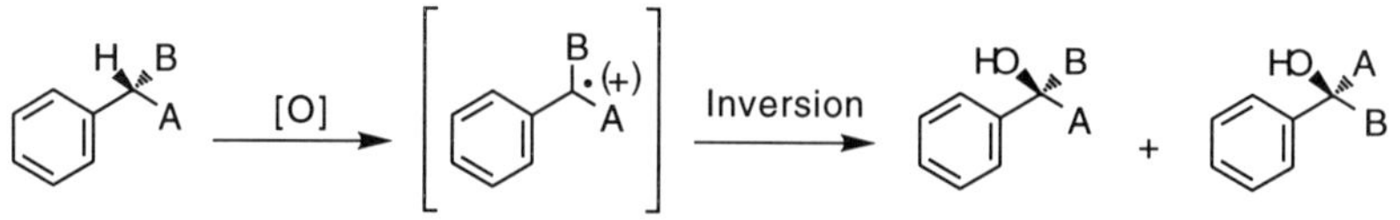

Fig. 6. Inversion Mediated by Putative Unbound Planar Intermediate of AMO.

7 PHENYLACETYLENE AS A SELECTIVE INHIBITOR OF AMO AND sMMO VS pMMO.

Unambiguously specific inhibitors are not available for evaluation of the relative contribution of various bacterial species to the transformation of ammonia or methane in nature. We find that phenylacetylene at a concentration of 100 μM completely inhibits the oxidation of ammonia by AMO in *Nitrosomonas*. Soluble MMO, which actively oxidizes ethyl benzene and styrene (Burrows et al., 1994), may be expected to be inactivated by phenylacetylene. Membrane MMO, which is notoriously inactive with aromatic substrates including ethyl benzene and styrene may be expected not to react with phenylacetylene. Phenyl acetylene has the advantage of an apparent high toxicity, water solubility and low volatility (as compared with acetylene). If the specificity is as great as predicted, it may prove useful for limited environmental studies.

8 ACKNOWLEDGEMENTS

(support from MN Sea Grant/NA90AA-D-SG149; USDA/92-3705-7811; NSF/MCB-9316906; DOE/DE-FG02-95ER20191).

9 REFERENCES
Abeliovich, A. and Vonshak, A. (1992) Arch. Microbiol. 158:267-270.
Aleem, M.I.H. (1966) Biochim. Biophys. Acta 113: 216-224.
Arciero, D., Balny, C. and Hooper, A.B. (1991) Biochemistry 30: 11466-11472.
Arciero, D.M. and Hooper, A.B. (1994) J. Biol. Chem. 269:11878-11886.

Bergmann, D., Arciero, D. and Hooper, A.B. (1994) J. Bacteriol. 176: 3148-3153.

Bergmann, D. and Hooper, A.B.(1994a) FEBS Lett 353: 324-326.

Bergmann, D. and Hooper, A.B.(1994b) Biochem. Biophys. Res. Comm. 204:759-762.

Berks, B.C., Richardson, D.J., Robinson, C., Reilly, A., Aplin, R.T. and
    Ferguson,S.J. (1994). GenBank accession number Z36773.

Bock, E. (personal communication).

Burrows, K.J., Cornish, A., Scott, D. and Higgins, I.J. (1984) J. Gen. Microbiol.130:
    3327-3333.

DiSpirito, A.A., Lipscomb, J.D., and Hooper, A.B. (1986) J. Biol. Chem. 261:17048-
    17056.

DiSpirito, A.A., L.R. Taaffe, J.D. Lipscomb, and A.B. Hooper (1985) Biochim.
    Biophys. Acta 827:320-326.

Ensign, S.A., Hyman, M.R. and Arp, D.J. (1993) J. Bacteriol. 175:1971-1980.

Fox, B.G.,Bourneman, J.G., Wackett, L.P. and Lipscomb, J.D. (1990) Biochem.
    263: 10553-10556.

Goreau, T.J., Kaplan, W.A., Wofsy,S.C., McElroy, M.B., Valois, F.W. and
    Watson, S.W. (1980) Appl. Envt. Microbiol. 40: 526-532.

Guenngerich, F.P. and MacDonald, T.L. (1990) FASEB J. 4:2453-2459.

Hooper, A.B. (1989) in Autotrophic Bacteria, Schlegel, H.G. and Bowein, B. (eds)
    pp239 - 265 Science Tech.Madison, WI.

Hyman, M.R., Murton, I.B. and Arp, D.J. (1988) Appl. Envt. Microbiol. 54:3187-
    3190.

Hyman, M.R. and Wood, P.M.1985. Biochem.J. 227: 779-725.

Jones,J.T., Taylor, W.R. and Thornton, J.M.(1994) Biochem. 33: 3038-3049.

Jungst, A., Wakayabashi, S., Matsubara, H. and Zumft, W.G. (1991) FEBS Lett.
    279:205-209.

Keener, W.K. and Arp, D.J. (1994) Appl. Envt. Microbiol. 60:1914-1920.

Klotz,M.G. and Norton,J.M.(1995a) Genbank accession number U20644.

Klotz,M.G. and Norton,J.M.(1995b) Genbank accession number U15733.

McTavish,H. (1992) PhD Thesis, University of Minnesota,

McTavish, H., Fuchs, J. and Hooper, A.B. (1993) J. Bacteriol. 175:2436-2444.

McTavish, H., Arciero, D.M. and Hooper,A.B. (1995) Arch. Biochem. Biophys. in press

Mejean,, V., Lobbi-Nivol, C., Lepelletier, M., Giordano, G., Chippaux, M. and   Pascal,
    M. (1994) Molec. Microbiol. 11:1169-79.

Priestly, N.D., Floss, H.D., Froland, W.A.,Lipscomb, J.D., Williams, P.G. and
    Morimoto, H. (1992) J.Am.Chem.Soc. 114: 7561-7562.

Semrau,J.D., Chistoserdov,A., Lebron,J., Costello, A., Davagnino, J., Kenna, E.,
    Holmes, A.J., Finch, R., Murrell, J.C. and Lidstrom, M.E. (1995) J.Bacteriol.
    177:3071-3079.

Suzuki,I., Dular, U. and Kwok, S.C. (1974) J. Bacteriol. 120: 556-558.

Tsang, D.C.Y. and Suzuki, I. 1982. Can. J. Biochem. 60: 1018-1024.

Vannelli, T. and Hooper, A.B.(1993) Appl. Envt. Microbiol. 59:3597-3601.

Vannelli, T. and Hooper, A.B.(1995a) Biochemistry 34, in press.

Vannelli, T. and Hooper, A.B. (1995b) FASEB J. 9: A1493.

Vannelli, T. and Hooper, A.B. (1995c) manuscript in preparation.

Wallace, W. and Nicholas, D.J.D. (1968) Biochem.J. 109: 763-773.

White, R.E., Miller, J.P., Favreau,L.V, and Battacharyya, A. (1986) J. Am. Chem. Soc.
    108: 6024-6031.

Yamanaka, T., and Shinra,M. (1974) J. Biochem. 75: 1265-1273.

Zahn, J. A., Duncan, C. and DiSpirito, A.A. (1994) J. Bacteriol. 176,5879-5887.

Genes related to carbon dioxide fixation in *Hydrogenovibrio marinus* and
Pseudomonas hydrogenothermophila

Yasuo Igarashi and Tohru Kodama

Department of Biotechnology, Division of Agriculture and Life Sciences,
University of Tokyo, Bunkyo-ku, Tokyo 113, Japan.

Introduction

Carbon dioxide fixation is probably the most important biological reaction to sustain
lives of all living maters on this planet, because organic matters indispensable for life are
mostly produced from inorganic carbon dioxide by this biological reaction. Hydrogen
bacteria are able to grow on carbon dioxide as the sole carbon source by using molecular
hydrogen as an energy source. Some kinds of hydrogen bacteria, such as *Pseudomonas
hydrogenothermophila* (Goto et al., 1978) and *Hydrogenovibrio marinus* (Nishihara et
al., 1989, 1991a), have been known to grow extremely fast by using carbon dioxide as
the sole carbon source. The specific growth rate for the autotrophic growth of these
bacteria are around 0.65-0.70. These bacteria fixes carbon dioxide via the reductive
pentose phosphate cycle (Calvin cycle). The key enzyme of the cycle is ribulose-1,5-
bisphosphate carboxylase/oxygenase (RubisCO, Ec 4.1.1.39). In this article, we describe
our recent results on the structural and related genes for RubisCO from *P.
hydrogenothermophila* and *H. marinus*.

Structure of three RubisCO genes from *H. marinus*.

Hydrogenovibrio marinus strain MH-110 is a gram-negative aerobic hydrogen
oxidizing bacterium isolated from sea water (Nishihara et al, 1989,1991a). The strain is a
comma-shaped rod, and motile with a polar flagellum. The optimal growth temperature
and NaCl concentration for growth are 37C and 0.5M, respectively. The most remarkable
feature of the strain is obligate chemolithoautotrophy, namely, the strain can utilize only
molecular hydrogen or reduced forms of sulfur compounds as an energy source for its
growth.

H. marinus fixes CO_2 via the reductive pentose phosphate cycle. We have shown that
the strain possesses three RubisCO genes, which code for two L8S8-type or Form I, and
one L2-type or Form II RubisCOs (Igarashi et al., 1992). Two Form I RubisCO genes,
cbbLS-1 and *cbbLS-2* were cloned by using the gene for the large subunit of RubisCO
from *Anacystis nidulans*. A gene for Form II RubisCO, *cbbM*, was cloned by the PCR
method by referring the N-terminal amino acid sequence of purified CbbM (Chung et al.,
1993). To our knowledge, a bacterium possessing three types of RubisCOs has not been
known.

M. E. Lidstrom and F. R. Tabita (eds.), Microbial Growth on C$_1$ Compounds, 88–93.

All of these three RubisCO genes were sequenced. The *cbbLS-1* gene coded for a large subunit (1416bp, 472aa and Mr=52,487) and a small subunit (351bp, 117aa and Mr=13,306). The *cbbLS-2* gene coded for another large subunit (1413bp,471aa and Mr=52,004) and small subunit (366bp,122aa and Mr=14133), respectively. The deduced amino acid sequences showed 78% identity between CbbL-1 and CbbL-2, and 62% homology between CbbS-1 and CbbS-2.

The *cbbM* gene from *H. marinus* (Yaguchi et al, 1994) coded for form II RubisCO (1389bp, 463aa and Mr=50,665). The deduced amino acid sequence of *cbbM* had high homology with those of the form II RubisCO of *Rhodospirillum rubrum* (63% identity) and *Rhodobacter sphaeroides* (62%), but low similarity to those of CbbL-1 and CbbL-2 from *H. marinus* (less than 30%).

Figure 1 shows the dendrogram of RubisCO large subunits of RubisCOs from various origins. A purple sulfur photosynthetic bacterium, *Chromatium vinosum* has been also known to possess two form I RubiSCOs. Surprisingly, CbbL-1 showed the highest homology to one of the two RubiSCO large subunits of *C. vinosum* (RbcA, Viale et al., 1989), and CbbL-2 showed the highest homology to the other (RbcL, Kobayashi et al., 1991).

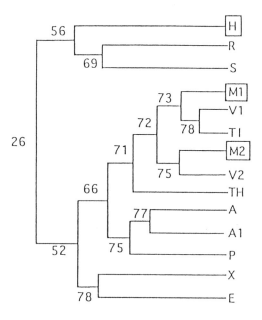

Fig.1. Dendrograms showing RubisCO large subunit amino acid sequence relatedness. The total length of the horizontal lines connecting two proteins is inversely proportional to the extent of similarity between their modules. Numbers in the fugure indicate the similarity scores. H; *H marinus cbbM*, R; *R. rubrum*, S; *R. sphaeroides*, M1; *H. marinus cbbL-1*, M2; *H. marinus cbbL-2*, V1; *C. vinosum rbcA*, V2; *C.vinosum rbcL*, T1, *T. ferrooxidans*, TH; *P. hydrogenothermophila*, A; *A. nidulans*, A1; *Synechococcus* a-1, P; *S.oleracea*, X; *X. fluvus* and E; *A. eutrophus*.

The analysis of 16S-rRNA from *H. marinus* suggested that this bacterium falls into γ-subdivision of the purple (photosynthetic) bacteria (Nishihara, unpublished), just same as *C. vinosum*. The dendrogram obtained here strongly suggests *cbbLS-1* and *cbbLS-2* had diverged from each other, before *H. marinus* diverged from *C. vinosum*. On the other hand, aerobic chemolithoautotrophic bacteria were thought to have diverged from unicellar cyanobacteria (Glover, 1989). But this dendrogram clearly indicated that *H. marinus* have diverged from anaerobic photosynthetic bacteria.We believe that more precise analyses on the three RubisCO genes and the phylogenic position of *H. marinus* will give us more information on the evolutionary aspects of RubisCO and autotrophic bacteria.

Expression of RubisCO genes under different CO_2 concentrations.

We examined the expression of these three RubisCO genes in *H. marinus* cells under different culture conditions. *Rhodobacter sphaeroides* has been known to possess two sets of RubisCO genes (Tabita et al, 1993). One is form I and the other is form II. Jouanneau and Tabita (1987) showed that form I RubisCO was inactivated, but Form II RubisCO was not affected, when *R. sphaeroides* was cultivated in high CO_2 concentration. So we are especially interested in the expression of three RubisCO genes in *H. marinus* under different CO_2 concentration.

When gas mixture with the composition of $H_2:O_2:CO_2=7:1:1$ was continuously introduced into a culture vessel, the strain can grow very fast at the doubling time of about 1 hour by using CO_2 as the sole carbon source (Nishihara et al., 1991b). By this culture condition, CbbM, Form II RubisCO, was mainly produced in the cells (Chung, 1993). In this condition, about 80-90% of total RubisCO activity was shown to be derived from CbbM (Igarashi et al., 1992 and Chung, unpublished).

We try to reduce the CO_2 concentration without the drastic decrease in the growth rate of the strain. At 2% CO_2 concentration in the inlet gas, *H. marinus* can grow rather well. The RubisCO activity in the cell free extract was measured after the cells were harvested at the logarithmic phase during the growth at 10% ($H_2:O_2:CO_2 =70:20:10$) and 2% (78:20:2) CO_2 concentration. The RubisCO activity was 2.5 fold higher when grown at 2% than 10% CO_2 concentration.

Western blotting by using an antibody for form I RubisCO of *P. hydrogeno-thermophila* showed that CbbM was detected from the cells grown under 10% CO_2 concentration, and form I RubisCO was detected when grown under 2% CO_2 concentration.This result agreed with our former experiment of RubisCO purification mentioned above.Our preliminary Northern blot analysis by using antisense RNAs for *cbbM* , *cbbS-1* and *cbbS-2* genes suggested that *cbbM* is mainly transcribed under 10% CO_2 concentration, and *cbbLS-2* seems to be mainly transcribed under 2% CO_2 concentration.

The specificity factor (τ; VcKo/VoKc) is used to determine the efficiency of carboxylation reaction relative to oxygenation reaction of RubisCO (Laing et al., 1974). We have determined the specificity factors of three RubisCOs by measuring the products of carboxylation and oxygenation reactions of RubisCO by ion chromatography (Yaguchi, unpublished). Each RubisCO gene was expressed in *E. coli* cells by using an expression vector, and then RubisCO was purified from the cells. The value for CbbLS-1, CbbLS-2 and CbbM were 26.6, 33.1 and 14.8, respectively.

In conclusion, *H. marinus* mainly transcribed *cbbM* under 10% CO_2 concentration, and mainly transcribed *cbbLS-2* with the highest τ-value under the 2% CO_2 concentration. This can be explained as follows. In a high CO_2 concentration, low efficiency but simple form II RubisCO is enough to support the autotrophic growth of this fast-growing bacterium, but in a low CO_2 concentration, form II RubisCO cannot support the growth, and high efficiency but complicated form I RubisCO, CbbLS-2 becomes necessary for the fast cell growth. The reason for the existence of another form I RubisCO, CbbLS-1, which is probably weakly transcribed in 10% CO_2 concentration, is still unclear.There may exist much more complicated regulatory mechanism for three RubisCO genes in *H. marinus*.

Structure and role of the *cbbQ* gene located at the downstream region of *cbbLS* from *P. hydrogenothermophila*

Pseudomonas hydrogenothermophila strain TH-1 (Goto et al., 1977, 1978) is a thermophilic hydrogen oxidizing bacterium with the optimal temperature of 52-53C. This strain, isolated from a hot spring in Japan, also fixes CO_2 via Calvin Cycle. RubisCO from the strain has been purified and shown to be Form I (Chung et al., 1994). The RubisCO gene (*cbbLS*) was cloned and sequenced (Yokoyama et al., 1995). The deduced amino acid sequences of large and small subunits of RubisCO from *P. hydrogenothermophila* showed the highest identity with those of *C. vinosum* RbcAB, *Alvinoconcha hessleri* symbiotic bacterium RubisCO, and *H. marinus* cbbLS-1. It is interesting that *P. hydrogenothermophila* and all of these three strains were obtained from aquatic environment.

In the case of *Alcaligenes eutrophus* (Windhovel, Bowien, 1991) and *Xanthobacter fluvus* (Van den Bergh et al., 1993), *cbbX* encoding an ATP-binding protein is located at the downstream region of *cbbS*. We also found a novel gene (ORF O, and later designated as *cbbQ*) in this region. Although the translated sequence of *cbbQ* had a putative ATP binding motif, but *cbbQ* was not homologous to *cbbX* from *A. eutrophus* or *X. fluvus*.

Surprisingly, we found a sequence similar to *cbbQ* in the gene cluster for denitrification from *Pseudomonas aeruginosa* (Fig. 2, Yokoyama et al., 1995). The translated amino acid sequence of *cbbQ* had 53% identity to *nirQ* gene product from *P. aeruginosa* (Arai et al.,1994), and 49% to *nirQ* from *Pseudomonas stutzeri* (Jungst, Zumft, 1992) . We also found a putative ORF similar to *cbbQ* downstream from *rbcAB* genes from *C. vinosum* (Viale et al., 1989). The deduced amino acid sequence of the incomplete ORF starting 153 downstream from *rbcB* showed 73% identity to the N-terminal portion of the *cbbQ* gene product. It is interesting that the two chemo- and photo-autotrophs having the similar RubisCO-encoding genes also possess the similar downstream ORF. Recently, we also found a similar DNA sequence to *cbbQ* at the downstream region of *cbbM* from *H. marinus*, though sequencing of the region has not been completed (Hayashi, unpublished).

In the cases for *P. aeruginosa* and *P. stutzeri*, the *nirQ* gene is located in the denitrification gene cluster, and has been shown to be essential for denitrification. A *nirQ* mutant of *P. stutzeri* simultaneously lost the activity of both of nitrite reductase(Nir) and nitric oxide reductase(Nor) in vivo, though both reductase proteins were synthesized (Jungst, Zumft, 1992). So, the role of the *nirQ* gene product is believed to be the post-translational activation of Nor and/or stabilization of the Nir/Nor complex.

```
Phy          MDLRNQYLVRSEPYYHAVGDEIERFEAAYANRIPMMLKGPTG    42
Cvi          MSDIDRNQFLIDHEPYYRPVSNEVALYEAAYAARMPVMLKGPTG  44
Pae             MRDATPFYEATGHEIEVEERAWRHGLPVLLKGPTG       35
Pst   MRYLPVNAIEIPTTAGTPDAPFYQPLGNEEQLFQQAWQHGMPVLIKGPTG    50


Phy   CGKSRFVEYMAWKLGKPLITVACNEDMTAADLVGRFLLDKEGTRWQDGPL    92
Cvi   CGKTRFVEYMAWKLGKPLITVACNEDMTAS                        74
Pae   CGKTRFVQYMARRLELPLYSVACHDDLGAADLLGRHLIGADGTWWQDGPL    85
Pst   CGKTRFVQHMAHRLNLPLYTVACHDDLSAADLVGRHLIGAQGTWWQDGPL   100


Phy   TTAARIGAICYLDEVVEARQDTTVVIHPLTDHRRILPLDKKGEVVEAHPD   142
Pae   TRAVREGGICYLDEVVEARQDTTVAIHPLADDRRELYLERTGETLQAPPS   135
Pst   TRAVREGGICYLDEVVEARQDTAVVLHPLADDRRELFIERTGEALKAPPG   150


Phy   FQIVISYNPGYQSAMKDLKTSTKQRFAAMDFDYPAPEVESEIVAHESGVD   192
Pae   FMLVVSYNPGYQNLLKGLKPSTRQRFVALRFDYPAAQQEARILVGESGCA   185
Pst   FMLVVSYNPGYQNLLKGMKPSTRQRFVAMRFDYPPTAEEERIVANEAQVD   200


Phy   AATAKKLVEVAIRSRHLKGHGLDEGISTRLLVYAGSLITKGIAPLIACEM   242
Pae   ETLAQRLVQLGQALRRLEQHDLEEVASTRLLIFAARLIGDGMDPREACRV   235
Pst   AALAAQVVKLGQALRRLEQHDLEEVASTRLLIFTARMIRSGMTPRQACLA   250


Phy   ALICPITDDPDLRDALRAAAQTLFA                            267
Pae   ALAEPLSDDPATVAALMDIVDLHVA                            260
Pst   CLAEPLSDDPQTVAALMDVVYVHFG                            275
```

Fig. 2. Comparison of amino acid sequence of NirQ type proteins. Phy; *Pseudomonas hydrogenothermophila* CbbQ, Cvi; *Chromatium vinosum* ORF, Pae; *Pseudomonas aeruginosa* NirQ, and Pst; *Pseudomonas stutzeri* NirQ. See Yokoyama et al., 1995.

Considering the position of *cbbQ* in the CO2 fixation gene cluster, the CbbQ protein may affect the post-translational activation and/or stabilization of RubisCO. Our preliminary experiment showed that *E. coli* cells harboring both *cbbLS* and *cbbQ* genes showed about 3 fold higher RubisCO activity than the cells harboring only *cbbLS* genes. Transcriptional activation of *cbbLS* genes by CbbQ could not occur in this experiment. Western blotting analysis showed that there existed no significant difference in the amount of RubisCO protein regardless that *E. coli* cells harbored the *cbbQ* gene or not. This observation indicated that CbbQ functioned post-translatioally to increase the RubisCO activity in *E. coli* cells. Recently we found another ORF at the downstream region of *cbbQ* of *P. hydrogenothermophila* (Hayashi, unpublished). The function of the ORF is still unknown, but it may have a role related to RubisCO activation/stabilization in cooperation with *cbbQ*.

Conclusions

(1) *Hydrogenovibrio marinus* contains three RubisCO genes (*cbbLS-1*, *cbbLS-2* and *cbbM*), and transcribes different RubisCO genes under different CO2 concentrations.
(2) The *cbbQ* gene, similar to *nirQ* of the denitrification gene cluster of *Pseudomonas* species, locates at the downstream region of cbbLS of *Pseudomonas hydrogenothermophila*. The *cbbQ* gene product may have an important role to activate and/or stabilize the RubisCO protein.

References

Arai H et al (1994) Biosci. Biotech. Biochem. 58, 1286-1291.
Chung SY et al (1993) FEMS Microbil. Lett. 109, 49-54.
Chung SY et al (1994) J. Ferment. Bioeng. 78, 469-471.
Glover HE (1989) Int. Rev.Cytol. 115, 67-138.
Goto E et al (1977) Agric. Biol. Chem. 41, 685-690.
Goto E et al (1978) Agric. Biol. Chem. 42, 1305-1308.
Igarashi Y et al (1992) Abstracts of 7th International Symposium on Microbial Growth on C1 compounds, C131.
Jouanneau Y, Tabita FR (1987) Arch. Biochem. Biophys. 254, 290-303.
Jungst A, Zumft WG (1992) FEBS Lett. 314, 308-314.
Kobayashi H et al (1991) Gene 97, 55-62.
Laing WA et al (1974) Plant Physiol. 54, 678-685.
Nishihara H et al (1989) Arch. Microbiol. 152, 39-43.
Nishihara H et al (1991a) Int. J. Syst. Bacteriol. 41, 130-133.
Nishihara H et al (1991b) J. Ferment. Bioeng. 72, 358-361.
Tabita FR et al (1993) In Murrell JC and Kelly DP, eds, Microbial Growth on C1 Compounds, pp469-479, Intercept Ltd, Andover, England.
Van den Bergh ERE et al (1993) J. Bacteriol. 175, 6097-6104.
Viale AM et al (1989) J. Bacteriol. 171, 2391-2400.
Windehovel U, Bowien B (1991) Mol. Microbiol. 5, 2695-2705.
Yaguchi T et al (1994) Biosci. Biotech. Biochem. 58, 1733-1737.
Yokoyama K et al (1995) Gene 153, 75-79.

Molecular Control and Biochemistry of CO_2 Fixation in Photosynthetic Bacteria

Janet L. Gibson, Yilei Qian, George C. Paoli, James M. Dubbs, H. Howard Xu, Hemalata V. Modak, Kempton M. Horken, Thomas M. Wahlund, Gregory M.F. Watson, and F. Robert Tabita

Department of Microbiology, The Ohio State University, Columbus, Ohio

INTRODUCTION

Photosynthetic bacteria are capable of diverse modes of CO_2 assimilation, employing either the Calvin reductive pentose phosphate, reductive tricarboxylic acid, or hydroxyproprionate cycles, or in some cases various ancillary reactions. In purple bacteria and cyanobacteria, the Calvin reductive pentose phosphate pathway predominates (Tabita 1988, 1994, 1995), while green sulfur bacteria use the reductive TCA cycle and green nonsulfur bacteria employ the hydroxypropionate path (Sirevåg 1995). Biochemical studies in our laboratory have focused on the enzymology of ribulose 1,5-bisphosphate carboxylase/oxygenase (RubisCO) and phosphoribulokinase (PRK), the key enzymes of the Calvin cycle, and ATP-citrate lyase and the ferrodoxin-linked reactions of the reductive TCA cycle. The molecular control of CO_2 fixation in nonsulfur purple bacteria has been a continuous interest (Gibson 1995), while the organism *Chlorobium tepidum* (Wahlund et al. 1991) has recently provided us with a useful and genetically tractable model system to study regulation of the reductive TCA cycle.

RESULTS

RubisCO and ATP-citrate lyase. Recently, four classes of form I (L_8S_8) RubisCO were grouped according to sequence relatedness (Tabita 1995). Interestingly, the primary structure of RubisCO from the phycoerythrin-containing marine cyanobacterium *Synechococcus* sp. strain WH 7803 was shown to more resemble the enzyme from various chemoautotrophic bacteria than previously determined sequences from cyanobacteria (G.M.F. Watson, unpublished). In as much as we are predominantly interested in the basis for differences in CO_2/O_2 specificity of RubisCO, we have sought to compare and contrast the ability of diverse sources of RubisCO to discriminate between CO_2 and O_2. Interestingly,

M. E. Lidstrom and F. R. Tabita (eds.), Microbial Growth on C₁ Compounds, 94–101.

neither the known x-ray structures nor sequence comparisons of Types IA, IB, IC, ID form I RubisCO provide obvious clues to the basis of RubisCO specificity. Indeed, homologous Type IC RubisCO from *Rhodobacter sphaeroides*, *Alcaligenes eutrophus*, and *Xanthobacter flavus* or homologous plant and cyanobacterial (*Synechococcus* sp. strain PCC 6301) enzymes of Type IB have widely diverse specificities. In an attempt to ascertain the basis for CO_2/O_2 discrimination of various Type IC and IB enzymes, we have made site-directed changes in regions of the molecule that show some variance and are known to influence specificity. We have also developed a method to employ biological selection, using a RubisCO-deficient host of *R. sphaeroides*, to assist in these studies. With respect to our studies in *C. tepidum*, we have isolated and highly purified the ATP-citrate lyase. This enzyme appears to have a native molecular weight of about 260,000 and is either comprised of a single 65,000 kD subunit that readily proteolyzes to a 45 kD product or is composed of two distinct polypeptides (Wahlund, unpublished). Several mutants of *C. tepidum* defective in photoautotrophic growth have been isolated to further our studies on the regulation of the reductive TCA cycle.

Organization of Calvin cycle genes in nonsulfur purple bacteria. Three organisms have been studied extensively, *R. sphaeroides* (Gibson 1995), *R. capsulatus* (Paoli et al. 1995), and *Rhodospirillum rubrum* (Falcone, Tabita 1993). In each of these organisms, structural (*cbb*) genes are clustered into distinct operons (Fig. 1), with the LysR-type transcriptional activator protein (CbbR), the product of *cbbR*, positively regulating transcription (Gibson, Tabita 1993), much like the situation in other autotrophic bacteria.

Chemoautotrophic growth of R. sphaeroides. Nonsulfur purple bacteria are capable of growing under as many as five different growth regimens including, in some cases, an aerobic chemoautotrophic growth mode. *R. capsulatus* has been shown to exhibit relatively good chemoautotrophic growth in an atmosphere of 10% O_2/10% CO_2/80% H_2, whereas the closely related organism *R. sphaeroides* was previously found to grow poorly under these conditions. Recently, through a serendipitous set of circumstances, a naturally occuring variant of *R. sphaeroides*, strain CAC (chemoautotrophic competent), was isolated that was capable of chemoautotrophic growth (G.C. Paoli, unpublished). Cultures of *R. sphaeroides* were shown to acquire the chemoautotrophic competent phenotype following incubation in an atmosphere of oxygen, carbon dioxide, and hydrogen for approximately three weeks. RubisCO activity and the relative amounts of form I and form II RubisCO protein, determined by immunological means, were not very different from that observed in photoautotrophic cultures. In addition, analysis of the ability of various Calvin cycle mutants to grow under these conditions allowed certain requirements for chemoautotrophic growth to be defined. Whereas neither form I nor form II RubisCO was found to be exclusively necessary for chemoautotrophic growth, RubisCO was required since a mutant lacking both forms of RubisCO could not become competent for chemoautotrophic growth. Strain 1312, which carries a mutation in the gene encoding CbbR, a positive regulator of

96

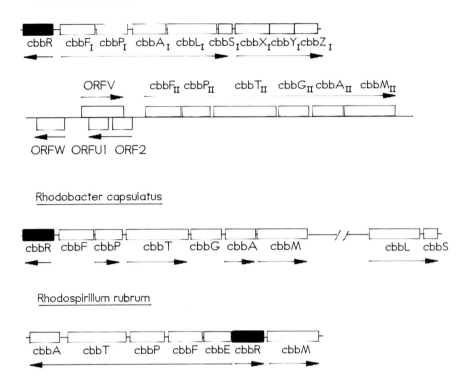

Fig. 1. Organization of *cbb* genes in three nonsulfur purple photosynthetic bacteria

both Calvin cycle operons, was also incapable of making the switch to chemoautotrophic growth.

Regulation of Calvin cycle gene expression in R. sphaeroides. As mentioned previously, nonsulfur purple bacteria are capable of growing under a wide range of culture conditions, in which the need for CO_2 fixation varies greatly. Synthesis of Calvin cycle enzymes, RubisCO and PRK, parallels this demand. During autotrophic growth, all of the cellular carbon is derived from CO_2 and maximal expression of RubisCO is observed. However, during photoheterotrophic growth on organic carbon compounds, CO_2 fixation functions primarily as an electron sink, and lower levels of RubisCO are present. Obviously, *cbb* gene expression is subject to stringent regulation that responds to a variety of environmental signals. In *R. sphaeroides*, two different forms of RubisCO are synthesized that are situated within distinct operons, cbb_I and cbb_{II} (Fig. 1). Expression of the two operons, though independent in response to various signals, is coordinated through a common regulatory protein (CbbR) of the LysR family, that is divergently transcribed from the cbb_I operon (Gibson, Tabita 1993). Mutations within one operon elicit a corresponding increase in transcription from the second operon, an effect mediated by the $cbbR$ gene product. Although it was previously shown that both the cbb_I and cbb_{II} operons were positively regulated by CbbR, nothing is known about the regulatory sequences involved in this control. Recent studies have been directed toward an elucidation of the cbb_I and cbb_{II} promoters and other factors or sequences that may be involved in the complex regulation that has been observed.

In order to define sequences required for cbb_I gene expression, $lacZ$ translational fusions to $cbbF_I$ were constructed that contained varying amounts of DNA upstream of $cbbF_I$. R. sphaeroides, containing the fusion constructs, was grown under a variety of growth conditions known to affect synthesis of Calvin cycle enzymes (J.M. Dubbs, unpublished). Although only 307 bp upstream of $cbbF_I$ were found to be required for regulated expression of ß-galactosidase, high level expression was observed only when 663 bp of upstream DNA were present. These results suggest that DNA sequences outside of the $cbbR$-$cbbF_I$ intergenic region, that actually extend into the $cbbR$ coding sequence, are involved in control of cbb_I expression. This observation is intriguing, because in most LysR-regulated systems, the entire complement of regulatory sequence is contained within the intergenic region between the $lysR$ coding sequence and the divergently oriented genes it regulates.

A similar line of experimentation was used to examine the cbb_{II} control region. A restriction fragment containing the putative cbb_{II} promoter sequence was cloned into a $xylE$ based promoter probe vector. A set of nested deletions obtained for both orientations of the 2.1 kb fragment was examined for promoter activity in R. sphaeroides grown under different culture conditions (Xu, Tabita, 1994). In addition to the $cbbF_{II}$ promoter, three additional promoter activities were revealed, U_p, V_p and W_p, that were found by sequence analysis to be associated with open reading frames of unknown function (Fig. 1). With the exception of W_p, all promoters were repressed by oxygen. In addition, when the promoter activities were examined in $cbbR$ and $cbbF_I$ mutants, backgrounds that have been shown to reduce and enhance cbb_{II} transcription, respectively, both positive and negative effects on $xylE$ expression were observed. Further studies are underway to elucidate the function of these genes, since the expression patterns indicate an association with cbb gene expression.

Global control of CO$_2$ fixation. Earlier studies in this lab were directed toward understanding the role of different forms of RubisCO in R. sphaeroides. To address this issue, strains were constructed in which genes encoding one or both forms of RubisCO had been inactivated. Mutants lacking either form I or form II RubisCO exhibited wild type growth rates in both photoheterotrophic and photoautotrophic growth modes. Strain 16, however, which lacks both forms of RubisCO, was unable to grow under conditions where CO_2 is used as an electron acceptor. This strain could not grow photoheterotrophically unless DMSO, an alternate electron acceptor, was supplied. However, upon incubation of strain 16 under photoheterotrophic conditions in the absence of DMSO for several weeks, growth was observed, and the resultant strain, 16PHC (photoheterotrophic competent), was shown to contain no RubisCO protein (Wang et al. 1993). RubisCO-independent growth was especially intriguing, since the electron acceptor pathway allowing growth to occur could represent an alternate CO_2 fixation pathway. Initial attempts at identifying the gene product(s) responsible for this conversion to photoheterotrophic competency involved mutagenesis of strain 16PHC with a gentamicin resistant derivative of Tn5. One mutant, strain A25, was obtained that grew poorly photoheterotrophically in the absence of DMSO. The transposon-containing DNA fragment of strain A25 was used as a hybridization probe to obtain a plasmid from a genomic library of strain 16PHC. Sequence analysis revealed three open reading frames, $regA$, $regB$, and $regC$ (Qian, Tabita 1995). Database searches showed the sequences to be identical to previously identified genes in $R.$ $capsulatus$ and $R.$ $sphaeroides$. Inactivation of the second gene, $regB$, was found to be responsible for the growth defect in strain A25. The deduced amino acid sequence of this gene shared 100% identity with $prrB$ from $R.$ $sphaeroides$ (Eraso, Kaplan 1995) and 57% identity to $regB$ from $R.$ $capsulatus$ (Mosely et al 1994). In both cases, the gene product is a protein related to a class of bacterial sensor kinases that have been shown to regulate anaerobic synthesis of light harvesting and reaction center proteins.

In order to assess the effect of the mutation in a wild type background, the $regB$ gene was insertionally inactivated in wild type strain HR (Qian, Tabita 1995). HRΩ, the $regB$ mutant of strain HR, exhibited normal growth rates during photoheterotrophic growth on malate but grew very poorly in an atmosphere of 1.5% CO_2/98.5% H_2. In addition, carbon starvation conditions failed to elicit the derepression of the cbb operons in strain HRΩ that is normally observed in the wild type. Since the expression of cbb genes seemed to be affected by the $regB$ mutation, plasmids containing the cbb_I and cbb_{II} promoters fused to $lacZ$ and $xylE$ reporter genes, respectively, were mated into strains HR and HRΩ and the respective enzyme activities monitored in the two strains under different growth conditions. In the wild type, RubisCO, ß-galactosidase [$lacZ$ fusion to the cbb_I promoter] and catechol 2,3-dioxygenase [$xylE$ fusion to the cbb_{II} promoter] were elevated upon incubation under photosynthetic growth conditions. However, in strain HRΩ, only low levels of RubisCO and reporter gene enzyme activities were found under photoheterotrophic and photoautotrophic conditions (1.5% CO_2/98.5% H_2 atmosphere), suggesting that the $regB$ mutation affected transcription of both the cbb_I and cbb_{II} operons.

Since $regB$ has been implicated in the anaerobic regulation of light harvesting components, it was of interest to examine the effect of the mutation on dark aerobic synthesis of RubisCO. To determine the effect of the $regB$ mutation on dark, aerobic expression of the cbb operons, $regB$ was inactivated in strain CAC, yielding strain CACΩ (Qian, Tabita 1995). Interestingly, strain CACΩ was found to be incapable of chemoautotrophic growth and was impaired in photoautotrophic growth. Following incubation of strain HRΩ under chemoautotrophic condtions, no derepression of RubisCO was observed. The cbb_I and cbb_{II} promoter fusion constructs were mated into strains CAC and CACΩ and reporter enzyme activities measured under chemoheterotrophic and chemoautotrophic conditions. ß-galactosidase and 2,3-catechol dioxygenase activities, indicative of cbb_I and cbb_{II} expression, respectively, were dramatically increased in strain CAC when switched from chemoheterotrophic to chemoautotrophic conditions. In contrast, the increase in reporter enzyme activity was not observed in strain CACΩ. Therefore, $regB$ appears to influence transcription of the cbb operons in the presence of oxygen in the dark, as well as in the absence of oxygen in the light.

In the course of analyzing strain 16PHC and the $regB$ mutant, strain A25, certain polypeptides were observed on SDS gels that were not present in the wild type strain HR or either of the mutant strains 16 and A25. The molecular weights of these proteins corresponded to those of the subunits of nitrogenase, which was unexpected since the cultures were grown in the presence of ammonia, a condition that normally represses synthesis of nitrogenase. Antibodies directed against the subunits of nitrogenase reacted with the suspect proteins in extracts of 16PHC, confirming the identity of the proteins as nitrogenase constituents (H.V. Modak, unpublished). RegB thus appears to be global in the regulation of diverse systems, including Calvin cycle gene expression, the alternate electron acceptor pathway in strain 16PHC, nitrogen fixation, and synthesis of light harvesting components for photosynthesis.

cbbXYZ operon in R. sphaeroides. Recently, three new genes, *cbbX, cbbY* and *cbbZ* have been identified, based on sequence comparisons, downstream from the cbb_I operon. Although the function of *cbbX* and *cbbY* has not been determined, it is known that *cbbZ* encodes phosphoglycolate phosphatase (PGP)(Shäferjohann et al. 1993). *R. sphaeroides* strains were constructed in which *cbbX, cbbYZ* or *cbbZ* were inactivated. None of the mutants exhibited PGP activity, suggesting the three genes form part of an operon. In addition, the presence of PGP activity in cbb_I insertion mutants implies the presence of a promoter separate from the cbb_I promoter. Only the strain carrying the mutation in *cbbX* exhibited a discernible phenotype. This strain was initially incapable of growing photoautotrophically, however, after prolonged incubation under photoautotrophic conditions, growth did occur. Of obvious interest, was the effect of mutations in these genes on aerobic RubisCO-dependent CO_2 fixation, since phosphoglycolate, the substrate of PGP, is the product of the RubisCO oxygenase reaction. Interestingly, the *cbbX* mutant was able to adapt to chemoautotrophic growth as well as the wild type strain, demonstrating a lack of a requirement for the gene products of the *cbbXYZ* operon during chemoautotrophic growth under the conditions described.

CONCLUSIONS

Photosynthetic prokaryotes are excellent systems for studying aspects of both the biochemistry and molecular biology of CO_2 fixation. Progress has been rapid relative to the nonsulfur purple bacteria and, to some extent, in various aspects of cyanobacterial CO_2 fixation (Tabita 1994, Kaplan et al. 1994). For organisms that use the reductive TCA pathway, *C. tepidum* promises to be an excellent model system.

Support from the NIH, DOE, and NSF is gratefully acknowledged.

REFERENCES

Eraso JM, Kaplan S (1995) J. Bacteriol. 177, 2695-2706.
Falcone DL, Tabita FR (1993) J. Bacteriol. 175, 5066-5077.
Gibson JL, Tabita FR (1993) J. Bacteriol. 175, 5778-5784.
Gibson JL (1995) In Blankenship, RE, Madigan MT, and Bauer CE, eds, Anoxygenic Photosynthetic Bacteria, pp. 1107-1124, Kluwer Acad. Publ., Dordrecht, The Netherlands.
Kaplan A et al (1994) In Bryant DA, ed, The Molecular Biology of Cyanobacteria, pp. 469-485, Kluwer Acad. Publ., Dordrecht, The Netherlands.
Mosely CF et al (1994) J. Bacteriol. 176, 7566-7573.
Paoli GC et al (1995) Arch. Microbiol. In press.
Qian Y, Tabita FR (1995) J. Bacteriol. In press.
Shäferjohann Y et al (1993) J. Bacteriol. 175, 7329-7340.
Sirevåg R (1985) In Blankenship RE, Madigan MT, and Bauer CE, eds, Anoxgenic Photosynthetic Bacteria, pp. 871-883, Kluwer Acad. Publ., Dordrecht, The Netherlands.
Tabita FR (1988) Microbiol. Rev. 52,155-189.
Tabita FR (1994) In Bryant DA, ed, The Molecular Biology of Cyanobacteria, pp. 437-467, Kluwer Acad. Publ., Dordrecht, The Netherlands.
Tabita FR (1995) In Blankenship RE, Madigan MT, and Bauer CE, eds, Anoxygenic Photosynthetic Bacteria, pp. 885-914, Kluwer Acad. Publ., Dordrecht, The Netherlands.

Wahlund TM et al (1991) Arch. Microbiol. 156, 81-90.
Wang X et al (1993) J. Bacteriol. 175, 3372-3379.
Xu HH, Tabita FR (1994) J. Bacteriol. 176, 7299-7308.

Biochemistry and genetics of organoautotrophy in *Alcaligenes eutrophus*

B. Bowien, J. Friedebold, B. Kusian, D. Bömmer and J. Schäferjohann

Institut für Mikrobiologie, Georg-August-Universität Göttingen, Göttingen, Germany

INTRODUCTION

Different metabolic capabilities have led to the distinction of two major groups of autotrophic organisms: (i) obligate and (ii) facultative autotrophs. The former are confined to the assimilation of CO_2 as the main source of cell carbon, whereas the latter are able to utilize many organic compounds as alternative or additional carbon and/or energy sources. The nutritional versatility varies considerably among the different subgroups of aerobic, facultatively autotrophic bacteria. Simultaneous utilization of inorganic and organic carbon and/or energy substrates for growth is designated mixotrophy and a rather common feature of these organisms (Barton et al. 1991). In this context the use of C_1 and related compounds such as methanol, formate and methylated sulfides are of particular interest. These compounds or their carbon moieties are oxidized to CO_2 for conservation of energy and generation of reducing power required for CO_2 assimilation (Quayle 1972; Kelly, Baker 1990). Such special type of mixotrophy is termed organoautotrophy as opposed to the conventional lithoautotrophic metabolism in these organisms that is based on the oxidation of inorganic energy-yielding substrates. In all bacteria studied so far for this property, CO_2 fixation involves the Calvin-Benson-Bassham carbon reduction pathway (Calvin cycle).

The Gram-negative bacterium *Alcaligenes eutrophus* is an aerobic facultative chemoautotroph able to grow either lithoautotrophically with hydrogen or organoautotrophically with formate as energy substrates (Bowien, Schlegel 1981). During growth on formate the organism forms two types of formate dehydrogenases (FDH), a soluble, NAD^+-reducing enzyme (S-FDH; EC 1.2.1.2) and a membrane-associated enzyme (M-FDH, formate:acceptor oxidoreductase) that is directly coupled to the respiratory chain. Although the metabolic requirements suggest a close regulatory relation between the FDH and the Calvin cycle, the the control of the syntheses of both enzyme systems does not appear to be directly linked (Friedrich et al. 1979). The same holds true for the regulation of the hydrogenase system in *A. eutrophus* (Friedrich, Schwartz 1993). Therefore, the study of biochemical and genetic aspects is essential for the elucidation of the regulatory network(s) controlling the autotrophic energy and carbon metabolism in

M. E. Lidstrom and F. R. Tabita (eds.), Microbial Growth on C_1 Compounds, 102–109.

this organism. The present work provides current molecular information about organoautrophy in *A. eutrophus*.

RESULTS

Formation of S-FDH. The synthesis of active S-FDH in *A. eutrophus* was strictly dependent on the presence of formate in the growth medium, suggesting that this compound might act as the inducer of the corresponding genes (Friedebold, Bowien 1993). It also required molybdate. Tungstate, a well-known antagonist of molybdate in biological systems, inhibited both the organoautotrophic growth and the formation of active enzyme. However, supplementation with molybdate or tungstate did not affect the synthesis of inactive S-FDH protein as disclosed by immunoblotting using polyclonal S-FDH antibodies. Rocket immunoelectrophoresis showed that S-FDH protein constituted about 0.65 % of the soluble protein in formate-grown cells (Friedebold et al. 1995). The pattern of M-FDH formation is presumably different, since lithoautotrophically grown cells also contained significant M-FDH activity. The synthesis of the two FDH is thus unlikely to be under common control (Friedrich et al. 1979). Surprisingly, M-FDH activity was not significantly influenced by tungstate, emphasizing the essential role of the S-FDH for growth of the organism on formate. The fact that wild-type strain H16, harboring the 450-kb plasmid pHG1 (Friedrich et al. 1981a), as well as plasmid-cured mutants utilize formate indicates a location of the FDH genes on the chromosome. Genetic studies on the formate-oxidizing system of *A. eutrophus* are currently being performed in our laboratory.

Structural and functional properties of S-FDH. The activity of S-FDH in cell extracts of *A. eutrophus* was unstable except in the presence of nitrate or azide (Friedebold et al. 1993). Both anions, which inhibited the enzyme, are thought to be transition-state analogs of formate during the catalysis by FDH (Blanchard, Cleland 1980). Purification of the stabilized enzyme resulted in an apparently homogeneous protein of high specific activity (80-100 U mg^{-1}). The molecular mass of the native enzyme was determined to be 197 kDa. Four nonidentical subunits of 110 (α), 57 (β), 19.4 (γ), and 11.6 kDa (δ), respectively, form the heterotetrameric S-FDH molecule (Friedebold, Bowien 1993). NH$_2$-terminal amino acid sequence analyses of the subunits (Friedebold et al. 1995) revealed significant similarities to the corresponding subunits of the FDH from the obligate methanotroph *Methylosinus trichosporium* (Jollie, Lipscomb 1991). The latter enzyme together with S-FDH from *A. eutrophus* and the FDH from another chemoautotroph, *Pseudomonas oxalaticus* (Müller et al. 1978) seems to belong to a subgroup of complex structured, NAD$^+$-reducing FDH occurring in aerobic bacteria (Table 1). Immunological studies confirmed this conclusion and suggested that other autotrophic bacteria might also contain this type of FDH (Friedebold et al. 1995).

An electron microscopical analysis provided the first direct image of an FDH molecule and confirmed the heterotetrameric quaternary structure of S-FDH. Four submasses of different sizes were observed that formed a nonglobular particle similar in appearance to a clover leaf (Friedebold et al. 1995). S-FDH is a molybdo-iron-sulfur-flavo protein (Table 1). Its molybdenum atom is bound to a molybdopterin guanine dinucleotide

(MGD) cofactor also found in FDH from other organisms (Rajagopalan, Johnson 1992). The flavin mononucleotide (FMN) component tended to get released from the enzyme during catalysis in vitro, causing inactivation. The high content of iron and acid-labile sulfur suggests the presence of several Fe-S clusters that is also indicated by the absorption spectrum of the enzyme (Friedebold, Bowien 1993). Despite of the significant content of nickel and zinc, these two metals are not considered to be redoxactive components of S-FDH (Friedebold et al. 1995). Since nickel is not required for growth of *A. eutrophus* on formate (Friedrich et al. 1981b), a functional role of the metal is unlikely.

Table 1. Some properties of heteromeric, NAD$^+$-dependent formate dehydrogenases from three aerobic bacteria

Property	*Alcaligenes eutrophus*	*Methylosinus trichosporium*	*Pseudomonas oxalaticus*
Mass of Holoenzyme (kDa)	197	400	315
Masses of Subunits (kDa)	α: 110 β: 56 γ: 19.4 δ: 11.6	α: 98 β: 56 γ: 20 δ: 11.5	α: 100 β: 59
Quaternary Structure	$\alpha\beta\gamma\delta$	$(\alpha\beta\gamma\delta)_2$	$(\alpha\beta)_2$
Cofactors (mol mol^{-1})	0.91 FMN 0.72 MGD Fe-S	1.8 Flavin Mo cofactor Fe-S	2 FMN Fe-S
Metals (g-atom mol^{-1})	21.6 Fe 0.76 Mo 2.18 Ni 0.59 Zn	46 Fe 1.5 Mo	18-25 Fe
S^{2-} (g-atom mol^{-1})	20±1	38	15-20
pH Optimum	7.7-7.9	n.d.[a]	7.4
K$_m$ Formate (mM)	3.3	0.1	0.14
K$_m$ NAD$^+$ (mM)	0.09	0.08	0.11
Reverse Reaction	no	no	yes

[a]n.d., not determined

Initial electron paramagnetic resonance (EPR) spectroscopic studies on the S-FDH reduced with either formate or NADH showed at least three distinct signals. The corresponding spin systems could be attributed to two [2Fe-2S] and one [4Fe-4S] centers. No signals common to molybdenum (V) or a flavin semiquinone radical were detected (Friedebold et al. 1995). Surprisingly, the EPR signals did not correlate to the high total iron content of the enzyme. The deficit of signals might have been due to spin-spin interactions between centers, incomplete reduction of the enzyme, lack of involvement of some of the [Fe-S] centers in the intramolecular electron tranport or/and partial damage of the purified S-FDH. The fact that the reduction with formate only allowed to address one of the [2Fe-2S] centers corresponded to the observed inactivation of S-FDH by formate in the absence of the elctron acceptor NAD^+, a probable consequence of FMN release (Friedebold, Bowien 1993). Based on the presently available information on its redoxactive components, a hypothetical scheme of the electron transport in the enzyme can be proposed (Fig. 1).

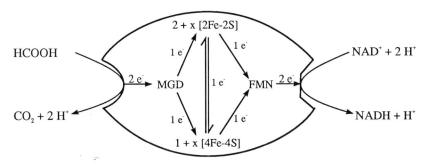

Fig. 1. Hypothetical scheme for organization of redox centers and for intramolecular electron transport in the S-FDH from *A. eutrophus*.

Organization of the cbb *operons.* Most enzymes of the Calvin cycle in *A. eutrophus* H16 are encoded by *cbb* genes (Tabita et al. 1992) organized in an unusual manner. The strain possesses two large, highly homologous *cbb* operons that are simultaneously expressed (Bowien et al. 1993). One operon is located on the chromosome, the other on plasmid pHG1, which also carries the genes of the hydrogenase system adjacent to the *cbb* genes (Friedrich, Schwartz 1993). The overall sequence identity of the two operons approaches about 95 %, comparing the regions between the 5'-terminal *cbbL* and the distal *cbbA* genes (Fig. 2). A gene duplication event is the most likely cause for this genetic setup. Several strains of *A. eutrophus*, with the exception of the type strain, have been shown to contain chromosomally as well as plasmid-borne *cbb* genes (Andersen, Wilke-Douglas 1984; Bowien 1989). Thus, chromosomal and plasmid locations of *cbb* genes might be considered as a characteristic feature of this species.

Except for pentose-5-phosphate isomerase and triosephosphate isomerase, all enzymes of the Calvin cycle are encoded within the *cbb* operons (Fig. 2). The functions of two genes, *cbbX* and *cbbY*, remain to be elucidated (Kusian et al. 1992), whereas *cbbZ* codes for 2-phosphoglycolate phosphatase (Schäferjohann et al. 1993). Although the latter enzyme is not involved in the Calvin cycle, it is necessary for the degradation of 2-phosphoglycolate that is formed by the action of ribulose-1,5-bisphosphate carboxylase oxygenase (RuBisCO) in the presence of molecular oxygen (Lorimer et al. 1973). The inclusion of *cbbZ* in the operons might, therefore, be the genetic manifestation of a metabolic interlock between the Calvin cycle and the oxidative glycolate pathway in autotrophically growing cells of *A. eutrophus*. Recently, an additional gene, designated *cbbB*, was detected downstream of *cbbA* in the chromosomal operon (Fig. 2). It does not

106

have a counterpart in the plasmid-borne operon and formally encodes an FDH-like product (D. Bömmer, B. Bowien: unpublished results).

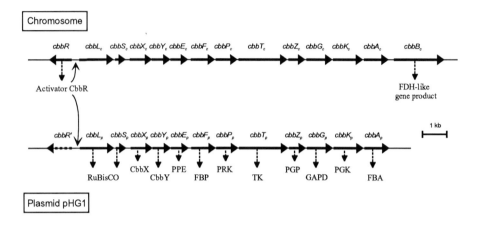

Fig. 2. Organization of the chromosomal and plasmid-borne *cbb* gene clusters of *A. eutrophus* H16. Abbreviations of gene products: FBA, fructose-1,6-/sedoheptulose-1,7-bisphosphate aldolase; FBP, fructose-1,6-/sedoheptulose-1,7-bisphosphatase; FDH, formate dehydrogenase; GAPD, glyceraldehyde-3-phosphate dehydrogenase; PGK, 3-phosphoglycerate kinase; PGP, 2-phosphoglycolate phosphatase; PPE, pentose-5-phosphate 3-epimerase; PRK, phosphoribulokinase; TK, transketolase; RuBisCO, ribulose-1,5-bisphosphate carboxylase oxygenase. *cbbR'* denotes the defective *cbbR* gene on phG1. The target sites of activator protein CbbR are indicated by arrows.

The formation of isoenzymes operating in the central carbon metabolism of *A. eutrophus* H16 is a consequence of the expression of the *cbb* operons. Since the functions of most of these enzymes are also required under heterotrophic conditions for glycolysis, gluconeogenesis and/or pentose biosynthesis, when the operons are fully repressed (see below), the organism needs another set of genes coding for the corresponding enzymes. Genes for glyceraldehyde-3-phosphate dehydrogenase (*gap*) and fructose-1,6-bisphosphatase (*fbp*) have already been detected on the chromosome of the organism (Windhövel, Bowien 1990a; Yoo, Bowien 1995).

Regulation of the cbb *genes.* The two *cbb* operons of *A. eutrophus* H16 form a regulon controlled by the transcriptional activator protein CbbR, the product of the chromosomal *cbbR* gene. This regulatory gene is divergently oriented to and separated by 167 bp from the chromosomal *cbb* operon (Fig. 2; Windhövel, Bowien 1991). The intergenic space between *cbbR* and *cbbL* comprises the *cbb* control region containing the target site (operator) of CbbR and the pertaining promoters (Kusian et al. 1995; Kusian, Bowien 1995). CbbR belongs to the LysR family of bacterial regulators. Mutational inactivation of the *cbbR* gene abolished expression of the operons and resulted in the inability of the mutant to grow autotrophically. Although the phG1-borne copy of the gene is defective due to a short deletion within its 5'-end (Windhövel, Bowien 1991), it is transcribed like the functional gene (Kusian et al. 1995).

Autotrophic growth of *A. eutrophus* led to strong expression of the *cbb* operons that was further increased by CO_2 limitation with hydrogen as energy source. Complete repression prevailed during heterotrophic growth on organic substrates such as acetate, pyruvate, or succinate. However, growth on gluconate, fructose, or glycerol caused a partial derepression which, for still unknown reasons, depended on the presence of the pHG1-borne *cbb* operon (Bowien et al. 1990; Windhövel, Bowien 1990b). Studies on transcriptional fusions with the operon promoters confirmed the observed expression pattern (Kusian et al. 1995). These findings suggest a repression/derepression mechanism to control the expression of the operons with an intracellular signal compound to act as an effector of the CbbR activator.

A transcriptional analysis of the *cbb* control region revealed an overlapping arrangement of the σ^{70}-type promoters for the *cbb* operon and *cbbR* (Fig. 3). The transcription of *cbbR* was constitutive at a low level (Kusian et al. 1995). Moreover, two different *cbbR* promoters were found that were alternatively used in response to autotrophic versus heterotrophic growth of *A. eutrophus*. The 'heterotrophic' *cbbR* promoter is likely to be blocked during activated transcription from the operon promoter.

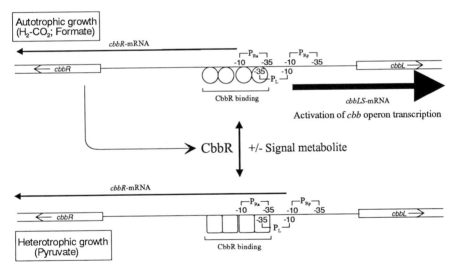

Fig. 3. Model of transcriptional regulation exerted in the cbb control regions of *A. eutrophus* H16. The chromosomal region is shown with the overlapping σ^{70}-type promoters of the *cbb* operon (P_L) and *cbbR* (P_{Ra} and P_{Rp}) as well as the CbbR-binding site. Depending on effector (signal metabolite) binding, two different conformations of CbbR are believed to exist, one of which promotes activation of the operon during autotrophic growth. The relative strengths of transcriptions are represented by the thickness of the arrows symbolizing the corresponding mRNA.

DNA footprintings showed that CbbR binds to an operator region comprising about 50 bp immediately upstream of the -35 region of the operon promoter. Dissection of this binding region allowed the differentiation of two subsites, the distal R (recognition)- and the proximal A (activation)-site. Insertion of one or two bp near the interface of the subsites did not influence CbbR binding, but prevented activation of the operon promoter (Kusian, Bowien 1995). Thus, a correct alignment of the subsites seems to be critical for the activation process. As proposed for other systems (Ishihama 1993), this process is thought to involve proper contacts between RNA polymerase and CbbR. Removal of the R-site not only abolished activation, but also strongly increased *cbbR* promoter activity, most probably to lack of CbbR binding at the A-site. The data confirmed the anticipated negative autoregulation of *cbbR* transcription. A working model of transcriptional *cbb* gene regulation in *A. eutrophus* summarizes our present knowledge of the system (Fig. 3).

CONCLUSIONS

The main difference between litho- and organoautotrophic metabolism of *A. eutrophus* concerns the operation of the energy-converting systems. Our studies have provided evidence that growth of *A. eutrophus* on formate depends on the function of the S-FDH, although the organism also synthesizes an M-FDH. The physiological role of the latter enzyme activity is still unclear. Unlike the simple, homodimeric enzymes occurring in some aerobic bacteria and in yeasts (Ferry 1990), S-FDH belongs to a subgroup of FDH that is characterized by a heteromeric subunit structure and redoxactive cofactors. Examination of further bacteria might reveal that this complex type of FDH is more widely distributed than presently known.

Phenotypic observations suggest no direct linkage in the genetic regulation of the FDH system and the *cbb* CO_2 assimilation operons of *A. eutrophus* H16. A thorough genetic analysis should provide the necessary information on the FDH control. Among the organisms studied so far, both the structure and size of the *cbb* operon are unique in *A. eutrophus* as is its almost complete duplication. However, there are partial similarities to the organization of *cbb* operons in other organisms (Meijer et al. 1991; Strecker et al. 1994; Tabita et al. 1993). The operons are apparently regulated by a *cbbR* gene, even in the obligate autotroph *Thiobacillus ferrooxidans* exhibiting constitutive expression of the *cbb* genes (Kusano, Sugawara 1993). It is tempting to speculate that in autotrophic bacteria using the Calvin cycle for CO_2 assimilation, with the possible exception of the cyanobacteria, control by a *cbbR* gene is the principal mode of *cbb* operon regulation. Whether a protein factor in addition to the activator protein CbbR might be necessary for the expression of the *cbb* operons in *A. eutrophus* is an open question. The search for (an) effector(s) of CbbR, most probably a metabolite acting as coactivator, remains a major task in the elucidation of *cbb* gene control.

ACKNOWLEDGEMENT

This work was supported by grants from the Deutsche Forschungsgemeinschaft.

REFERENCES

Andersen K, Wilke-Douglas M (1984) J. Bacteriol. 159, 973-978.

Barton LL et al (1991) In Shively JM and Barton LL, eds, Variations in Autotrophic Life, pp 1-23, Academic Press, London, UK.

Blanchard JS, Cleland WW (1980) Biochemistry 19, 3543-3550.

Bowien B (1989) In Schlegel HG and Bowien B, eds, Autotrophic Bacteria, pp 437-460, Science Tech, Madison, WI, USA.

Bowien B, Schlegel HG (1981) Annu. Rev. Microbiol. 35, 405-452.

Bowien B et al (1990) FEMS Microbiol. Rev. 87, 445-450.

Bowien B et al (1993) In Murrell JC and Kelly DP, eds, Microbial Growth on C_1 Compounds, pp 481-491, Intercept, Andover, UK.

Ferry JG (1990) FEMS Microbiol. Rev. 87, 377-382.

Friedebold J, Bowien B (1993) J. Bacteriol. 175, 4719-4728.

Friedebold J et al (1995) Biol. Chem. Hoppe-Seyler 376, in press.

Friedrich B, Schwartz E (1993) Annu. Rev. Microbiol. 47, 351-383.

Friedrich B et al (1981a) J. Bacteriol. 147, 198-205.

Friedrich B et al (1981b) J. Bacteriol. 145, 1144-1149.

Friedrich CG et al (1979) J. Gen. Microbiol. 115, 185-192.

Ishihama A (1993) J. Bacteriol. 175, 2483-2489.

Jollie DR, Lipscomb JD (1991) J. Biol. Chem. 266, 21853-21863.

Kelly DP, Baker SC (1990) FEMS Microbiol. Rev. 87, 421-246.

Kusano T, Sugawara K (1993) J. Bacteriol. 175, 1019-1025.

Kusian B, Bowien B (1995) J. Bacteriol. 177, in press.

Kusian B et al (1992) J. Bacteriol. 174, 7337-7344.

Kusian B et al (1995) J. Bacteriol. 177, 4442-4450.

Lorimer GH et al (1973) Biochemistry 12, 18-23.

Meijer WG et al (1991) Mol. Gen. Genet. 225, 320-330.

Müller U et al (1978) Eur. J. Biochem. 83, 485-498.

Quayle JR (1972) Adv. Microb. Physiol. 7, 119-203.

Rajagopalan KV, Johnson JL (1992) J. Biol. Chem. 267, 10199-10202.

Schäferjohann J et al (1993) J. Bacteriol. 175, 7329-7340.

Strecker M et al (1994) FEMS Microbiol. Lett. 120, 45-50.

Tabita FR et al (1992) FEMS Microbiol. Lett. 99, 107-110.

Tabita FR et al (1993) In Murrell JC and Kelly DP, eds, Microbial Growth on C_1 Compounds, pp 469-479, Intercept, Andover, UK.

Windhövel U, Bowien B (1990a) FEMS Microbiol. Lett. 66, 29-34.

Windhövel U, Bowien B (1990b) Arch. Microbiol. 154, 85-91.

Windhövel U, Bowien B (1991) Mol. Microbiol. 5, 2695-2705.

Yoo J-G, Bowien B (1995) Curr. Microbiol. 31, 55-61.

Hydrogen Oxidation by *Alcaligenes*

B. Friedrich, M. Bernhard, J. Dernedde, T. Eitinger, O. Lenz, C. Massanz, E. Schwartz
Institute of Biology, Humboldt-University of Berlin, Invalidenstrasse 43, D-10115 Berlin, Germany.

INTRODUCTION
       Detailed molecular studies of hydrogen metabolism have progressed rapidly thanks to the development of genetic systems, which are now available for selected proteobacteria, including species of lithoautotrophs, nitrogen-fixing and photosynthetic microorganisms (reviewed by Friedrich, Schwartz 1993 and Vignais, Toussaint 1994). These facultative hydrogen-oxidizers are abundant in soil and water. Adaptation to shortage of nutrients may explain the great metabolic versatility of these organisms. The ability to utilize a wide range of organic and inorganic substrates and to adjust rapidly to changing environmental conditions demands a rather flexible, balanced metabolic regulation. This article focuses on the hydrogen-oxidizing enzyme system of *Alcaligenes eutrophus,* the prototype of the so-called hydrogen (knallgas) bacteria, presenting our current knowledge of the arrangement, structure and function of genes involved in the biosynthesis of two nickel-iron-containing hydrogenases and summarizing the mechanisms that regulate hydrogenase expression.

RESULTS
       Organization of the hydrogenase genes. The hydrogenase genes of *A. eutrophus* H16, designated *hox* and *hyp,* are located on the 450-kb megaplasmid pHG1 adjacent to a set of genes coding for carbon dioxide-fixing enzymes. The order of the hydrogenase genes in different hydrogenotrophs is not constant but a conserved pattern is obvious (Friedrich , Schwartz 1993). In *A. eutrophus* the structural genes encoding a membrane-bound (MBH) and a cytoplasmic NiFe-hydrogenase (SH) form the borders of the *hox* cluster, which occupies approx. 80 kb of DNA (Fig. 1). Both, the MBH and SH structural genes are accompanied by sets of accessory genes whose products appear to play a role in hydrogenase maturation. The *hyp* genes, named after their *E.coli* homologs (Lutz et al. 1991), encode proteins with a more general function in nickel metabolism. The central region of the hydrogenase gene cluster is occupied by five regulatory genes (*hoxX, A, B, C, J*). These genes are flanked by *hypE* on one side and *hoxN* on the other side. *hoxN* codes for a high affinity nickel permease (Fig. 1).

       Molecular structure and function of the two NiFe-hydrogenases of A. eutrophus. Autotrophic carbon dioxide fixation via the ribulose-bisphosphate cycle is an energy demanding process. To synthesize one molecule of hexose from six molecules of carbon dioxide requires 18 ATP and 12 NAD(P)H. The majority of hydrogen bacteria

110

M. E. Lidstrom and F. R. Tabita (eds.), Microbial Growth on C$_1$ Compounds, 110–117.
© 1996 Kluwer Academic Publishers. Printed in the Netherlands.

contain a single membrane-bound hydrogenase, whose redox reaction is coupled to electron transport phosphorylation. Reducing power in these bacteria is generated via reverse electron transport. Strains of *Alcaligenes* harbor, in addition to a membrane-bound hydrogenase (MBH) of the common type, a second cytoplasmic NiFe-hydrogenase (SH), which directly interacts with NAD as electron acceptor (reviewed by Bowien, Schlegel 1981). Unlike MBH⁻ mutants, SH⁻ strains show a severe decrease in growth rate (from td 3.6 h to >12h, Hogrefe et al. 1984). This observation clearly shows that the acquisition of an NAD-reducing hydrogenase is extremely beneficial for *A. eutrophus*.

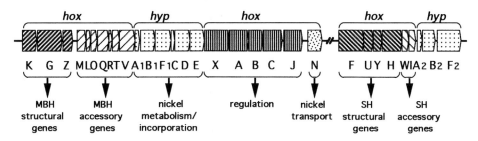

Fig. 1. The hydrogenase gene complex of *Alcaligenes eutrophus* strain H16

NiFe-hydrogenases have been the subject of several reviews (Przybyla et al. 1992, Voordouw 1992, Albracht 1994). Recently the first crystal structure of a periplasmic NiFe-hydrogenase from the anaerobic sulfate-reducer *Desulfovibrio gigas* was published (Volbeda et al. 1995). The active site is located in the large (60-kDa) subunit of the heterodimeric protein and contains, in addition to nickel, a second metal ion, probably iron. The small (28-kDa) subunit is considered to be the electron-transferring polypeptide, which coordinates one [3Fe-4S] and two [4Fe-4S] clusters. The sequence data for the *A. eutrophus* MBH (Fig. 2, Kortlüke et al.1992) indicate an extensive similarity with the *D. gigas* hydrogenase.

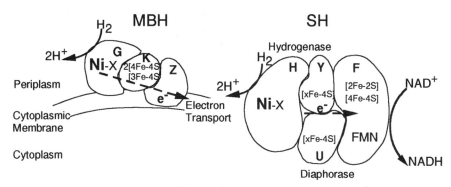

Fig. 2. Structural model of two NiFe-hydrogenases from *A. eutrophus*

Our own efforts focused on the structural analysis of the *A. eutrophus* SH, a representative of the less widespread multimeric NiFe-hydrogenases (Fig. 2). The SH consists of four dissimilar subunits which form two dimeric moieties. The large dimer is composed of a 67- and a 26-kDa polypeptide (sizes as deduced from the nucleotide sequences) encoded by the genes *hoxF* and *hoxU*. This portion of the SH reveals NADH oxidoreductase (diaphorase) activity. Based on biochemical data (Schneider, Schlegel 1976), spectroscopic investigations (Zaborosch et al. 1991) and sequence analysis (Tran-Betcke et al. 1990) we propose that two glycine-rich segments of the HoxF subunit provide the binding sites for NAD(H) and FMN. Two cysteine-containing motifs could be involved in the liganding of a tetranuclear and a binuclear iron-sulfur cluster. An additional Fe-S center may be buried in the small HoxU subunit (Fig. 2). Homology searches uncovered striking similarities between the diaphorase of the SH and three peripheral subunits of the mitochondrial NADH ubiquinone oxidoreductase, indicating that the two enzymes have a common ancestor (reviewed by Walker 1992).

The 55- and 23-kDa products of the *hoxH* and *hoxY* genes form the so-called hydrogenase dimer of the SH with the active site in the HoxH subunit. The large subunits of all NiFe- hydrogenases share 19 strictly conserved amino acid residues located in four homologous elements (Voordouw 1992). We have substituted ten mostly conserved residues of the HoxH polypeptide by site-directed mutagenesis and examined the resulting mutant enzymes for diaphorase activity, hydrogen-mediated NAD reduction, acceptor-independent deuterium exchange (H/D) and nickel content (Fig. 3; C. Massanz , V. Fernandez, B. Friedrich , unpublished). As expected the diaphorase activity was not significantly affected in these mutants. Substitution of the paired amino-terminal or paired carboxyl-terminal cysteine residues, however, led to a complete loss of hydrogen-activating activity. Furthermore, three of the corresponding mutant enzymes were devoid of nickel. This result correlates with the x-ray structure of the *D. gigas* hydrogenase, which shows four thiol groups as ligands for the active Ni ion (Volbeda et al. 1995). The replacement of Asp456 by Val in HoxH led to the same mutant phenotype as exchanges of Cys62, Cys65 and Cys458 (Fig. 3). This indicates an important function of this residue either in metal coordination and/or proton transfer. Exchanging a non-conserved residue (Cys72) only reduced the catalytic activities slightly (Fig. 3). Mutants with substitutions at the C-terminal end of HoxH will be discussed below.

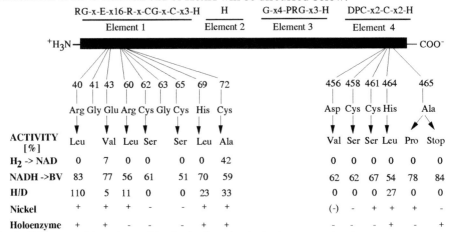

Fig. 3. Mutants with amino acid substitutions in the hydrogen-activating subunit HoxH

Another important aspect is the intramolecular channelling of protons either released during hydrogen oxidation or consumed during hydrogen evolution. Volbeda et al. (1995) proposed a proton-translocating pathway consisting of several histidines and a glutamate residue. It is interesting to note that the exchange of Glu43 and His69 in the

HoxH subunit disrupted hydrogen-activating activity substantially, while nickel was still detectable in the mutant enzymes (Fig. 3).

The small subunit (HoxY) of the SH (Fig. 2) is supposed to bridge the electron transfer between the active site and the diaphorase moiety. HoxY is considerably smaller than the electron-transferring subunits of other NiFe-hydrogenases. The nine cysteines present in HoxY could coordinate at least one Fe-S center. It is still a matter of debate whether all three iron sulfur clusters present in a variety of NiFe-hydrogenases including the *D. gigas* enzyme participate in electron transfer. It has been suggested (Volbeda et al. 1995) that alternative electron pathways exist in different classes of nickel-containing hydrogenases. In fact, its pattern may correlate with the electron-accepting site, which is a soluble cytochrome c_3 in the case of the periplasmic *D. gigas* enzyme and a membrane-bound cytochrome b -like protein in the case of the MBH family hydrogenases (Dross et al. 1992). In *A. eutrophus* the cytochrome b is encoded by *hoxZ* , the third gene of the MBH operon (Fig. 1).

Steps of MBH and SH synthesis involving nickel. The active site of both hydrogenases of *A. eutrophus* contain nickel and possibly a second metal ion (Fig. 2). Thus the biosynthesis of SH and MBH involves a series of common reactions leading to the active site assembly, in addition to a number of specific steps related to subunit processing, oligomerization and membrane translocation. Most of our present knowledge is based on the analysis of well-defined mutants of *A. eutrophus* carrying nonpolar mutations in the hydrogenase accessory genes.

Since nickel is an essential component of hydrogenase biosynthesis it is not surprising that *A. eutrophus* has acquired a very specific, high affinity nickel permease (K_T range: 20 nM $NiCl_2$), which is encoded by *hoxN* (Fig. 1; Eitinger, Friedrich 1991). HoxN is an integral membrane protein with eight transmembrane helices (Eitinger, Friedrich 1994). Under nickel starvation the metal permease does not only enhance hydrogenase activity but also urease, another nickel-containing enzyme (Wolfram et al. 1995).

Insertion mutations in the *A. eutrophus hyp* gene region are pleiotropic. Both the MBH and SH are catalytically inactive containing only traces of nickel (Dernedde et al. 1993). A recent analysis of in-frame deletions in the six *hyp* genes (Fig. 1) revealed two classes of mutants (J. Dernedde, T. Eitinger, N. Patenge, B. Friedrich, submitted). To our initial surprise, mutations in *hypA1, B1* and *F1* were silent. This unexpected result may be explained by the existence of a second copy of each gene located downstream of the SH operon (Fig. 1; A. Pohlmann, B. Friedrich, unpublished). The assignment of specific functions of the respective gene products must await the analysis of double mutants. On the basis of sequence comparisons and studies on other hydrogen-metabolizing bacteria, we assume that the product of *hypB* plays a role in nickel storage and/or insertion of the metal ion into the target proteins. *E. coli* HypB was classified as a guanine nucleotide-binding protein with intrinsic GTPase activity. A mutation eliminating this activity prevented nickel insertion (Maier et al. 1995). The HypB protein of *A. eutrophus* contains in addition to the GTP-binding domains extended clusters of histidines which could chelate nickel. In fact, HypB purified from *Rhizobium leguminosarum* (Rey et al. 1994) and *Bradyrhizobium japonicum* (Fu et al. 1995) bound significant amounts of nickel.

The second class of mutants with deletions in *hypC, D,* or *E* showed a clear phenotype (Table 1). On the basis of these results we conclude that the products of *hypC, D* and *E* are essential for the catalytic activity of the SH and MBH but are not involved in *hox* gene expression. The absence of nickel in the enzymes points to a general role of these Hyp proteins in nickel supply and/or insertion into the hydrogenase apoprotein.

Their action appears to be required for the subsequent maturation of the SH and MBH as discussed below.

Table 1. Properties of mutants defective in *hypC, D* or *F*

Character tested	*hypC*Δ	*hypD*Δ	*hypE*Δ
Growth on hydrogen	-	-	-
Enzymatic activity of MBH and SH	-	-	-
Presence of MBH and SH antigen	+	+	+
Nickel content in SH and MBH	-	-	-
Subunit assembly of SH	-	-	-
Membrane location of MBH	-	-	-
C-terminal proteolysis of HoxH and HoxG	-	-	-

Maturation of the MBH and SH. The analysis of MBH-deficient mutants of *A. eutrophus* demonstrated that both subunits of the enzyme exist in two electrophoretically distinct conformers. The conversion of the small subunit HoxK to the faster moving form is undoubtedly the result of proteolytic removal of a 43-amino-acid signal peptide (Kortlüke, Friedrich 1992). Little is known about the proteolytic mechanism. Studies with a hybrid gene showed that the leader peptide of the *Desulfovibrio vulgaris* small subunit directs the translocation of a ß-lactamase fusion protein (Nivière et al. 1992). Maturation of the *A. eutrophus* MBH to the membrane-associated form was shown to require in addition to the *hyp* genes specific functions encoded downstream of the MBH operon (Kortlüke et al. 1992). A nonpolar mutant with an internal deletion in the *hoxM* gene was blocked in the modification of the HoxG subunit and accumulated the precursor in the cytoplasm (Fig. 4; M. Bernhard, E. Schwartz, J. Rietdorf, B. Friedrich, unpublished). The deduced product of *hoxM* is homologous to HycI of *E. coli,* HoxM of *Azotobacter vinelandii,* HupD of *R. leguminosarum, B. japonicum* and *Rhodobacter capsulatus,* HynC of *Desulvovibrio fructosovorans* and the deduced product of ORF4 of *Wolinella succinogenes* (Rossmann et al. 1995 and references therein). Rossmann et al. (1995) presented evidence that HycI acts as a carboxyl-terminal processing protease.

Recently we isolated a mutant with an in-frame deletion in the first SH accessory gene, *hoxW* (Fig. 1). This mutant was devoid of SH activity but unaffected in MBH activity and accumulated HoxH in the unassembled non-processed form (S. Thiemermann, B. Friedrich, unpublished; Fig. 4). Thus, if HoxM and HoxW are the carboxyl-terminal proteases for the MBH and SH, respectively, their catalytic mechanism must be highly specific, since they apparently can not compensate for each other .

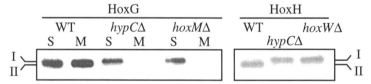

Fig. 4. Mutants defective in C-terminal proteolysis of the the hydrogen-activating subunits HoxG (MBH) and HoxH (SH). S, soluble extract; M, membrane fraction; unprocessed; II, processed.

C-terminal processing reactions have been reported for a number of NiFe- and NiFeSe-containing hydrogenases, including the enzymes of *E. coli, Methanococcus voltae* (reviewed by Albracht 1994) and *D. gigas* (Volbeda et al. 1995). Processing takes place at a histidine or an arginine residue three amino acids away from a cysteine (element 4 in Fig. 3) which is thought to be a ligand of the nickel ion. Replacement of His464 by Val in *A. eutrophus* HoxH disrupted NAD-reducing activity and strongly inhibited proteolytic cleavage (Fig. 3). Proteolysis was completely blocked by exchanging the adjacent Ala, the first residue of the cleaved polypetide, indicating that this amino acid is important for the interaction with the putative protease HoxW. The introduction of a stop codon at this position destroyed hydrogen-activating activity completely indicating that the proteolytic cleavage is an obligate step in the formation of the metal-containing active site.

The hydrogenase regulon. Both hydrogenases of *A. eutrophus* H16 are expressed coordinately in response to the energy status of the cell and growth temperature. On growth rate-limiting carbon sources such as glycerol, the SH and MBH are formed at a fully derepressed level. Repression occurs on preferentially utilized substrates and temperatures above 33°C (Friedrich, Schwartz 1993). *Alcaligenes hydrogenophilus,* a close relative of *A. eutrophus* H16, shows a different regulatory pattern. Synthesis of the MBH and SH in this strain is strictly hydrogen-dependent and not temperature-sensitive (Friedrich et al. 1984).

The molecular analysis of the *A. eutrophus hox* system uncovered a complex regulon. The expression of both the SH and MBH operons, relies on several transcriptional components: An alternate sigma factor (σ^{54}) is encoded by the chromosomal gene *rpoN*. The megaplasmid-borne *hoxA* gene encodes a transcriptional activator belonging to the response regulator family (Friedrich, Schwartz 1993). More recently a modulatory function was assigned to the neighboring gene *hoxX* (Fig. 1), whose product is necessary for complete derepression (Lenz et al. 1994).

Gel retardation assays provided the first evidence that HoxA and the *E. coli* integration host factor bind to target sequences in the SH promoter region *in vitro*. Deletion analysis pointed out a tandem palindrome (CAAG-N9-10-CTTG) which may well be the binding site for the hydrogenase regulator. Interestingly HoxA-containing extracts prepared from a temperature-resistant mutant of *A. eutrophus* gave a much stronger retardation than those from the wild type. Sequence analysis of the mutant *hoxA* allele revealed a base pair exchange determining a Gly --> Val replacement in the hypothetical DNA recognition helix of HoxA. Moreover, the corresponding *hoxA* allele from the temperature-tolerant strain *A. hydrogenophilus* codes for a valine in the same position as in the temperature-resistant *A. eutrophus* mutant (Zimmer et al. 1995). Thus the structure of the C-terminal segment of HoxA appears to be crucial for its regulatory activity.

Sequencing of the *hoxA* downstream region revealed three additional open reading frames (*hoxB, C* and *J;* O. Lenz, E. Schwartz, B. Friedrich, unpublished). HoxJ shows extensive homology to the histidine protein kinases. HoxB and HoxC are homologous to HysB and HysA, the small and large subunits of the *Desulfovibrio baculatus* periplasmic NiFeSe-hydrogenase, repectively, (Menon et al. 1987) and to *hupU* and *hupV* of *B. japonicum* (Black et al. 1994). Unexpectedly, in-frame deletions in *hoxB, hoxC* and *hoxJ* of *A. eutrophus* did not affect hydrogenase activity (Table 2).

Sequencing of a regulatory locus from the hydrogen-inducible strain *A. hydrogenophilus* revealed the same arrangement of genes found in *A. eutrophus* (*(hoxX, A, B, C* and *J ;* Fig. 1). Replacement of the *A. eutrophus hoxJ* gene by its *A. hydrogenophilus* counterpart generated a strictly hydrogen-inducible recombinant (Table

2). Moreover, an isogenic *A. eutrophus* mutant harboring a hybrid gene (*hoxJ**) encoding the N-terminal part of *A.e.* HoxJ and the C-terminal part of *A.h.* HoxJ showed total repression of the MBH and SH in the absence of hydrogen (Table 2). This observation points to functional differentiation of the C-terminal domains of the two HoxJ proteins. Moreover, the analysis of double mutants clearly showed that the products of *hoxB* and *hoxC* are instrumental in induction by hydrogen (Table 2). Based on these results we predict a hydrogen-sensing apparatus consisting of a dimeric hydrogenase-like protein and a two component regulatory system consisting of the transcriptional activator HoxA and the histidine protein kinase HoxJ (O. Lenz, A. Strack, B. Friedrich, unpublished).

Table 2. SH and MBH expression in regulatory mutants under heterotrophic (FGN) and lithoautototrophic (H_2 CO_2 O_2) growth conditions.

Strain[a]	FGN	H_2 CO_2 O_2
A. eutrophus H16	+	+
*A. eutrophus hoxJ*Δ	+	+
*A. eutrophus hoxB*Δ	+	+
*A. eutrophus hoxC*Δ	+	+
A. hydrogenophilus M50	-	+
A. eutrophus hoxJ M50	-	+
A. eutrophus hoxJ *	-	+
*A. eutrophus hoxB*Δ/*hoxJ* *	+	+
*A. eutrophus hoxC*Δ/*hoxJ* *	+	+

[a] *hoxJ* * is a gene fusion consisting of the 5´ part of *A. eutrophus hoxJ* and the 3´ part of *A. hydrogenophilus* M50 *hoxJ* .

CONCLUSIONS

Hydrogen is probably the simplest substrate available on earth. The complexity of the *A. eutrophus* system consisting of 30 hydrogenase-related genes so far identified indicates that the organism has to invest much effort to exploit this energy source. Major advancements in the understanding of the structure and function of the hydrogen-oxidizing enzyme system were made possible by genetic investigations and the recent solution of the first three-dimensional structure of a NiFe-hydogenase. This fascinating view of the Ni-active site has greatly facilitated the interpretation of our own investigations by site-directed mutagenesis of *A. eutrophus* HoxH, the hydrogen-activating subunit of the multimeric SH. Our results support the model for the coordination of nickel by four cysteine ligands that has emerged from the *D. gigas* crystal structure. Furthermore, we have obtained evidence that an aspartate residue is essential for catalytic activity. Since the crystallized *D. gigas* enzyme was in the catalytically "unready" state, further studies will be needed to establish the hydrogen-activating site.

The biosynthesis of the two hydrogenases of *A. eutrophus* depends on the activity of a series of common and specific ancillary proteins. The first steps include the uptake, storage and insertion of nickel into SH and MBH mediated by the products of *hoxN* and the *hyp* genes. These reactions are succeeded by maturation of the hydrogen-activating subunits HoxG (MBH) and HoxH (SH). This process depends on the presence of nickel and Hyp functions as well as the products of *hoxM* and *hoxW*, which appear to cleave a 15- or 24-amino-acid peptide from the carboxy terminus of the respective subunit. The proteolytic step is considered a prerequisite for locking the nickel inside the protein.

The coordinate function of the complex system is guaranteed by a multicomponent regulatory apparatus. Its major transcriptional factor is HoxA, a response regulator which activates σ^{54} RNA polymerase to form an open complex. We presented evidence that induction of *hox* gene expression by hydrogen is transmitted by a sensor composed of a hydrogenase-like protein (HoxB, C) and a histidine protein kinase (HoxJ).

Our results show that biological hydrogen oxidation not only provides an interesting system to study the mechanism of energy generation but it also provides an excellent model for the study of such basic cellular processes as metalloenzyme biosynthesis, protein maturation and signal transduction.

REFERENCES
Albracht SPJ (1994) Biochim. Biophys. Acta 1188, 167-204.
Aragno M (1994) In Balows et al, eds. The Prokaryotes, pp 3917-3933, Springer-Verlag New York Inc, USA.
Black LK et al (1994) J. Bacteriol. 176, 7102-7106.
Bowien B, Schlegel HG (1981) Annu. Rev. Microbiol. 35, 405-452.
Dernedde et al (1993) Arch. Microbiol. 159, 545-553.
Dross F et al (1992) Eur. J. Biochem. 206, 93-102.
Eitinger T, Friedrich B (1991) J. Biol. Chem. 266, 3222-3227.
Eitinger T, Friedrich B (1994) Mol. Microbiol. 12. 1025-1032.
Friedrich B et al (1984) J. Bacteriol. 158, 331-333.
Friedrich B, Schwartz E (1993) Annu. Rev. Microbiol. 47, 351-383.
Fu C et al (1995) Proc. Natl. Acad. Sci. USA 92, 2333-2337.
Hogrefe C et al (1984) J. Bacteriol. 158, 43-48.
Kortlüke C et al (1992) J. Bacteriol. 174, 6290-6293.
Kortlüke C, Friedrich B (1992) J. Bacteriol. 174, 6277-6289.
Lenz O et al (1994) J. Bacteriol. 176, 4385-4393.
Lutz S et al (1991) Mol. Microbiol. 5, 123-135.
Maier T et al (1995) Eur. J. Biochem. 230, 133-138.
Menon et al (1987) J. Bacteriol. 169, 5401-5407.
Niviere et al (1992) J. Gen. Microbiol. 138, 2173-2183.
Pryzbyla AE et al (1992) FEMS Microbiol. Rev. 88, 109-136.
Rey L et al (1994) J. Bacteriol. 176, 6066-6073.
Rossmann R. et al (1995) Eur. J. Biochem. 227, 545-550.
Schneider K, Schlegel HG (1976) Biochim. Biophys. Acta 452, 66-80.
Tran-Betcke A (1990) J. Bacteriol. 172, 2920-2929.
Vignais M, Toussaint B (1994) Arch Microbiol. 161, 1-10.
Volbeda A (1995) Nature 373, 580-587.
Voordouw G. (1992) Adv. Inorg. Chem. 38, 397-422.
WalkerJE (1992) Biophysics 25, 253-324.
Wolfram et al (1995) J. Bacteriol. 177, 1840-1843.
Zaborosch C et al (1991) Int. Conf. Molec. Biol. of Hydrogenases, 3rd, Troia, Portugal, pp. 133-134 (Abstr.).
Zimmer D et al (1995) J. Bacteriol. 177, 2373-2380.

Genetics of CO_2 fixation in methylotrophs

W.G. Meijer

Department of Microbiology, Groningen Biomolecular Sciences and
Biotechnology Institute (GBB), University of Groningen; Kerklaan 30,
9751 NN Haren, The Netherlands

INTRODUCTION

In contrast to photoautotrophic bacteria which use light energy for the assimilation of carbon
dioxide via the Calvin cycle, methylotrophic bacteria oxidize reduced C_1 compounds for this
purpose. Those methylotrophs which belong to the genus *Xanthobacter* are yellow
pigmented Gram-negative, obligate aerobes (Wiegel, Schlegel, 1984). They employ the
Calvin cycle for the fixation of carbon dioxide during autotrophic growth, which is driven
by the oxidation of methanol via a PQQ-dependent methanol dehydrogenase and an NAD-
dependent formate dehydrogenase (Janssen, Keuning et al. 1987; Weaver, Lidstrom, 1985).
Formaldehyde dehydrogenase activities could only be detected using activity staining in non-
denaturing acryl amide gels (Weaver, Lidstrom, 1985). An inducible NAD-independent
hydrogenase is used for the oxidation of hydrogen (Schink, 1982). This paper reviews the
recent advances on the genetics and physiology of carbon dioxide fixation of *X. flavus*, the
best studied representative of this genus.

RESULTS

*Transition from heterotrophic to autotrophic growth: a simple induction of two
enzymes?* *X. flavus* uses a form I (L_8S_8) ribulosebisphosphate carboxylase oxygenase
(RuBisCO) which catalyses the carboxylation and oxygenation of ribulosebisphosphate
(Meijer, Arnberg et al. 1991). This enzyme and phosphoribulokinase (PRK), which uses ATP
to phosphorylate ribulose-5-phosphate, are the characteristic enzymes of the Calvin cycle
(Fig. 1). Facultative autotrophic bacteria, such as *X. flavus*, only synthesize these enzymes
under autotrophic growth conditions (Croes, Meijer et al. 1991). The other enzymes of the
Calvin cycle are also required for glycolysis and the pentose phosphate cycle, and are
produced consitutively. It could therefore be concluded that the transition from heterotrophic
to autotrophic growth conditions only involves the induction of two specific enzymes,
RuBisCO and PRK, since the other enzymes required for carbon dioxide fixation are already
present.

     That this is an oversimplification is evident from the observation that the activity of
the Calvin cycle enzymes other than PRK and RuBisCO increase significantly following

M. E. Lidstrom and F. R. Tabita (eds.), Microbial Growth on C₁ Compounds, 118–125.
© 1996 *Kluwer Academic Publishers. Printed in the Netherlands.*

transition to autotrophic growth. Characterization of glycolytic mutants of *Escherichia coli* showed that the gluconeogenic requirement of growing cells accounts for only approximately 5% of the total catabolic and anabolic needs (Irani, Maitra, 1977). In contrast, all cellular carbon has to pass through the pool of triosephosphates during autotrophic growth via the Calvin cycle. The high activity of these Calvin cycle enzymes is therefore probably necessary to sustain the high rate of carbon dioxide fixation required for autotrophic growth. This is exemplified by a mutation in phosphoglycerate kinase of *X. flavus* which causes only a 13% decrease in growth rate of *X. flavus* on succinate; autotrophic growth however is no longer possible (Meijer, 1994).

In addition to the fact that the enzymes participating in both heterotrophic and autotrophic growth significantly increase in activity, isoenzymes of some of these proteins have been identified. *X. flavus* employs a class I fructosebisphosphate (FBP) aldolase duringheterotrophic growth. Interestingly, a class II FBP aldolase is induced following a shift from heterotrophic to autotrophic growth conditions (Meijer, unpublished). Class I aldolases form a Schiff-base between the substrate and a lysine residue during catalysis, whereas class II enzymes depend on a divalent cation as electrophile in the catalytic cycle. It remains unclear why *X. flavus* uses two different FBP aldolases during heterotrophic and autotrophic growth.

Fructosebisphosphatase (FBPase) catalyses the dephosphorylation of fructosebisphosphate and sedoheptulosebisphosphate. The former reaction is important in both gluconeogenesis and the Calvin cycle; the latter is only required for autotrophic growth. Extracts of heterotrophically grown cells contained FBPase with a 1:1 ratio of both dephosphorylating activities. Cells growing autotrophically contained an additional FBPase which was twice as active with sedoheptulosebisphoshate as substrate compared with fructosebisphosphate. In addition to the different substrate specificities, they also display differences in the regulation of their activity. ATP causes a two-fold stimulation of the inducible enzyme; the constitutive enzyme uneffected (van den Bergh, van der Kooij et al. 1995). Two FBPase enzymes have also been purified from *Nocardia opaca* (Amachi, Bowien, 1979). This Gram-positive bacterium produces a FBPase with a high sedoheptulosebisphosphatase (SBPase) activity during autotrophic growth which has different regulatory properties than the enzyme which is synthesized constitutively; the latter has a low SBPase activity.

These results clearly demonstrate that the induction of the Calvin cycle is not the result of a simple induction of two enzymes. The transition from heterotrophic to autotrophic growth requires the reorganization of central metabolism which demands not only a strong increase in activity of the Calvin cycle enzymes but also the replacement of some of these enzymes with proteins uniquely adapted to a role in Calvin cycle.

Cbb operon. Mutants of *X. flavus* which were defective in carbon dioxide fixation could be complemented with cosmid from a genomic library (Lehmicke, Lidstrom, 1985). Subsequent subcloning and nucleotide sequencing (Fig. 1) revealed the presence of the *cbb* genes, previously referred to as *cfx* (Tabita, Gibson et al. 1992), encoding the large and small subunits of RuBisCO, and genes encoding FBPase, PRK, transketolase, class II fructosebisphosphate aldolase and pentose-5-phosphate epimerase (Meijer, Enequist et al. 1990b; Meijer, Arnberg et al. 1991; Meijer, unpublished). The identity of these genes was confirmed by heterologous expression in *E. coli*. A gene of unknown function (*cbbX*) is located between *cbbS* and *cbbF*. This gene has since been identified in other autotrophic bacteria, which may indicate that it fulfills an important function in carbon dioxide fixation (Kusian, Yoo et al. 1992). The protein encoded by *cbbX* is similar to SpoVJ, which is involved in sporulation in *Bacillus subtilis* (Foulger, Errington, 1991). Both SpoVJ and CbbX contain an ATP-binding motif (GxxGxGKT/S), suggesting that the protein encoded by *cbbX* is a kinase.

The distance between the first five genes of the *cbb* cluster is small, varying between 11 and 91 base pairs, indicating that these genes are co-transcribed. Total RNA isolated from autotrophically grown *X. flavus* was hybridized to a plasmid containing *cbbLSXFP*. Messenger RNA hybridizing to DNA displaces the non-coding strand of the DNA, causing it to loop out; this can be visualized by electron microcopy. Using this technique it was shown that the *cbbLSX* genes and the *cbbFP* are cotranscribed (Meijer, Arnberg et al. 1991).

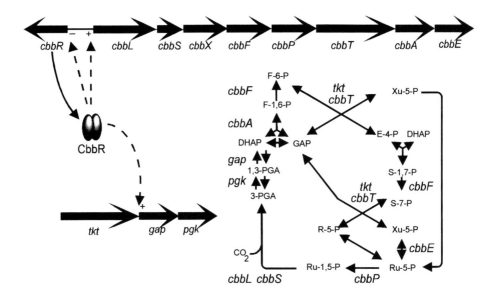

Figure 1: The Calvin cycle and *cbb* genes. Abbreviations: DHAP, dihydroxyacetonephosphate; E-4-P, erythrose-4-phosphate; F-6-P, fructose-6-phosphate; F-1,6-P, fructose-1,6-bisphosphate; GAP, glyceraldehyde-3-phosphate; 3-PGA, 3-phosphoglycerate; 1,3-PGA, 1,3-diphosphoglycerate; R-5-P, ribose-5-phosphate; Ru-5-P, ribulose-5-phosphate; Ru-1,5-P, ribulose-1,5-bisphosphate; S-7-P, seduheptulose-7-phosphate; S-1,7-P, sedoheptulose-1,7-bisphosphate; Xu-5-P, xylulose-5-phosphate. Genes: cbbA, class II FBP aldolase; *cbbE*, pentose-5-phosphate epimerase; *cbbF*, fructosebisphosphatase; *cbbLS*, RuBisCO; *cbbP*, phosphoribulokinase; *cbbR*, LysR-type transcriptional activator; *cbbT*, transketolase; *cbbX*, unknown; *gap*, glyceraldehyde-3-phosphate dehydrogenase; *pgk*, phosphoglycerate kinase; *tkt*, transketolase

A gene encoding kanamycin resistance was introduced in the *cbbX* gene located on a plasmid which is unable to replicate in *X. flavus*. Subsequent mobilization of this plasmid to *X. flavus* resulted in the integration of this plasmid into the chromosome. The resulting *X. flavus*

X. flavus strain	RuBisCO	PRK
Wild type	39	96
Q77	25	3

Table I: RuBisCO and PRK activities in *X. flavus* wild type and strain Q77 following growth on a mixture of formate (30 mM) and gluconate (5 mM). Enzyme activities are in nmoles per milligram protein per minute.

strain, Q77, was grown on a mixture of gluconate and formate in order to analyze the effect of the *cbbX* insertion on the expression of *cbbP*, a gene located downstream from *cbbX* (Table I; Fig. 1). The insertion in the *cbbX* gene had a strong polar effect on the expression of *cbbP*, showing that the transcription of *cbbFP* is dependent on a promoter upstream from *cbbX*. It was therefore concluded that the *cbbLSXFP* genes form an operon. Primer extension analysis demonstrated that the transcriptional start site of the *cbb* operon is located 22 base pairs upstream from *cbbL* (Meijer, Arnberg et al. 1991). A fusion between *cbbL* and *lacZ* on a broad-host range plasmid was subsequently mobilized to *X. flavus*, in order to examine the regulation of the promoter of the *cbb* operon. The *cbbL-lacZ* fusion was not expressed during heterotrophic growth on gluconate. Induction of the Calvin cycle by the addition of formate, resulted in the rapid increase in β-galactosidase activity and the appearance of RuBisCO protein and activity (Meijer, Arnberg et al. 1991). These results demonstrate that the *cbb* promoter is not active during heterotrophic growth, and is induced following a transition to autotrophic growth.

The mRNA analysis further showed that mRNA hybridizing to the *cbbLSX* genes was six times as abundant as *cbbFP* mRNA (Meijer, Arnberg et al. 1991). This indicates that the *cbbLSXFP* transcript is processed into a stable *cbbLSX* product and a less stable *cbbFP* RNA. Because RuBisCO is a poor catalyst it needs to be present in large amounts; this could be accomplished by means of a relatively stable mRNA.

The distance between the *cbbP* and the genes encoding transketolase, FBPaldolase and 5-pentose-epimerase is much larger (219 to 267 base pairs) than between the first five *cbb* genes (Meijer, unpublished). These large intergenic region can easily accommodate promoters for the *cbbT*, *cbbA*, and *cbbE* genes. Current research aims to elucidate if these genes are part of the *cbb* operon or if they are transcribed from separate promoters.

Phosphoglycerate kinase. In a study aiming to investigate the regulation of the promoter of the *cbb* operon, a mutant was isolated which failed to grow autotrophically. Further characterization of this mutant revealed that it had a defective phosphoglycerate kinase gene (*pgk*). Since the mutant had no phosphoglycerate kinase activity under heterotrophic or autotrophic growth conditions it was concluded that *X. flavus* employs a

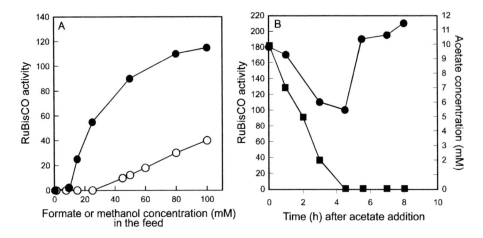

Figure 2: Metabolic regulation in *Xanthobacter* strain 25a (Croes, Meijer et al). A:, Acetate (30 mM) limited continuous culture (D=0.1 h^{-1}) with increasing concentrations of methanol (●) or formate (○)in the feed. The activities of RuBisCO were determined when the culture reached steady state. B, Methanol limited (100 mM) continuous culture (D=0.1 h^{-1}). A pulse of acetate (10 mM) was added to the culture vessel at t=0 hours. ■, acetate concentration; ●, RuBisCO activity. Enzyme activities are nmoles per minute per milligram of protein.

single phosphoglycerate kinase gene. In contrast *A. eutrophus* has three phosphoglycerate kinase genes, two of which (*cbbK*) are located within the duplicate *cbb* operons (Schäferjohann, Yoo et al. 1995). The *pgk* gene is not part of the *cbb* operon (Fig. 1). Instead, it is clustered with genes encoding transketolase (*tkt*) and glyceraldehyde-3-phosphate dehydrogenase (Meijer, unpublished). Since *cbbK* genes have not been described for other autotrophic bacteria, it may well be that phosphoglycerate kinase in these bacteria is also encoded outside the *cbb* operon (Gibson, Falcone et al. 1991; Chen, Gibson et al. 1991; Falcone, Tabita, 1993; Strecker, Sickinger et al. 1994; Gibson, Chen et al. 1990).

Since phosphoglycerate kinase is required during heterotropic and autotrophic growth, it not surprising that the *pgk* gene is constitutively expressed. However, induction of the *cbb* operon in *X. flavus* resulted in the simultaneous and parallel increase of both phosphoglycerate kinase and RuBisCO (Meijer, 1994). This indicates that the expression of the *pgk* gene and the *cbb* operon is under co-ordinate control. The results of recent experiments suggest that CbbR, the transcriptional activator of the *cbb* operon (see below), also plays a role in the regulation of *pgk* expression (Meijer, unpublished).

Regulation of the Calvin cycle. In order to understand how the expression of the Calvin cycle is regulated in response to the growth conditions, detailed physiological studies were carried out with *X. flavus* and a related strain, *Xanthobacter* 25a (Meijer, Croes et al. 1990a; Croes, Meijer et al. 1991; Dijkhuizen, Croes et al. 1989). The C$_1$ substrates methanol and formate are oxidized via the same linear pathway to carbon dioxide. In order to investigate whether both substrates are equally strong inducers of the Calvin cycle, increasing concentrations of these substrates were added to the feed of an acetate (30 mM) limited continuous culture (Fig. 2A). The activity of RuBisCO, indicative of the expression of the *cbb* operon, was measured when the culture was in a steady state. Cultures growing on mixtures of acetate and up to 100 mM formate utilized both substrates to completion. RuBisCO activity was only present when the formate concentration in the feed exceeded 25 mM. Methanol was found to be a far more potent inducer than formate. Not only were lower concentrations (5 mM) of methanol needed to induce the *cbb* operon, the levels of RuBisCO were higher when cells were growing on mixtures of acetate and methanol than was observed when formate replaced methanol at the same concentration. Substrates as chemically diverse as thiosulphate, methanol and molecular hydrogen can act as inducers of the *cbb* operon in *Xanthobacter* species and are capable of supporting autotrophic growth. Although it is possible that sensors for each of the autotrophic substrates are present in the cell, it is more likely that a common element in the metabolism of these substrates, e.g., NAD(P)H, acts as the true inducer. This provides an attractive explanation for the observed differences in induction of the *cbb* operon by formate and methanol under otherwise identical growth conditions. The oxidation of formate yields only one reducing equivalent whereas methanol oxidation results in the production of three reducing equivalents.

The ability of autotrophic substrates to induce the Calvin cycle genes is not only dependent on the number of reducing equivalents generated during their oxidation but also on the presence and identity of substrates supporting heterotrophic growth. Cells growing in a batch fermenter on a mixture of succinate and formate did not induce the Calvin cycle enzymes; gluconate however was not able to repress the synthesis of these enzymes when formate or hydrogen was present in the growth medium. The repression of the Calvin cycle genes by heterotrophic substrates was further demonstrated by the addition of acetate to a methanol limited chemostat (Fig. 2B). Immediately following the addition of acetate, the activity of RuBisCO started to decline and returned to the original steady state level after the completion of acetate. The results of these physiological experiments showed that expression of the *cbb* operon is regulated by both the energy and the carbon status of the cell. Full induction of the Calvin cycle only takes place in the absence of organic carbon sources and in the presence of sufficient reducing power generated by the oxidation of, for example, hydrogen, methanol or formate.

CbbR. A gene (*cbbR*) was identified 94 base pairs upstream from *cbbL*, the first gene of the *cbb* operon (Meijer, Arnberg et al. 1991; van den Bergh, Dijkhuizen et al. 1993). The *cbbR* gene displayed significant similarities to transcriptional regulators belonging to the LysR-family of transcriptional activators. These proteins are characterized by an amino-terminal DNA-binding domain containing a helix-turn-helix motif. The central domain of these proteins is believed to be involved in binding of effector molecules (Schell, 1993). This domain is poorly conserved because the various LysR-type proteins recognize different effector molecules. However, it is highly conserved amongst CbbR proteins, indicating that the same effector molecule is recognized by CbbR proteins from chemoautotrophic and photoautotrophic bacteria (Windhövel, Bowien, 1991; Viale, Kobayashi et al. 1991; Meijer, Arnberg et al. 1991; Kusano, Sugawara, 1993; Falcone, Tabita, 1993; Gibson, Tabita, 1993; van den Bergh, Dijkhuizen et al. 1993; Strecker, Sickinger et al. 1994).

A *cbbR* disruption mutant of *X. flavus* was constructed via the insertion of a gene specifying kanamycin resistance. The induction of the *cbb* operon by the addition of formate to cells growing heterotrophically on gluconate no longer took place in this mutant. However, the *cbbR* mutant strain could oxidize formate, demonstrating that the failure to induce the *cbb* operon was not caused by an inability to oxidize autotrophic substrates. The *cbbR* gene was subsequently expressed in *E. coli* by optimizing the translational signals and placing it under the control of the IPTG inducible *tac* promoter. Using a band-shift assay it was shown that CbbR binds to the *cbb* operon promoter. Inspection of the *cbb* promoter revealed the presence of an inverted repeat (TTCAG-N_5-CTGAA; IR_I) at position -65 with respect to *cbb* transcriptional start site. A second imperfect repeat similar to IR_I and located on the same side of the DNA helix (TTCAT-N_5-TTAAA; IR_{II}) is present at position -43 (van den Bergh, Dijkhuizen et al. 1993). DNA fragments containing one or both repeats were therefore used in band-shift assays to show that IR_I and IR_{II} represent the CbbR binding sites (van den Bergh and Meijer, unpublished). These experiments clearly demonstrate that CbbR acts as a transcriptional activator of the *cbb* operon. Current research aims to unravel the molecular mechanism by which CbbR achieves this.

CONCLUSION

A superficial examination of the physiology and genetics of the Calvin cycle in autotrophic bacteria could easily lead to the conclusion that the acquisition and expression of just two enzymes, RuBisCO and PRK, is sufficient to enable autotrophic growth. A more detailed study revealed that autotrophic growth, however, is the result of a far more complicated process, which we are only now beginning to understand. The activity of enzymes other than those unique to the Calvin cycle need to be increased in order achieve sufficiently high rates of carbon dioxide fixation required for autotrophic growth. In addition, isoenzymes of FBPase, FBP aldolase and transketolase are employed in the Calvin cycle. Physiological studies showed that the expression of the Calvin cycle genes is regulated by the carbon and energy status of the cell. Although it is firmly established that CbbR plays an important role in mediating these signals to the transcription apparatus, it remains unclear by which mechanism this is accomplished.

REFERENCES

Amachi T et al (1979) J Gen Microbiol 113, 347-356.
Chen J-H et al (1991) J Biol Chem 266, 20447-20452.
Croes LM et al (1991) Arch Microbiol 155, 159-163.
Dijkhuizen L et al (1989) In Hamer G, Egli T, Snozzi M eds Mixed and multiple substrates and feedstocks, pp 21-32. Hartung-Gorre verlag, Konstanz,
Falcone DL et al (1993) J Bacteriol 175, 5066-5077.
Foulger D et al (1991) Mol Microbiol 5, 1363-1373.
Gibson JL et al (1990) Biochemistry 29, 8085-8093.
Gibson JL et al (1991) J Biol Chem 266, 14646-14653.
Gibson JL et al (1993) J Bacteriol 175, 5778-5784.
Irani MH et al (1977) J Bacteriol 132, 398-410.
Janssen DB et al (1987) J Gen Microbiol 133, 85-92.

Kusano T et al (1993) J Bacteriol 175, 1019-1025.
Kusian B et al (1992) J Bacteriol 174, 7337-7344.
Lehmicke LG et al (1985) J Bacteriol 162, 1244-1249.
Meijer WG et al (1990a) Arch Microbiol 153, 360-367.
Meijer WG et al (1990b) J Gen Microbiol 136, 2225-2230.
Meijer WG et al (1991) Mol Gen Genet 225, 320-330.
Meijer WG (1994) J Bacteriol 176, 6120-6126.
Schäferjohann J et al (1995) Arch Microbiol 163, 291-299.
Schell MA (1993) Ann Rev Microbiol 47, 597-626.
Schink B (1982) FEMS Microbiol Lett 13, 289-293.
Strecker M et al (1994) FEMS Microbiol Lett 120, 45-50.
Tabita FR et al (1992) FEMS Microbiol Lett 99, 107-110.
van den Bergh ERE et al (1993) J Bacteriol 175, 6097-6104.
van den Bergh ERE et al (1995) J Bacteriol, In press
Viale AM et al (1991) J Bacteriol 173, 5224-5229.
Weaver CA et al (1985) J Gen Microbiol 131, 2183-2197.
Wiegel JKW et al (1984) In Krieg NR, Holt JG eds Bergey's manual of systematic bacteriology, pp 325-333. Williams and Wilkins, Baltimore/London,
Windhövel U et al (1991) Mol Microbiol 5, 2695-2705.

Genetics of C_1 metabolism regulation in *Paracoccus denitrificans*

N. Harms, J. Ras, S. Koning, W.N.M Reijnders, A.H. Stouthamer, R.J.M. van Spanning.
Department of Microbial Physiology, Biology Faculty, Vrije Universiteit, de Boelelaan 1087, 1081 HV Amsterdam, The Netherlands
tel: 31 20 4447179; fax: 31 20 4447123; e-mail: harms@bio.vu.nl

INTRODUCTION

Paracoccus denitrificans is a facultative anaerobic bacterium that can be found in soil, sewage or sludge. The readily changing composition of these habitats forces this bacterium to adapt its metabolism frequently to the available carbon and free-energy sources. In addition to heterotrophic growth, *P. denitrificans* is able to grow autotrophically with either hydrogen, thiosulphate or reduced C_1 compounds (methanol, methylamine or formate) as electron donors. To adjust smoothly to the changing environment, unicellular organisms have evolved signal transduction systems that report to the cytoplasm aspects of the changes of the extracellular conditions. Our research focuses on the question: through which signal transduction routes does *P. denitrificans* adapt its C_1 metabolism and how do these routes communicate with each other.

Methanol and methylamine are oxidized by the periplasmically located quinoproteins methanol dehydrogenase (MDH) and methylamine dehydrogenase (MADH), respectively (Fig 1.). Formaldehyde, formed in either reaction, is transported to the cytoplasm and is coupled non-enzymatically to reduced glutathione (GSH). The S-hydroxy methyl glutathione is converted by NAD-GSH-dependent formaldehyde dehydrogenase (GD-FALDH) to S-formylglutathione. The latter compound is hydrolyzed by S-formylglutathione hydrolase (FGH) to formate and GSH. The homo-tetrameric GD-FALDH has been partially purified from *P. denitrificans* (Van Ophem, Duine 1994). The enzyme is a member of group I subclass III of the long chain NAD-dependent zinc-containing alcohol dehydrogenases (Reid, Fewson 1994; Van Ophem, Duine 1994). FGH has not been purified from *P. denitrificans*, but isolates from human liver and the methylotrophic yeast *Kloeckera* sp No.2201 evidence a homo-dimeric enzyme with a molecular mass of 58-kDa (Kato et al. 1980; Uotila, Koivusalo 1974). Genes encoding MDH (*mxa* genes) or MADH (*mau* genes) and GD-FALDH (*flhA*) of *P. denitrificans* have been isolated (Chistoserdov et al. 1992; Harms et al. 1987; Ras et al. 1995; Van Spanning et al. 1991; Van Spanning et al. 1990b). The *flhA* gene is located upstream of *cycB*, encoding cytochrome c_{553i}, which is involved in C_1 metabolism as well (Ras et al. 1991).

M. E. Lidstrom and F. R. Tabita (eds.), Microbial Growth on C_1 Compounds, 126–132.
© 1996 Kluwer Academic Publishers. Printed in the Netherlands.

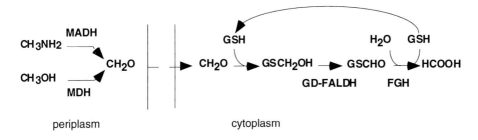

Fig 1. Reaction scheme of C_1 metabolism in *P. denitrificans*

C_1 metabolism in *P. denitrificans* is strictly regulated. Both MDH and MADH are repressed during heterotrophic growth. MADH is induced during growth on methylamine but not choline or methanol and this induction is not repressed by the simultaneous presence of choline or methanol (Page, Ferguson 1993; Van Spanning et al. 1994). In *P. denitrificans*, the methylamine-induced expression of the *mau* genes is under control of MauR, a LysR-type transcription activator (Van Spanning et al. 1994). MDH is induced during growth on methanol, methylamine or choline. Formaldehyde seems to be effector of this activation due to some product induction mechanism (De Vries et al. 1988). Expression of the structural *mxa* genes is regulated by a two-component regulatory system, MxaZYX (Harms et al. 1993). MxaY is the sensor of the system and MxaX the response regulator. The function of MxaZ is unknown but a role in the sensing system has been suggested (Harms et al. 1993). Since a deletion of both *mxaY* and *mxaYZ* results in a wild type phenotype with respect to methanol oxidation, a second regulatory system that is able to cross talk with MxaX has been hypothesized (Yang et al. 1995). The regulation of *flhA* expression is largely unknown. GD-FALDH can be found in small amounts in succinate grown cells, but larger amounts have been detected in methanol, methylamine and choline grown cells (Van Ophem, Duine 1994). The induction pattern of this enzyme is therefore comparable with that of MDH. The same holds for the induction pattern of cytochrome c_{553i}. Both GD-FALDH and cytochrome c_{553i} are present in mutants with a defect in *mxaX*, indicating that the two-component regulatory system MxaZYX is not controlling the expression of *flhA* and *cycB* (Harms et al. 1993). We hypothesize that the regulatory system that activates *flhA* and *cycB* expression is able to cross talk with the two-component regulatory system MxaYX. Following this hypothesis we initiated the isolation of a gene that encodes the protein that is able to activate MxaX in the $\Delta mxaY$ mutant. In addition we here report the isolation of the gene encoding FGH and of the genes encoding an additional PQQ-dependent dehydrogenase.

RESULTS
Analysis of the flhA, clpP, fghA, xoxF, cycB, xoxJ and xoxI genes. A 9.5 kbp DNA fragment was isolated from the genome of *P. denitrificans*. The nucleotide sequence of this fragment was determined and taking into account the codon preference of *P. denitrificans* genes, nine open reading frames could be recognized in this fragment (see Fig 2).

Fig 2. Physical map of the chromosomal DNA fragment containing *orf2*, *flhA*, *clpP*, *orf3*, *fghA*, *xoxF*, *cycB*, *xoxJ*, and *xoxI*. Open bars indicate open reading frames and the signal sequences are indicate by a hatched box. Promoters are indicated by an arrow.

       The genes and their products were analyzed and mutations were generated in some of them. The fysiological consequences of the mutations were studied by determining the growth of the wild type and the mutant strains in batch cultures and on solid media with different carbon and energy sources (see Table 1.). The *flhA* gene encoded GD-FALDH. A mutation in *flhA* resulted in a mutant that was unable to grow on methanol, methylamine and choline, indicating that GD-FALDH is essential for the oxidation of formaldehyde produced during methylotrophic growth. The *orf2* and *3* and their deduced amino acid sequences showed no homology with any DNA or protein sequences present in data banks. An *orf3* mutant was still able to grow on all C_1 substrates and a role in C_1 metabolism for this *orf* is therefore questionable. The deduced amino sequence of the *orf* downstream of *flhA* was for 37-41% identical to the caseinolytic protease proteolytic subunit (ClpP) of *E. coli* (Maurizi et al. 1990) and *Streptococcus salivarius* (Giffard et al. 1993). A mutation in this *clpP* gene did not affect C_1 metabolism, indicating that this gene is not necessary for growth on C_1 substrates. The role of this gene in *P. denitrificans* is still unknown. The deduced amino acid sequence of the *orf* downstream of *orf3* showed an identity of 52% with human esterase D (Lee, Lee 1986) and an identity of 58% with the deduced amino acid sequence of the *yeiG* gene, isolated from *E. coli* (Robison 1994). Nothing has been

Table 1. Characteristics of wild type and mutant strains of *P. denitrificans*

Strain	growth		on[a]		protein[b]
	CH_3OH	CH_3NH_2	choline	succinate	cyt c_{553i}
wild type	+	+	+	+	+
flhA::Km	-	-	-	+	nt
clpP::vector	+	+	+	+	nt
orf3::Km	+	+	+	+	+
fghA::Km	-	-	+	+	+
xoxF::Km	+/-	+	+	+	+
cycB::Km	+/-	+	+	+	-

[a] +, good growth; +/-, growth less than wild type; [b] presence of cytcochrome *c*553i was determined after SDS Polyacrylamide electrophoresis and staining for covalently bound heme.

reported about the *yeiG* gene product, but human esterase D was found to be identical to S-formylglutathione hydrolase (Eiberg, Mohr 1986). It was concluded, therefore, that this *orf,* designated *fghA,* was the gene encoding FGH. An *fghA* mutant is unable to grow on methanol and methylamine, while growth on choline was possible be it at half the wild type growth rate. This suggests that during growth on choline another enzyme was expressed that could hydrolyze S-formylglutathione.

Downstream of the *fghA* gene a cluster of four genes was identified that was similar in terms of sequence to the *mxaFJGI* gene cluster, encoding MDH. The deduced amino acid sequence of *xoxF* showed homology with several quinoproteins, such as MDH of both *P. denitrificans* (Harms et al. 1987) and *Methylobacterium extorquens* AM1(Anderson et al. 1990), quino-haemo ethanol dehydrogenase type II of *Acetobacter* species (Inoue et al. 1989; Takemura et al. 1993; Tamaki et al. 1991) and glucose dehydrogenase of *Acinetobacter calcoaceticus* (Cleton-Jansen et al. 1988) and *E. coli* (Cleton-Jansen et al. 1990). The homology studies revealed an overall identity of 50% with the MDHs, 30% with QH-EDHs and 16% with the GDHs. Recently, the three dimensional structure of MDH from *M. extorquens* AM1 has been determined (Anthony et al. 1994; Ghosh et al. 1995). In this structure several conserved amino acid residues were identified that appear to be involved in the special tryptophan docking motif, in PQQ binding or in calcium binding. All these residues were also found in the *xoxF* gene product at exactly the same spacing. A remarkable feature of the outer face of MDH is a funnel mouth consisting of hydrophobic residues leading to the active-site chamber. These surface-accessible residues are identical in all MDHs, but completely different in QH-EDHs type II and GDHs. Most of the surface-accessible residues from MDH were also present in XoxF. A mutation in *xoxF* reduced growth on methanol by some 50%. Downstream of *xoxF*, the gene *cycB*, encoding cytochrome c_{553i} is present (Ras et al. 1991). A *cycB* mutant had a decreased maximal specific growth rate on methanol, while the growth yield was unaffected. Immediately downstream of *cycB*, the gene *xoxJ* is located. The gene product of thie latter gene is an identity of 19% identical to MxaJ both of *P. denitrificans* (Van Spanning et al. 1991) and of *M. extorquens* AM1 (Anderson et al. 1990). The *xoxI* gene is located downstream of *xoxJ*. The gene product of *xoxI* shows no homology with any protein in the data banks. On basis of the similarity of the primary structure of XoxF with other quinoproteins it was concluded that the *xoxF* gene encodes a periplasmically located PQQ-dependent dehydrogenase (XDH) using cytochrome c_{553i} as the electron acceptor. Whether the *xoxI* gene encodes a second subunit of the XDH is not clear. Although the β-subunit is necessary for MDH activity, its function is unclear. In the PQQ-dependent enzymes QH-EDH type II and GDH no sequence homology was found with the β-subunit of MDH. The substrate for XDH is still unknown, but since mutations in either *xoxF* or *cycB* affected growth on methanol, a role in C_1 metabolism has to be considered.

Cytochrome c_{553i} was synthesized in the *orf3, fghA* and *xoxF* mutants, while the protein was absent in the *cycB* mutant strain (see also Table 1.). This indicated that mutations in *orf3, fghA* and *xoxF* had no downstream effects on the expression of the *cycB* gene and that directly upstream of *cycB* a promoter was present.

Isolation of a regulatory mutant. DNA fragments containing the promoter regions of either the *mxaF* gene or the *cycB* gene were cloned in front of the reporter gene *lacZ*. These transcriptional fusions were transferred in single copy to the genome of a Δ*mxaY* mutant. Both strains showed blue colonies on methanol, methylamine or choline minimal agar plates in the presence of X-gal, while colonies on succinate agar plates remained very light blue. This indicated that not only the promoter region of the *mxaF* gene but also the one of the *cycB* gene contained a C_1 inducible promoter. To both strains a plasmid was conjugated that harboured an *flhA* gene under control of the *cycA* promoter. *cycA* gene encoding cytochrome c_{550} is constitutively expressed in *P. denitrificans* (Van Spanning et al. 1990a). The presence of this plasmid ensured that GD-FALDH is always present and consequently that formaldehyde could not accumulate in the cell. Transposon mutagenesis was subsequently performed and mutants were selected that showed light blue colonies on choline agar plates in the presence of X-gal. With this selection procedure, it should be possible to obtain mutants that were affected in the activation of the C_1 inducible promoters. And since the background strain contained a deletion in *mxaY* we hypothesized that mutants could be obtained that were mutated in the system that could cross talk with MxaX. One mutant was obtained from the strain carrying the p*cycB*::*lacZ* fusion that was interesting to characterize further.

Table 2. Growth characteristics of wild type and regulatory mutants of *P. denitrificans*

Strain	growth		on			
	CH_3OH	CH_3NH_2	choline	succinate	formate	H_2/CO_2
wild type	+	+	+	+	+	+
Δ*mxaY*	+	+	+	+	+	+
mutant + p*cycA*::*flhA*	-	-	+	+	+	+
mutant - p*cycA*::*flhA*	-	-	-	+	+	+

As is shown in Table 2, the mutant harbouring the plasmid was unable to grow on methanol or methylamine. Growth on choline just as the wild type was still possible. Curing of the plasmid, however, resulted in a strain that was unable to grow on choline. This indicated that expression of the chromosomal *flhA* gene is impaired in the mutant. Apparently the constitutively expressed *flhA* gene could complement the growth on choline, but not the growth on either methanol or methylamine. Consequently the expression of other genes is abolished as well. Blocked expression of the *fghA* gene would give this phenotype, but it is also possible that *mxa* and *mau* expression was impaired. To investigate this, cell extracts were made from choline+methylamine grown cells. Protein analysis revealed that the mutant was synthesized neither cytochrome c_{553i}, nor MDH nor MADH. This indicated that the expression of *mxa* , *mau flhA* and *cycB* was abolished. Whether the expression of the *fghA* gene was also blocked is not known yet. The mutation appears to be highly pleiotropic. Since autotrophic growth with hydrogen and growth on formate was unaffected, the mutation is not controlling C_1 metabolism beyond formate. Experiments are in progress to clone and characterize the gene that is mutated in this mutant. This will certainly give more information about the regulatory network involved in C_1 metabolism.

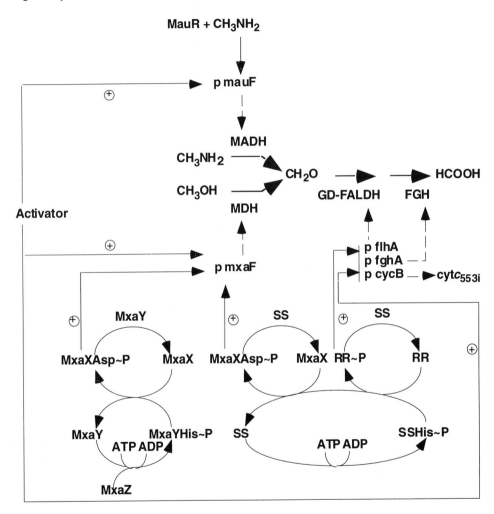

Fig 3. Model of the regulatory network controlling the expression of the structural genes encoding MDH, MADH, GD-FALDH, FGH and cytochrome c_{553i}.

CONCLUSIONS

The studies summarized here indicate that the structural genes of *P. denitrificans* involved in the oxidation of methanol, methylamine and formaldehyde are located on three separate loci on the genome. The locus encoding the enzymes involved in the oxidation of formaldehyde (*flhA* and *fghA*), comprises genes encoding a PQQ-dependent dehydrogenase of unknown function. Somehow this dehydrogenase is involved in the oxidation of methanol, since mutants in the gene encoding the large subunit showed a reduced growth on this substrate. A cytochrome c_{553i} is most probably the electron acceptor of this dehydrogenase. Upstream of the gene encoding this cytochrome a C_1 inducible promoter was identified.

A mutant was isolated that was unable to activate the expression of the genes encoding MDH, MADH, GD-FALDH and cytochrome c_{553i}. These results led to our current working model for the network regulation of C_1 metabolism in *P. denitrificans* (See Fig 3).

The sequence data will appear in Nucleotide Sequence Data Libraries under the accession number U34346.

REFERENCES

Anderson DJ et al (1990) Gene 90, 173-176.

Anthony C et al (1994) Biochem. J. 304, 665-674.

Chistoserdov AY et al (1992) Biochem. Biophys. Res. Comm. 184, 1226-1234.

Cleton-Jansen A-M et al (1988) Nucleid Acids Research 16, 6228.

Cleton-Jansen A-M et al (1990) J. Bacteriol. 172, 6308-6315.

De Vries GE et al (1988) J. Bacteriol. 170, 3731-3737.

Eiberg H, Mohr J (1986) Hum. Genet. 74, 174-175.

Ghosh M et al (1995) Structure 3, 177-187.

Giffard PM et al (1993) Journal General Microbiology 139, 913-920.

Harms N et al (1987) J. Bacteriol. 169, 3969-3975.

Harms N et al (1993) Mol. Micobiol. 8, 457-470.

Inoue T et al (1989) J. Bacteriol. 171, 3115-3122.

Kato N et al (1980) Biochim. Biophys. Acta 611, 323-332.

Lee EY-HP, Lee W-H (1986) Proc. Natl. Acad. Sci. USA 83, 6337-6341.

Maurizi MR et al (1990) J. Biol. Chem. 265, 12536-12545.

Page MD, Ferguson SJ (1993) Eur. J. Biochem. 218, 711-717.

Ras J et al (1995) J. Bacteriol. 177,

Ras J et al (1991) J. Bacteriol. 173, 6971-6979.

Reid MF, Fewson CA (1994) Critical Reviews in Microbiology 20, 13-56.

Robison K (1994) Genbank accession U00007

Takemura H et al (1993) J. Bacteriol. 175, 6857-6866.

Tamaki F et al (1991) Biochim. Biophys. Acta 1088, 292-300.

Uotila L, Koivusalo M (1974) J. Biol. Chem. 249, 7664-7672.

Van Ophem PW, Duine JA (1994) FEMS Microb. Let. 116, 87-94.

Van Spanning RJM et al (1994) Eur. J. Biochem. 226, 201-210.

Van Spanning RJM et al (1991) J. Bacteriol. 173, 6948-6961.

Van Spanning RJM et al (1990a) J. Bacteriol. 172, 986-996.

Van Spanning RJM et al (1990b) FEBS Lett. 275, 217-220.

Yang H et al (1995) Microbiology 141, 825-830.

Molecular Biology of Particulate Methane Monooxygenase

J. Colin Murrell and Andrew J. Holmes,   Department of Biological Sciences,
University of Warwick, Coventry, CV4 7AL, UK.

Introduction

Methane oxidizing bacteria (methanotrophs) grow on methane as their sole source of carbon and energy by oxidizing this one-carbon (C1) compound, via methanol, formaldehyde and formate, to carbon dioxide. Carbon is assimilated into biomass at the level of formaldehyde by either the serine or the ribulose monophosphate pathway. The first step in the methane oxidation pathway is catalysed by the enzyme methane monooxygenase (MMO). The MMO can exist in two forms, a soluble, cytoplasmic enzyme complex (sMMO) or a membrane-bound, particulate enzyme (pMMO). The sMMO is only found in methanotrophs such as *Methylococcus* spp., *Methylosinus* spp., and some *Methylocystis* and *Methylomonas* spp. The sMMO has been purified from several methanotrophs and the genes encoding this enzyme complex have been cloned and sequenced from *Methylococcus capsulatus* (Bath) and *Methylosinus trichosporium* OB3b. The biochemistry and molecular biology of sMMO has recently been reviewed extensively (Lipscomb 1994; Dalton et al. 1993; Murrell 1992, 1994) and will not be covered here. The pMMO appears to be present in all methanotrophs but is less-well characterised than sMMO. The enzyme has proved difficult to purify in active form, presumably because of instability in the pMMO polypeptides once they are removed from the membranes. In those methanotrophs that possess both sMMO and pMMO, eg *Mc. capsulatus* (Bath) and *Ms. trichosporium* OB3b, the copper-to-biomass ratio of cultures seems to regulate the switch between expression of sMMO or pMMO (Stanley et al. 1983). A high copper-to-biomass ratio favours expression of the pMMO and sMMO is repressed by copper ions (copper also appears to inhibit sMMO activity). Cells grown at low copper-to-biomass ratios express sMMO and there is now good evidence for transcriptional control of sMMO (and pMMO) expression by copper ions (A. Nielsen, J.C. Murrell, unpublished). This switch can be monitored by polyacrylamide gel electrophoresis of cell-free extracts of methanotrophs. sMMO has major characteristic cytoplasmic polypeptides of around 60, 45 and 20 kDa (Protein A components) whereas cells expressing pMMO have major membrane-bound polypeptides of approximately 45 and 27 kDa. Partially purified pMMO extracted from membranes (Smith, Dalton, 1989) and purified membrane preparations (Nguyen et al. 1994) of *Mc. capsulatus* (Bath) contained polypeptides of 45 and 27 kDa, suggesting that these were components of the pMMO enzyme. Acetylene, a potent inhibitor of methane oxidation, can be used to radiolabel the 27 kDa polypeptide, providing further evidence for this being a subunit of pMMO. EPR studies have also revealed a correlation between copper content of membranes of *Mc. capsulatus* (Bath) and pMMO activity and suggested that the active site of pMMO may contain unusual trinuclear copper centres which may be involved in methane oxidation by pMMO (Nguyen et al. 1994).

The pMMO shares many similarities with ammonia monooxygenase (AMO) found in the ammonia-oxidizing nitrifying bacteria. The two enzymes are both membrane-associated, they share broadly similar substrate profiles and although apparent Km values are substantially different, pMMO and

133

M. E. Lidstrom and F. R. Tabita (eds.), Microbial Growth on C₁ Compounds, 133–140.
© 1996 *Kluwer Academic Publishers. Printed in the Netherlands.*

AMO are both capable of oxidizing ammonia and methane. AMO is also inhibited by acetylene and ^{14}C-acetylene can be used to label a 27 kDa membrane-bound polypeptide that co-purifies with a 43 kDa polypeptide in *Nitrosomonas europaea* (McTavish et al. 1993). Copper ions also stabilize AMO activity *in vitro* and may be important in the active site of this enzyme (Ensign et al. 1993). Therefore, it appeared that the 45 and 27 kDa polypeptides of pMMO and the 43 and 27 kDa components of AMO were analogous.

Until recently, nothing was known about the molecular biology of pMMO. However, the genes encoding the putative 27 and 43 kDa polypeptides of AMO from *Nm. europaea*, *amoA* and *amoB* respectively, have been cloned and sequenced by McTavish et al. (1993). These genes were linked on the chromosome of this organism. They also reported that *Nm. europaea* probably contained two copies of these genes. The availability of DNA sequence information on *amo* genes allowed us to design oligonucleotide primers which we used to PCR amplify and clone a 693-bp internal DNA fragment of the *amoA* gene from *Nm. europaea*. When this was used as a hybridization probe against chromosomal DNA digests from a number of representative species of methanotrophs in our culture collection, specific DNA fragments from all of the methanotroph DNAs hybridized to this probe at relatively high stringency hybridization and washing conditions. No hybridization was observed with this *amoA* gene probe and DNA from methanol-utilizers or other negative control organisms DNA (E. Kenna, A. Holmes, J. C. Murrell, unpublished). This suggested that *amoA* homologues, ie putative *pmo* genes, were being identified by this probe. Independent to these studies, Professor Mary Lidstrom's group at Caltech had obtained N-terminal amino acid sequence information on the putative 45 kDa pMMO polypeptide from *Mc. capsulatus* (Bath) from which they designed a degenerate oligonucleotide probe (AC10). This was used to probe Southern blots containing restriction endonuclease digested chromosomal DNAs from *Mc. capsulatus* (Bath) and *Methylobacter albus* BG8 (formerly called *Methylomonas albus* BG8). A number of DNA fragments were identified in digests from both organisms which bound the AC10 probe. This enabled the identification and subsequent cloning of a 0.9 kbp *Pst*I DNA fragment of *Mc. capsulatus* (Bath) which when sequenced, showed considerable homology with the 5' two-thirds of *amoB* and was thus designated *pmoB*. The translated N-terminal region of this gene also corresponded exactly with the N-terminal sequence obtained by protein sequencing the 45 kDa pMMO polypeptide. When digests of *Mc. capsulatus* (Bath) and *Mb. albus* BG8 DNA were subsequently probed with the 693 bp amoA gene fragment, AC10 and the 0.9 kbp *Pst*I fragment containing *pmoB*, hybridization to specific DNA fragments was observed with all probes and in many cases, fragments were of the same size. Moreover, in most cases two DNA fragments were identified in each restriction digest, suggesting that duplicate copies of pmo genes may exist in methanotrophs (Semrau et al. 1995).

The 0.9 kbp *Pst*I fragment of *Mc. capsulatus* (Bath) containing part of *pmoB* was then used as a homologous probe to identify and clone overlapping chromosomal DNA fragments. A 2.1 kbp region was sequenced and was found to contain both *pmoB* and *pmoA*. The predicted amino acid sequences of these genes revealed approximately 43% and 47% identity with the corresponding *amoB* and *amoA* gene products respectively. Hydropathy plots of PmoB and PmoA also showed great similarity to those of AmoB and AmoA and were diagnostic of membrane-bound polypeptides (Semrau et al. 1995).

The high level of similarity between the *Mc. capsulatus pmo* genes and *Nm. europaea amo* genes led us to speculate that *amo* and *pmo* genes from other methanotrophs and nitrifiers may show similar conservation of sequence and comprise a gene family. This was supported by hybridization data for both *Mc. capsulatus pmoA* and *Nm. europaea amoA* probes with a diverse collection of methanotroph strains. As the *Mc. capsulatus* and *Nm. europaea* genes represent functional extremes (ie methane oxidation or ammonia oxidation) for this predicted gene family and belong to different phylogenetic groups, we reasoned that conserved peptide sequences between these two would also be conserved in other methane or ammonia oxidizing bacteria from the α, β, or γ-*Proteobacteria*. A suite of degenerate oligonucleotide primers targeting conserved peptide sequences at positions 46 - 71 and 217 - 222 of PmoA was synthesized, taking into account possible codon useage in methanotrophs and nitrifiers (Holmes et al. 1995[a]). We aimed to use these primers to generate sequence information from a diverse collection of methanotrophs and ammonia oxidizers to determine the evolutionary relationships between these groups, identify potential functionally important residues, and define group-specific sequences for probe design.

Results

Primers were tested in PCR against the following organisms: *Mc. capsulatus* (Bath), *Mb. albus* BG8, *Methylosinus trichosporium* OB3b, *Methylocystis parvus* OBBP, *Methylomonas methanica* S1, *Nitrosomonas europaea*, *Nitrosospira* sp. Np22, *Nitrosolobus multiformis* and *Nitrosococcus oceanus* (ammonia oxidizing nitrifier cultures were kindly supplied by Prof Jim Prosser, Aberdeen). Chromosomal DNA was prepared from these organisms and used as templates in PCR reactions with two degenerate oligonucleotide primers, A189 (GGNGACTGGACTTCTGG) and A682 (GAASGCNGAGAAGAASGC) which were designed from the aligned sequences of the *pmo* and *amo* gene clusters. Their location is shown in Figure 1. PCR products of the predicted size (525 bp) were checked for purity on a 1% agarose gel. The pMMO/AMO gene-specific PCR primers amplified a single DNA fragment of 525 bp from all methanotroph and nitrifier DNA samples. PCR primers were also tested against DNA samples from a range of bacteria which do not oxidize methane or ammonia, including *Methylobacterium extorquens* and *Paracoccus denitrificans*. No PCR products of the predicted size were obtained from any of these negative control organisms. The identity of the PCR products was confirmed by hybridization to a probe corresponding to the same portion of *pmoA* generated from *Mc. capsulatus*. The products were considered to be *pmoA* if they were obtained from methanotrophs and *amoA* if they were obtained from ammonia oxidizers. This distinction was based on the disparate physiologies of these two groups rather than biochemical characterization of the enzymes. The *pmoA* and *amoA* PCR products obtained from all DNA samples were cloned using the TA cloning kit (Invitrogen) and subsequently sequenced using Sequenase (United States Biochemicals). These DNA sequences have been deposited in Genbank under the Accession numbers U31649-U31655.

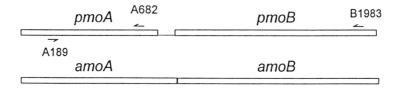

Figure 1. Schematic alignment of the particulate methane monooxygenase and ammonia monooxygenase gene clusters of *Methylococcus capsulatus* (Bath) and *Nitrosomonas europaea*, showing the approximate positions of PCR primer binding sites for A189, A682 and B1983.

The cloned PCR products obtained in this study showed strong conservation of nucleic acid sequence, inferred polypeptide sequence and predicted polypeptide structure. Pairwise comparisons of each of the predicted AmoA and PmoA sequences showed 44-99% identity (65-99% similarity) throughout these sequences. The alignment of each of these sequences is shown in Figure 2, with the amino acid alignment blocks starting at residue 52 and ending at residue 216 in accordance with the published PmoA sequence (Semrau et al. 1995). Methanotrophs include Ms.tri, *Methylosinus trichosporium*; Mc. par, *Methylocystis parvus* (both alpha-*Proteobacteria* and Type II methanotrophs); Mb. albus, *Methylobacter albus*; Mm. met, *Methylomonas methanica*; Mc. caps, *Methylococcus capsulatus* (Type I methanotrophs and gamma-

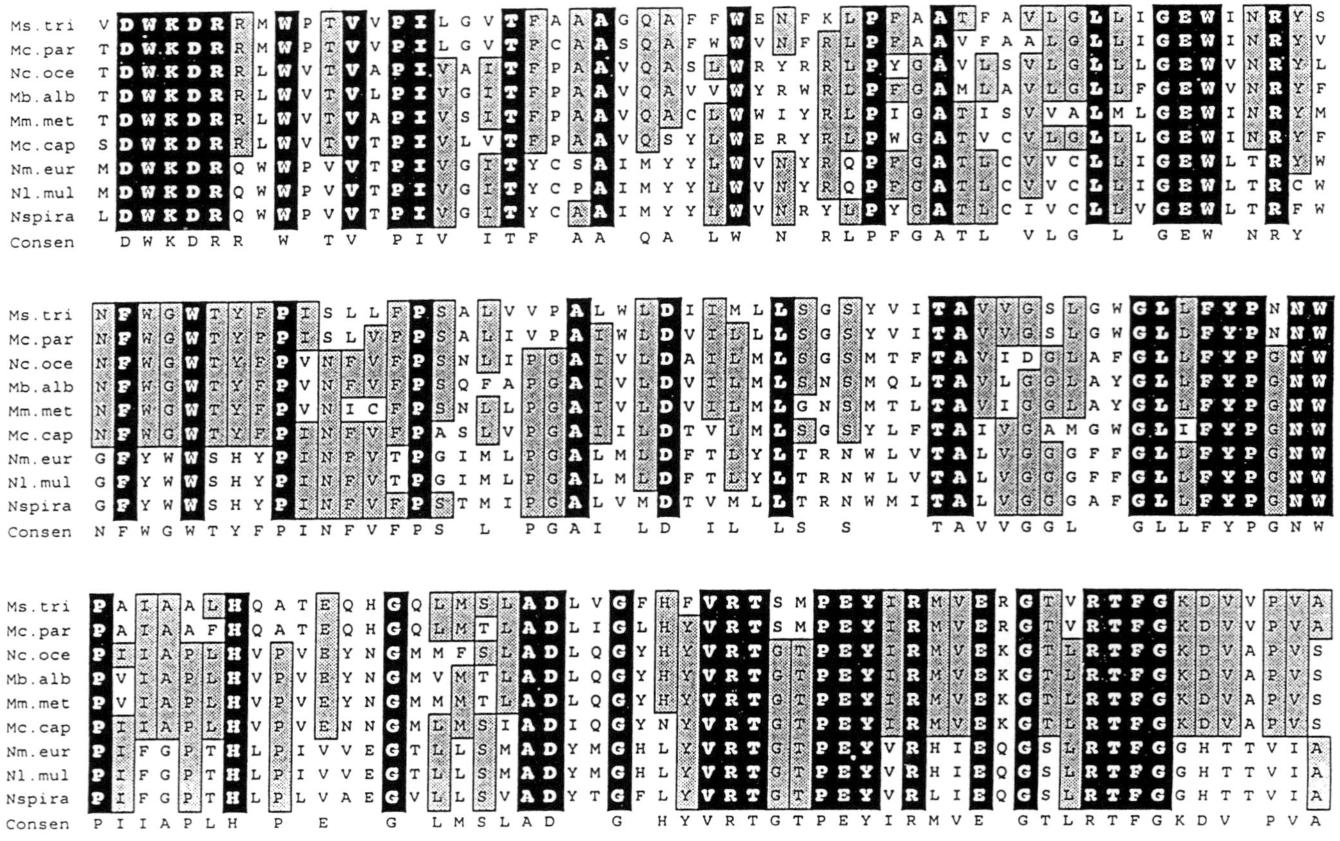

Figure 2. Alignment of the predicted PmoA and AmoA polypeptide sequences from methanotrophs and ammonia oxidizing nitrifyers. See text for details

Proteobacteria) (Hanson et al. 1993; Brusseau et al. 1994). Ammonia oxidizing nitrifiers are: Nc.oce, *Nitrosococcus oceanus* (gamma-*Proteobacteria*); Nm. eur, *Nitrosomonas europaea*; Nl.mul, *Nitrosolobus multiformis*; Nspira, *Nitrosospira* sp. Np22 (all beta-*Proteobacteria*) (Head et al. 1993). Consen., is the consensus sequence. A total of 54 out of 164 amino acid residues in the portion of the PmoA and AmoA examined were conserved in all sequences. These universally conserved residues are boxed in black and residues conserved in a majority of the species included in this study are shaded. These highly conserved residues are distributed throughout the alignment, suggesting that all of these proteins will show structural conservation. This is supported by hydropathy plots and protein structure predictions for these polypeptides which are all similar for each sequence.

The most notable feature of sequence comparisons was that the level of conservation showed a stronger correlation with the phylogenetic relatedness of the organisms than with the physiological function of the gene products. Three identity groups representing the alpha, beta and gamma subdivisions of the *Proteobacteria* were observed, rather than two identity groups representing AmoA and PmoA. Significantly, *Nc. oceanus* AmoA showed a higher identity (>75%) to all gamma-*Proteobacteria* methanotroph PmoA sequences than to any of the beta-*Proteobacteria* nitrifier AmoA sequences (<50%). The first identity group contained *Ms. trichosporium* and *Mcy. parvus* (88% identical), the second group contained *Nc. oceanus*, *Mb. albus*, *Mm. methanica* and *Mc. capsulatus* (75-85% identical) and the third contained *Mn. europaea*, *Nl. multiformis* and *Nitrosospira* sp. Np22 (83-99% identical). These data suggest the possibility that AMO and pMMO are evolutionarily homologous and that sequence differences are due partly to neutral changes as a result of evolutionary separation rather than solely to specialization to different functions.

To test this hypothesis, a more detailed phylogenetic analysis was performed using 16S rRNA and monooxygenase gene sequences. Identical clades, which corresponded to the identity groups, were found in trees constructed from PmoA/AmoA and those from 16S rRNA. The tree topography did however show some differences, most notably that the PmoA/AmoA tree showed the γ-*Proteobacteria* group to form a "supercluster" with the α-*Proteobacteria* group, whilst this was not found for the 16S rRNA tree (Figure 3). All putative PmoA sequences belonged to this "supercluster", together with the *Nc. oceanus* AmoA. The β-*Proteobacteria* group contained exclusively AmoA sequences. `Signature' amino acid sequence motifs could be defined for each of the identity groups, (*ie* α, β, and γ-*Proteobacteria*) and for the alpha/gamma "supercluster". However, such signatures could not be defined for all AmoA or PmoA sequences, assuming our designation of the *Nc. oceanus* sequence as AmoA is correct.

In a separate study we have also used another degenerate oligonucleotide primer (B1983) based on the aligned sequences of the *amo* and *pmo* gene clusters (see Figure 1). The target site for this primer is at the 3' end of pmoB and it has been used, together with the PCR primer A189, to PCR amplify a 1.8 kbp fragment of the genome of *Mb. albus* BG8. This was subsequently cloned using the TA cloning vector (Invitrogen) and sequenced (Murrell et al, Lidstrom et al, unpublished). This clone contains the majority of *pmoA* and *pmoB* from *Mb. albus* BG8 and when compared with the corresponding regions of the *pmoA* and *pmoB* genes from *Mc. capsulatus,* it was found to have a high identity with these genes (75 and 80% respectively).

Discussion

The importance of methanotrophs in the global methane cycle and their poor growth in laboratory conditions has led us to develop molecular probes for their detection. The pMMO genes were selected as targets for a general methanotroph probe as this enzyme has been found in all methanotrophs so far examined. The primers A189 and A682 proved to be highly efficient for amplification of pMMO genes but also amplify AMO genes. The strong conservation of pMMO and AMO genes suggests they are homologous, having evolved from a common ancestor. This hypothesis is supported by the identical groupings found in phylogenetic trees derived from PmoA/AmoA sequences and 16S rRNA sequences. The presence of Pmo and Amo genes in representatives of each of the α, β, and γ-*Proteobacteria* suggests that a homologue of these genes was present in the common ancestor of the *Proteobacteria*. This then

138

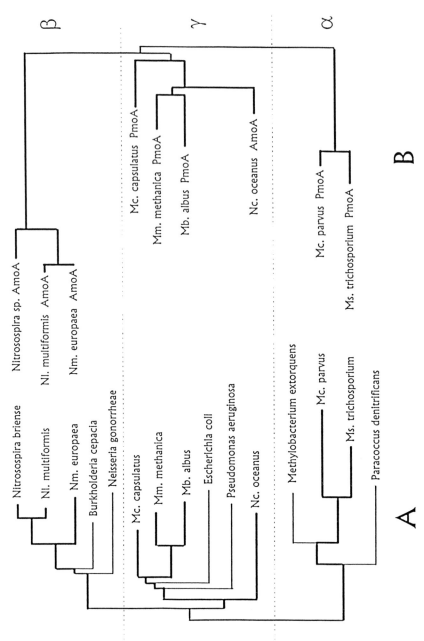

Figure 3. Evolutionary relationships of methane and ammonia oxidizing bacteria derived from 16S rRNA sequences (A) and monooxygenase peptide sequences (B). The alpha, beta and gamma-subdivisions are indicated by dotted lines. Phylogenetic trees were constructed from Kimura distance matrices using the neighbour-joining method.

strengthens the case for the hypothesis that all methanotrophic bacteria will contain the particulate form of methane monooxygenase. If pMMO genes can be distinguished from AMO genes by sequence information, then these genes may represent ideal "functional group-specific" probes for methane and ammonia oxidizing bacteria. On the basis of the phylogenetic analysis presented above it is possible to discriminate between β-subdivision ammonia oxidizers and methanotrophs on the basis of PmoA/AmoA sequence information. All PmoA sequences were found in a "supercluster" comprising members of the α and γ-*Proteobacteria*. We have designed primers targeting coding regions for amino acid sequences which were unique to this "supercluster". Experiments so far indicate these primers are diagnostic for pMMO, although they also recognise the putative *amoA* from *Nc. oceanus*.

Phylogenies inferred from rRNA genes, show that the β and γ-*Proteobacteria* branch together to the exclusion of the α-subdivision (Woese 1987). The phylogeny inferred from these monooxygenase genes differs in that it shows a closer relationship between α and γ-*Proteobacteria*, although the branch order could not be determined. This may reflect the different functions of the pMMO and AMO. It is not possible to define a root for the tree derived from monooxygenase gene sequences as no outlying sequences are available. If we assume that the ancestral monooxygenase gene was present in the Proteobacterial common ancestor then the root for the trees derived from 16S rRNA and monooxygenases should be the same, as is shown approximately in Figure 3. The long branch length for the β-*Proteobacteria* group resulting from this position of the root may be accounted for, either by adaptation to the pMMO function having led to convergent evolution in the α and γ-*Proteobacteria,* or that the β-*Proteobacteria* ammonia oxidizers represent a highly evolved group specialized for oxidation of ammonia. These hypotheses bring into question the status of *Nc. oceanus* as an ammonia oxidizing chemoautotroph and our putative identification of the 525 bp PCR product from this strain as a*moA*. It is interesting to note that this species has been reported to incorporate ^{14}C-labelled methane into cell carbon, although it could not be grown on methane as sole carbon and energy source (Ward 1987). In this context, it is noteworthy that of the strains tested in this study, methanol dehydrogenase was found to be present in all methanotrophs and absent from all ammonia oxidizers, including *Nc. oceanus*. This was tested by using each of the DNAs as template, with PCR primers targeting an approximately 500 bp DNA fragment of the methanol dehydrogenase large subunit structural gene *mxaF*. These PCR primers have previously been used to amplify this gene fragment from a number of methanotrophs and methylotrophs (Murrell et al. 1993; Holmes et al. 1995[b]; McDonald and Murrell, unpublished observations). PCR products of the correct (predicted) size were obtained with all methanotroph DNAs but no PCR product was obtained with any nitrifier DNA, including *Nc. oceanus*, suggesting the absence of methanol dehydrogenase in these nitrifiers. We also ensured the purity of *Nc. oceanus* DNA by PCR-amplifying and partial sequencing of the 16S rRNA gene from the DNA sample we obtained from this culture. It matched exactly the published 16S rRNA sequence for *Nc. oceanus*.

A major question in the global cycle of methane has been the relative importance of methanotrophs and ammonia oxidizers for methane oxidation in various habitats. These new probes and the developing sequence database allow us to directly address this question. We have successfully used the PCR primers A189 and A682 to amplify discrete PCR products from environmental DNA samples obtained from seawater. Gene libraries constructed from these PCR products are currently being analysed, with preliminary data confirming the amplification of *Nitrosospira* sp. *amoA* from British coastal waters.

In this study we have confirmed the presence of *pmoA* or *amoA* in a representative collection of methanotrophs and ammonia oxidizers. On the basis of the evidence presented above we tentatively consider probes targeting the α/γ -*Proteobacteria* monooxygenase "supercluster" to be pMMO-specific and universally applicable to methanotrophs. Similarly probes targeting the β-*Proteobacteria* monooxygenase group are considered to be AMO-specific. These probes are a useful tool for screening and analysis of gene libraries to assess the relative importance of these organisms in natural systems. These genes are likely to have been present in the ancestral *Proteobacteria* but amongst extant organisms appear to have been retained only in the methanotrophs and ammonia oxidizers. The possibility of homologues of *pmoA* being present in other *Proteobacteria* cannot be discounted and probes targeting either pMMO or AMO genes must therefore be carefully tested.

140

References

Brusseau GA et al (1994) Appl. Environ. Microbiol. 60, 626-636.

Dalton H et al (1993) In Murrell JC, Kelly DP, eds, Microbial Growth on C1 Compounds, pp 65-80, Intercept, Andover, UK.

Ensign SA et al (1993) J. Bacteriol. 175, 1971-1980.

Hanson RS et al (1993) In Murrell JC, Kelly DP, eds, Microbial Growth on C1 Compounds, pp 285-302, Intercept, Andover, UK.

Head IM et al (1993) J. Gen. Microbiol. 139, 1147-1153.

Holmes AJ et al (1995[a]) FEMS Microbiol. Lett. in press.

Holmes AJ et al (1995[b]) Microbiol. 141, 1947-1955.

Lipscomb JD (1994) Annu. Rev. Microbiol. 48, 371-399.

McTavish et al (1993) J. Bacteriol. 175, 2436-2444.

Murrell JC (1992) FEMS Microbiol. Rev. 88, 233-248.

Murrell JC (1994) Biodegradation 5, 145-159.

Murrell JC et al (1993) Chemosphere, 26, 1-11.

Nguyen H-H T et al (1994) J. Biol. Chem. 269, 14995-15005.

Semrau et al (1995) J. Bacteriol. 177, 3071-3079.

Smith DD, Dalton H (1989) Eur. J. Biochem. 182, 667-671.

Stanley SH et al (1983) Biotechnol. Lett. 5, 487-492.

Ward BB (1987) Arch. Microbiol. 147, 126-133.

Woese CR (1987) Microbiol. Rev. 51, 221-271.

Carboxylate Shifts in the Active Site of the Hydroxylase Component of Soluble Methane Monooxygenase from *Methylococcus capsulatus* (Bath)

Amy C. Rosenzweig[a], Christin A. Frederick,[a] and Stephen J. Lippard[b]
[a]Department of Biological Chemistry and Molecular Biology, Harvard Medical School and Dana Farber Cancer Institute, 44 Binney St., Boston, Massachusetts 02115, U. S. A. [b]Department of Chemistry, Massachusetts Institute of Technology, Cambridge, Massachusetts 02139, U. S. A.

Introduction

Carboxylate anions constitute an especially versatile group of ligands in inorganic chemistry. Numerous complexes in which the electron lone pairs of a carboxylate anion are coordinated to one or more metal ions have been reported (R. C. Mehrotra, R. Bohra, 1983). The carboxylate ligand can adopt a variety of different binding modes, which have been described as 'carboxylate shifts' (Figure 1) (R. L. Rardin et al., 1991). In

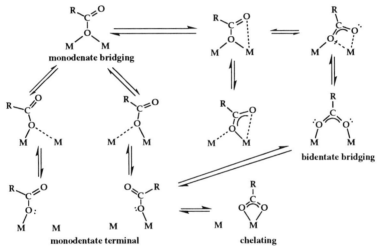

Figure 1. Carboxylate binding modes in metal complexes and the carboxylate shift (adapted from Rardin et al., 1991)

particular, polymetallic complexes containing a monodentate bridging carboxylate were classified based on the strength of interaction between the non-bridging or 'dangling' oxygen atom and one of the bridged metal ions. It was suggested that the monodentate bridging mode represents an intermediate between other coordination modes (Figure 1). Carboxylate ligands also occur in nature, as aspartate and glutamate amino acid residues coordinated to metal ions in metalloproteins, and a number of compounds having variable modes of carboxylate ligation have been prepared as models of metalloprotein active sites, particularly those containing iron or manganese ions (A. L. Feig, S. J. Lippard, 1994, D. M. Kurtz, 1990, L. Que, Jr., A. E. True, 1990). It was therefore hypothesized that carboxylate shifts may occur in metalloproteins as a means of facilitating changes in oxidation state and mechanisms of catalysis.

The hydroxylase component of soluble methane monooxygenase (sMMOH) from *Methylococcus capsulatus* (Bath) belongs to growing group of crystallographically

141

M. E. Lidstrom and F. R. Tabita (eds.), Microbial Growth on C₁ Compounds, 141–149.
© 1996 *Kluwer Academic Publishers. Printed in the Netherlands.*

142

characterized proteins that contain a catalytic carboxylate-bridged diiron center. Details of the active sites of hemerythrin (Hr) (M. A. Holmes, R. E. Stenkamp, 1991, M. A. Holmes et al., 1991), the R2 protein of ribonucleotide reductase (R2) (P. Nordlund, H. Eklund, 1993, P. Nordlund et al., 1990), and rubrerythrin (Rr) (P. Nordlund, personal communication) are now available. In the active site of each of these proteins, one or more ligands to the iron atoms is provided by a carboxylate derived from a glutamate or aspartate side chain. These carboxylates exhibit variable binding modes, including monodentate terminal, chelating, and bidentate bridging. The prevalence of carboxylate ligands in these diiron centers indicates that carboxylate shifts may indeed be involved in catalysis. The first evidence of a carboxylate shift in one of these proteins was observed upon reduction of the R2 diiron center to the diiron(II) state (A. Åberg, 1993). We have recently discovered that carboxylate shifts occur quite readily at the diiron center of MMOH.

Four different X-ray structures of MMOH have been determined: a 2.2 Å resolution structure of the diiron(III) hydroxylase (H_{ox}) at 4 °C, a 1.7 Å resolution structure of H_{ox} at -160 °C, a 1.7 Å resolution structure of the diiron(II) hydroxylase (H_{red}) at -160 °C, and a 1.7 Å resolution structure of H_{ox} in 10% DMSO (H_{DMSO}) at -160 °C. The overall protein structure of the MMOH $\alpha_2\beta_2\gamma_2$ dimer is the same in all four analyses, but the diiron centers differ significantly. The first three structures have been reported elsewhere (A. C. Rosenzweig et al., 1993, A. C. Rosenzweig, S. J. Lippard, 1994, A. C. Rosenzweig et al., 1995). The fourth structure, which has not yet appeared in the literature, is a result of attempts to bind DMSO to the iron atoms. Such an adduct has been detected by ENDOR spectroscopic studies of mixed valent Fe(II)Fe(III) solution samples treated with DMSO (M. P. Hendrich et al., 1992, V. J. DeRose et al., unpublished). Although no DMSO was found in the active site of H_{DMSO}, a new geometry for the diiron center was revealed. Here we describe the structure of H_{DMSO} in detail and compare the four MMOH active site structures with special emphasis on the observed carboxylate shifts and their possible significance in the catalytic mechanism.

Results

Structure of H_{ox} at 4 °C. Since MMOH is isolated in the diiron(III) oxidation state, our initial X-ray studies focused on this form of the enzyme. The first structure of H_{ox} was determined at 2.2 Å resolution, with all data collected at 4 °C (A. C. Rosenzweig et al., 1993). The active site is shown in Figure 2. Four carboxylate ligands

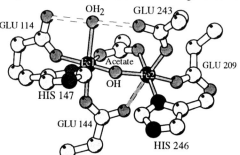

Figure 2. The diiron center in H_{ox} at 4 °C (Molscript, (P. Kraulis, 1991)). Hydrogen bonds are indicated by open dashed lines. The filled dashed line indicates a weak interaction.

from the protein coordinate the two Fe atoms: Glu 114, Glu 144, Glu 209, and Glu 243. Glu 114 is a monodentate terminal ligand to Fe1, and its dangling oxygen atom is hydrogen bonded to a terminal water molecule also coordinated to Fe1. We have called Glu 144 a 'semi-bridging' carboxylate ligand because, although it bridges the two Fe atoms, the Fe2-O distance is longer than a normal bonding distance. The two carboxylate ligands coordinated to Fe2, Glu 209 and Glu 243, are both monodentate, with the dangling oxygen atom of Glu 243 hydrogen bonded to the water molecule coordinated to Fe1. There is a fifth carboxylate present in the bidentate bridging position, which we assigned as an acetate ion derived from the crystallization buffer. Hence three difference carboxylate binding modes are observed: monodentate terminal, bidentate 'semi-bridging,' and bidentate bridging. The remainder of the coordination sphere comprises two histidine ligands coordinated by their δ-N atoms, His 147 and His 246, and a bridging hydroxide ion. Each Fe atom is therefore six-coordinate, and the Fe···Fe distance is 3.4 Å.

Structure of H_{ox} at -160 °C. In order to obtain a higher resolution structure of H_{ox}, we flash froze the crystals at -160 °C in the presence of 20% glycerol as a cryosolvent (A. C. Rosenzweig et al., 1995). The frozen crystals are far less sensitive to radiation damage, allowing us to collect data to 1.7 Å resolution. The resulting diiron center structure is shown in Figure 3. The 4 glutamic acid ligands are coordinated in the

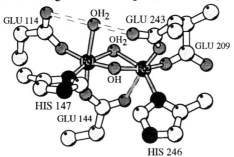

Figure 3. The diiron center in H_{ox} at -160 °C. Hydrogen bonds are indicated by open dashed lines. The filled dashed line indicates a weak interaction.

same manner as in the 4 °C, 2.2 Å structure, and the two histidine ligands, the hydroxide bridge, and terminal water molecule on Fe1 are also present. The bridging acetate, however, has been replaced with a monoatomic bridge, which we have assigned as a weakly coordinated water molecule from the long average $Fe-OH_2$ distance of 2.5 Å. Consequently, the Fe···Fe distance has been reduced from 3.4 Å to 3.1 Å. Electron density that may represent the acetate ion was found in the active site cavity, within hydrogen bonding distance of the water bridge. The acetate ion has therefore 'carboxylate-shifted' completely out of the coordination sphere.

Structure of H_{red} at -160 °C. The first step in the catalytic cycle of MMOH is reduction of H_{ox} to the diiron(II) state (H_{red}). Two electrons are transferred from NADH to H_{ox} by a 2Fe-2S flavoprotein, the reductase (R) (J. Lund, H. Dalton, 1985). The electron transfer is regulated by R and a third protein which has no cofactors, protein B (J. Green, H. Dalton, 1985, K. E. Liu, S. J. Lippard, 1991). H_{red} then reacts with dioxygen to form a diiron(III) peroxo intermediate (H_{peroxo}), which has been characterized by stopped-flow optical (K. E. Liu et al., 1995), freeze-quench Mössbauer (K. E. Liu et al., 1995, K. E. Liu et al., 1994) and resonance Raman spectroscopy (K. E. Liu et al., 1995). A second intermediate, Q, also characterized by stopped-flow optical and freeze-quench Mössbauer spectroscopy (K. E. Liu et al., 1995, K. E. Liu et al., 1994, S. K. Lee et al., 1993, S. K. Lee et al, 1993), then forms and reacts with substrate to form product and regenerate H_{ox}. In order to understand fully the hydroxylation mechanism, crystallographic information about each of these species is desirable. As a first step in this direction, we added chemical reductants to the H_{ox} crystals, imitating procedures used to generate H_{red} in solution. Upon addition of sodium dithionite and methyl viologen, the active site structure shown in Figure 4 was obtained in the A protomer of the MMOH dimer (A. C. Rosenzweig et al., 1995). This structure differs from

144

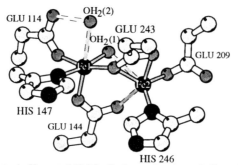

GLU 114

OH₂(2)

GLU 243

OH₂(1)

GLU 209

HIS 147

GLU 144

HIS 246

Figure 4. The diiron center in H_{red} at -160 °C. Hydrogen bonds are indicated by open dashed lines. The filled dashed lines indicate weak interactions.

that in H_{ox}, and we concluded that a change in oxidation state due to the addition of chemical reductants had occurred.

Many aspects of the H_{red} active site resemble that observed in H_{ox}. The two Fe atoms are coordinated to two histidine residues, two monodentate terminal carboxylates, Glu 114 and Glu 209, and a semi-bridging carboxylate, Glu 144. The binding mode of Glu 243 is dramatically different, however. In H_{red}, the oxygen atom of Glu 243 that was not bonded in H_{ox} bridges the two Fe atoms, rendering Glu 243 a monodentate bridging ligand and a bidentate, chelating ligand to Fe2. The bridging hydroxide is no longer present, and the bridging water molecule is only very weakly coordinated to Fe2. The terminal water molecule on Fe1 is also weakly coordinated. Each Fe atom is therefore five-coordinate, with a weak sixth interaction with a water molecule. The Fe···Fe distance is 3.3 Å. Carboxylate shifts resulting from a change in oxidation state are not unprecedented. In the diiron(II) model compound [$Fe_2(O_2CH)_4(BIPhMe)_2$], a monodentate bridging formate becomes a monodentate terminal ligand upon oxidation to the diiron(III) state (W. B. Tolman et al., 1991). In the R2 protein, a glutamate residue, Glu 238, shifts from bidentate bridging in the diiron(II) form to monodentate terminal in the diiron(III) form (A. Åberg, 1993). The fact that changes in carboxylate binding modes accompany oxidation in two biological diiron centers, MMOH and R2, suggests that carboxylate shifts are important in promoting reduction and thus key in the catalytic mechanisms. The R and protein B components raise the reduction potential of the MMOH diiron center (K. E. Liu, S. J. Lippard, 1991). One way that they could cause such a change would be to perturb the helix containing Glu 243 upon binding, causing a carboxylate shift. The results of EPR and MCD spectroscopic studies have demonstrated that the binding of protein B directly affects the diiron center (B. G. Fox et al., 1991, S. Pulver et al., 1993). No crystallographic information is currently available for intermediates H_{peroxo} and Q, but carboxylate shifts may also accompany their formation.

Structure of H_{ox} in 10% DMSO at -160 °C. In addition to H_{red}, H_{peroxo} and Q, the substrate and product complexes of H_{ox} are important species in the MMO reaction cycle. In order to gain structural information relevant to substrate and product binding, we utilized the inhibitor DMSO. According to recent ENDOR spectroscopic results on the mixed valent Fe(II)Fe(III) hydroxylase, the oxygen atom of DMSO binds to the Fe(III) ion with one methyl group ~4 Å from the Fe(III) ion (V. J. DeRose et al., unpublished). The ENDOR study also revealed that a water ligand coordinated to the Fe(II) ion and the bridging hydroxide ion remain intact in the presence of DMSO. We attempted to introduce DMSO into the H_{ox} crystals in two ways. Firstly, we crystallized H_{ox} as described previously (A. C. Rosenzweig et al., 1992) with the addition of 20 mM DMSO to the crystallization buffer. Secondly, we soaked crystals of H_{ox} in a synthetic mother liquor containing 25 mM Li_2SO_4, 50 mM NH_4OAc, 20% polyethylene glycol 4000, and 10% DMSO in 25 mM 3-[N-morpholino]propanesulfonic acid (MOPS) buffer, pH 7.0 for 28 days. Both types of crystals were transferred briefly to solutions having the same compositions but with the addition of 20% glycerol as a cryosolvent, and flash-frozen at -160 °C. A number of these crystals were transported to Stanford Synchrotron Radiation Laboratory (SSRL) for data collection, and 1.7 Å data sets were obtained by using a 300 mm diameter Marresearch imaging plate detector equipped with a 2theta vertical lift. The data set for the crystal co-crystallized with 20 mM DMSO resulted in a structure identical to H_{ox} at -160 °C, with no ordered DMSO present anywhere in the protein. The data set for the crystal soaked in 10% DMSO (H_{DMSO}) yielded an interesting result, however. The R_{sym} for this data set is 0.079 (0.170 between 1.8 and 1.7 Å), and the completeness is shown by resolution range in Table 1. The lack of completeness at very high

Table 1. Completeness of data in different resolution ranges for the 10% DMSO data set

Resolution (Å)	5.0	3.3	2.8	2.5	2.3	2.1	2.0	1.9	1.8	1.75	1.7
Completeness (%)	81	78	75	70	65	60	54	47	36	13	

resolution is due to the size of the imaging plate and the use of the 2theta lift. The structure was refined with X-PLOR (A. T. Brünger, 1993) by using the 1.7 Å -160 °C H_{ox} structure as a starting model. The program O was used for all model building (T. A. Jones et al., 1991). The final model contains the same residues as the other three structures with the addition of 919 water molecules, and the R-value to 1.7 Å resolution is 0.201. The root-mean-square deviations from ideal geometry are 0.013 Å for bond lengths and 1.58° for bond angles, and >99% of residues fall within the allowed regions of the Ramachandran plot. The average error in the refined coordinates is estimated by a Luzzati plot to be ~0.2 Å.

The active site structure is shown in Figure 5 and metrical details are presented in Table 2. No ordered DMSO was found coordinated to the Fe atoms, in the active site cavity, or elsewhere in the protein. Drastic changes have occurred in the diiron center in both protomers, however. In particular, Fe2 has moved by ~1.2 Å in a direction roughly perpendicular to the Fe⋯Fe vector, increasing the Fe⋯Fe distance to 3.64 Å in the A protomer and 3.68 Å in the B protomer (Table 2). One result of the new position of Fe2 is that Glu 144 is not within coordinating distance of Fe2, with an average Fe2-O distance of 3.9 Å. This semi-bridging carboxylate has therefore become a monodentate

Table 2. Interatomic distances (Å)[a] in H_{DMSO}[b]

Atom	Atom	Distance (Å)		
		Protomer A	Protomer B	Average
Fe1	Fe2	3.64	3.68	3.7
Fe1	Glu114 OE1	2.01	2.11	2.0
Fe1	His147 ND1	2.15	2.17	2.2
Fe1	Glu144 OE2	2.01	2.14	2.1
Fe1	μOH	2.30	2.48	2.4
Fe1	OH_2(1)	2.09	2.11	2.1
Fe2	Glu209 OE2	1.90	1.94	1.9
Fe2	His246 ND1	2.30	2.09	2.2
Fe2	Glu243 OE1	2.26	2.03	2.1
Fe2	Glu 243 OE2	2.49	2.26	2.4
Fe2	Glu144 OE1	3.81	3.93	3.9
Fe2	μOH	1.66	1.65	1.7
Fe2	OH_2(2)	2.36	2.57	2.5
OH_2(1)	Glu 114 OE2	2.81	2.82	2.8
OH_2(1)	OH_2(2)	2.82	2.95	2.9
Glu 144 OE1	μOH	2.45	2.44	2.4

[a]Estimated standard deviations in distances are ~0.2 Å. [b]Protomers A and B are the two crystallographically independent halves of the hydroxylase $\alpha_2\beta_2\gamma_2$ dimer (see A. C. Rosenzweig et al., 1993).

(a)

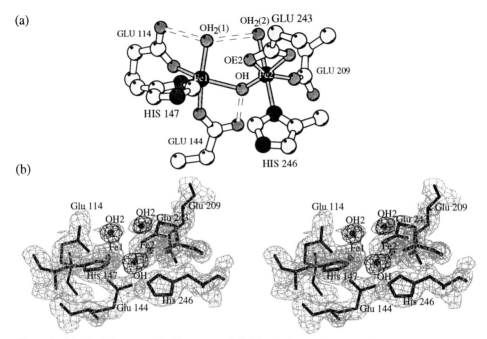

(b)

Figure 5. (a) The diiron center in H_{DMSO} at -160 °C. Hydrogen bonds are indicated by open dashed lines. (b) Stereo diagram of the final $2F_o$-F_c electron density at the diiron center in H_{DMSO} (O, (T. A. Jones et al., 1991)). The $2F_o$-F_c map is shown in gray and is contoured at the rms density. Superimposed in black is an F_o-F_c map showing the hydroxo bridge and water molecules coordinated to Fe1 and Fe2. The F_o-F_c map is contoured at three times the rms density.

terminal ligand to Fe1. Glu 243 has also adopted a new coordination mode, different from that seen in H_{ox} or H_{red}, chelating Fe2, with Fe2-O distances of ~2.1 and ~2.4 Å. Glu 209 is still monodentate, but has changed conformation slightly. His 147, His 246, Glu 114, and the terminal water ligand to Fe1 all remain in the same positions as in H_{ox}. A hydroxo bridge is also present. Instead of a bridging water molecule as in H_{ox} at -160 °C, there is a terminal water coordinated to Fe2. This molecule is hydrogen-bonded to the water molecule coordinated to Fe1, which is hydrogen bonded to Glu 114. The 2 Fe atoms are linked solely by the hydroxo bridge and by this hydrogen bonding interaction. The diiron center is therefore asymmetrical with Fe1 being five-coordinate and Fe2 being six-coordinate.

The bridging hydroxide ion is asymmetrical with Fe1-O longer than Fe2-O, and is hydrogen bonded to the dangling oxygen atom of Glu 144 with a short hydrogen bonding distance of ~2.4 Å. In both H_{ox} structures, the bridging hydroxide is also asymmetrical, but with Fe2-O longer than Fe1-O by 0.2-0.4 Å (A. C. Rosenzweig et al., 1993, A. C. Rosenzweig et al., 1995), a difference within experimental error. In this case, Fe1-O is longer than Fe2-O by ~0.7 Å, a difference which may be outside experimental error. If a small fraction of the diiron centers contain Fe2 in the position it occupies in H_{ox} at -160 °C, ~3.1 Å from Fe1, the hydroxide ion could refine to a position artificially closer to Fe2, leading to the observed asymmetry. The assignment of the exogenous ligands in H_{DMSO} as two water molecules and a hydroxide ion renders the active site charge positive, in contrast to the active sites in H_{ox} both at 4 °C and at -160 °C, in H_{red}, and in several forms of the R2 protein (A. Åberg, 1993). There is, however, a portion of poorly defined electron density in the active site cavity, similar to what was observed in the frozen crystal H_{ox} structure and postulated to be acetate. If acetate ion were also present in H_{DMSO}, the active site charge would remain neutral. Alternatively, one of the terminal ligands assigned as water may actually be a hydroxide ion.

Treatment of the H_{ox} crystals with 10% DMSO has thus resulted in several major changes: a new Fe···Fe distance of ~3.65 Å, different coordination modes for Glu 144 and Glu 243, and the replacement of the bridging water molecule with a terminal water molecule coordinated to Fe2. It is not obvious why DMSO should alter the diiron center in this way nor why it does not bind to an Fe atom as it does in the solution ENDOR samples (V. J. DeRose et al., unpublished). There is no indication that the extra electron density in the active site cavity suggested above to be acetate represents disordered DMSO, although the possibility cannot be excluded. According to the ENDOR study, the hydroxo bridge and terminal water molecule are not affected by DMSO binding. Similarly, in the crystal treated with 10% DMSO, the hydroxo bridge and terminal water molecule coordinated to Fe1 are not altered. Therefore, the coordination site on Fe2 occupied by a water molecule in the current structure could delineate the DMSO binding site. Perhaps this coordination site is sometimes occupied by water and sometimes by DMSO, and the water ligand predominates in the crystal or displaces DMSO as the temperature is lowered to -160 °C. Alternatively, the water molecule could actually represent the oxygen atom of a multiply disordered DMSO molecule. The observed structural variations could also be due to changing the solvent rather than to DMSO binding. Little information is currently available about the behavior of the hydroxylase either in solution or in the crystal in mixtures of water and aprotic solvents. Finally, the DMSO structure could represent another oxidation state of the protein core. We have previously considered the possibility that the crystals used for the H_{ox} structure determinations could have photoreduced to the mixed valent Fe(II)Fe(III) oxidation state in the synchrotron-generated X-ray beam (A. C. Rosenzweig, S. J. Lippard, 1994, A. C. Rosenzweig et al., 1995), a phenomenon observed in EXAFS studies of oxidized hydroxylase samples (J. G. DeWitt et al., 1991, J. G. DeWitt et al., 1995). A preliminary single crystal EXAFS experiment indicated that photoreduction does not occur in the crystalline state, however (H. Bufford et al., unpublished). A less plausible explanation for the present results is that DMSO somehow sensitizes the crystal to photoreduction, resulting in the Fe(II)Fe(III) oxidation state. The Fe···Fe distance of ~3.65 Å is longer than the 3.4 Å value obtained from EXAFS measurements on the photoreduced hydroxylase, however.

Conclusions

The four structures described herein reveal that the diiron center in MMOH is flexible and can accommodate changes in environment induced by freezing, addition of chemical reductants, and addition of DMSO. A number of carboxylate shifts are observed: the displacement of the bridging acetate ion, the shift of Glu 243 from monodentate terminal in H_{ox} to monodentate bridging and chelating in H_{red} to chelating in H_{DMSO}, and the shift of Glu 144 from semi-bridging in H_{ox} and H_{red} to monodentate terminal in H_{DMSO}. The different binding modes of Glu 243 are superimposed in Figure 6. These types of changes may be analogous to what happens during the catalytic cycle as H_{ox} is converted to H_{red} and then to H_{peroxo} and intermediate Q. In addition to demonstrating carboxylate shifts, the changes observed in the crystal structures reveal that the solvent molecules coordinated to the Fe atoms can be readily exchanged. The terminal water molecule coordinated to Fe1 is present in H_{ox} and H_{DMSO}, but only weakly coordinated in H_{red}. The bridging water molecule is present in H_{ox} at -160 °C and is weakly coordinated to Fe2 in H_{red}. In H_{DMSO}, the bridging water molecule has been replaced by a terminal water molecule on Fe2. Finally, the bridging hydroxide ion present in H_{ox} and H_{DMSO} is absent in H_{red}. The lability of these solvent molecules is essential if open coordination sites for the binding of dioxygen, substrates, and products are to be provided. Furthermore, the solvent molecules are all hydrogen bonded to the carboxylate ligands, suggesting that these hydrogen bonding interactions play a crucial role in stabilizing various carboxylate coordination modes. The ability to exchange solvent molecules may facilitate carboxylate shifts which in turn may accompany changes in oxidation state and the binding of protein B, the reductase, and substrate. Further crystallographic studies, especially in the presence of protein B and reductase, may reveal new carboxylate shifts in the MMOH active site.

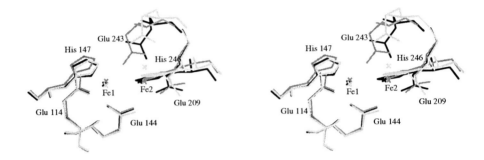

Figure 6. Stereo superposition of the diiron centers in H_{ox} (black), H_{red} (dark gray), and H_{DMSO} (light gray). All solvent molecules have been omitted for clarity.

Acknowledgements: This work was supported by grants from the National Institute of General Medical Sciences (GM32134 to S. J. L. and GM48388 to C. A. F. and S. J. L.). A. C. R. is the recipient of NIGMS postdoctoral fellowship GM15914. We thank D. A. Whittington for assistance with data collection and the staff at SSRL for support.

References

Åberg, A (1993), Ph. D. thesis, Stockholm University.
Brünger, AT (1993), X-PLOR Version 3.1 Manual, Yale University Press, New Haven, CT.
DeWitt, JG et al. (1991) J. Am. Chem. Soc. 113, 9219-9235.
DeWitt, JG et al. (1995) Inorg. Chem. 34, 2505-2515.
Feig, AL et al. (1994) Chem. Rev. 94, 759-805.
Fox, BG et al. (1991) J. Biol. Chem. 266, 540-550.
Green, J et al. (1985) J. Biol. Chem. 260, 15795-15801.
Hendrich, MP et al. (1992) J. Biol. Chem. 267, 261-269.
Holmes, MA et al. (1991) J. Mol. Biol. 220, 723-737.
Holmes, MA et al. (1991) J. Mol. Biol. 218, 583-593.
Jones, TA et al. (1991) Acta Crystallogr. Sect. A 47, 110-119.
Kraulis, P (1991) J. Appl. Crystallogr. 24, 946-950.
Kurtz, DM (1990) Chem. Rev. 90, 585-606.
Lee, S-K et al. (1993) J. Am. Chem. Soc. 115, 6450-6451.
Lee, S-K et al. (1993) J. Biol. Chem. 268, 21569-21577.
Liu, KE et al. (1991) J. Biol. Chem. 266, 12836-12839.
Liu, KE et al. (1995) J. Am. Chem. Soc. 117, 4987-4990.
Liu, KE et al. (1995), in press.
Liu, KE et al. (1994) J. Am. Chem. Soc. 116, 7465-7466.
Lund, J et al. (1985) Eur. J. Biochem. 147, 291-296.
Mehrotra, RC et al. (1983), Metal Carboxylates, Academic Press, New York.
Nordlund, P et al. (1993) J. Mol. Biol. 231, 123-164.
Nordlund, P et al. (1990) Nature 345, 593-598.
Pulver, S et al. (1993) J. Am. Chem. Soc. 115, 12409-12422.
Que, L, Jr. et al. (1990) Prog. Inorg. Chem. 38, 97-200.
Rardin, RL et al. (1991) New. J. Chem. 15, 417-430.
Rosenzweig, AC et al. (1992) J. Mol. Biol. 227, 283-285.

Rosenzweig, AC et al. (1993) Nature 366, 537-543.
Rosenzweig, AC et al. (1994) Acc. Chem. Res. 27, 229-236.
Rosenzweig, AC et al. (1995) Chem. & Biol. 2, 409-418.
Tolman, WB et al. (1991) J. Am. Chem. Soc. 113, 152-164.

THE BIOCHEMISTRY OF THE PARTICULATE METHANE MONOOXYGENASE

Hiep-Hoa A.T. Nguyen[†], Mei Zhu[†], Sean J. Elliott[†], Kent H. Nakagawa[§], Britt Hedman[¶¶], Andria M. Costello[¶], Tonya L. Peeples[¶], Barrie Wilkinson[‡], Hiromi Morimoto[*], Philip G. Williams[*], Heinz G. Floss[‡], Mary E. Lidstrom[¶], Keith O. Hodgson[§], Sunney I. Chan[†**]

[†]A.A. Noyes Laboratory of Chemical Physics and [¶]W.M. Keck Laboratory of Environmental Engineering, California Institute of Technology, Pasadena CA 91125. [§]Department of Chemistry and [¶¶]Stanford Synchrotron Radiation Laboratory, Stanford University, Stanford CA 94305. [‡]Department of Chemistry, University of Washington, Seattle, Washington 98195. [*]National Tritium Labeling Facility, Lawrence Berkeley National Laboratory, Berkeley, CA 94720
[**]Author to whom correspondence should be addressed.

1 SUMMARY

The particulate methane monooxygenase (pMMO) found in the intracytoplasmic membranes of methanotrophs has been known to be very difficult to study. Recent progress in our laboratory indicates that the pMMO is a novel copper-containing enzyme [1]. Metal/protein ratio data analysis clearly suggests that the pMMO is a multiple copper-containing enzyme. The pMMO-associated copper ions appear to be organized into trinuclear cluster units with rather defined-magnetic and redox properties. These copper clusters has been shown to be involved in dioxygen activation. The as-isolated pMMO enriched-membranes often contain a mixture of Cu(I) and Cu(II) ions in various proportions, depending on the history of the samples. The functional form of the enzyme has been found to be the reduced or partially reduced form.

2 ISOLATION & MOLECULAR STRUCTURE

One of the major obstacles in characterizing the pMMO is the lack of a purified enzyme system. Previous efforts in isolating the pMMO have been largely unsuccessful. Enzymatic activity is frequently lost upon cell lysis and detergent solubilization. However, we have made significant progress in this area. Methodology now has been developed to obtain highly active pMMO-enriched membranes with unusual stability. Activity now can be maintained for extended period of time (>7 days at 4 °C), stable with respect to freeze-thaw cycles, and prolonged storage at -80 °C (> 2 years). These membranes have been found to be the most suitable for pMMO isolation. The membranes were deoxygenated, treated with reductant(s), and solubilized by non-ionic detergent. The solubilized membranes were then centrifuged to remove unsolubilized materials. Upon detergent removal by dialysis and lipid reconstitution, activity can be recovered from these solubilized membranes. Furthermore, these solubilized fractions can be subjected to additional chromatographic and purification procedures to obtain partially purified pMMO with significant level of activity recovery.

M. E. Lidstrom and F. R. Tabita (eds.), Microbial Growth on C_1 Compounds, 150–158.

After subjecting the solubilized membranes to various methods of purification (ion-exchange, ammonium sulfate precipitation, gel filtration, and affinity chromatography), we have observed that three polypeptides were consistently co-purified together concomitant with recovered activity. In addition, they are also the most abundant polypeptides in the as-isolated membranes, lending evidence to support the notion that the pMMO is the major protein in these membranes. This result suggests that the pMMO or at least one form of the enzyme contains three subunits. SDS-PAGE analysis of partially purified pMMO fractions obtained from these experiments indicates an apparent molecular weight of 46 kDa, 23 kDa, 20 kDa, for these α, β, and γ subunits, respectively (Figure 1).

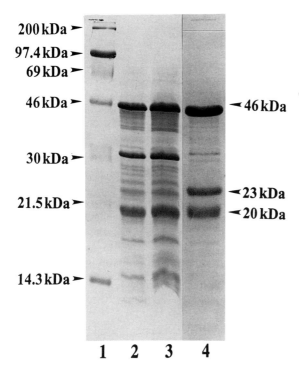

Figure 1. SDS-PAGE of the pMMO membranes and purified pMMO.
Lane 1. Proteins standards
Lane 2&3. Membrane fractions.
Lane 4. Purified pMMO (The three subunit form)

The large 46 kDa subunit has been implicated previously as part of the pMMO complex. When the organisms switch from expressing sMMO to pMMO, this 46 kDa polypeptide appears in the SDS-PA gel of the membranes. The 23 kDa polypeptide apparently is the acetylene-binding subunit. The presence of a third unit in the pMMO complex was not obvious from earlier works. At the moment, its involvement in the pMMO complex is not known. In the present study, the N-terminus of all three subunits have been sequenced using Edman degradation (Figure 2). Using oligonucleotides derived from these N-terminal sequences, we have identified two copies of the genes encoding these polypeptides in _Methylococcus capsulatus_ (Bath) (Figure 3). The translated sequences from the two different copies are highly homologous. Thus, the evidence is converging toward the notion that the pMMO hydroxylase is a three subunit enzyme.

N-Terminal Sequence of the α Subunit
HGEKSOAAFMRMRTIHWYDLSWSKEKVKINGTV
N-Terminal Sequence of the β Subunit
SAVRSHAEAVQSRTIDWMALFVVFFVI
N-Terminal Sequence of the γ Subunit
AAAEAPLKDKKWLTFA

Figure 2. The N-Terminal Sequence of the pMMO hydroxylase subunits

The *pmo* genes in *Methylococcus capsulatus* Bath

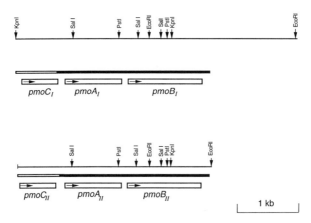

Figure 3. Physical map of the two restriction fragments containing pMMO hydroxylase genes. The region sequenced is indicated by the stippled box.

3 THE COPPER IONS OF THE pMMO

Our characterization of the pMMO-associated copper ions indicate that these copper ions exist in both Cu(I) and Cu(II) oxidation states in the membranes. The presence of significant level of Cu(I) ions is a surprising revelation since for most of redox-active copper-containing enzymes, the oxidized form is most often obtained. X-ray absorption edge data provide unequivocal evidence for the presence of Cu(I) ions in the as-isolated pMMO membranes [2]. The absorption feature at ~8984 eV in the Cu K-edge X-ray absorption spectra offers a diagnostic signature for Cu(I) ions [3]. The weak absorption at ~8984 eV has been attributed to the 1s -> 4p transition according to ligand field analysis. This diagnostic band has been observed in all of Cu K-edge X-ray absorption spectra of the as-isolated pMMO membranes obtained to date, albeit with various degrees of intensity, hence indicative of different level of reduction (Figure 4). This result strongly indicates that the membrane-bound copper ions are redox active toward dioxygen. In another word, these copper ions are sensitive to dioxygen tension, and can be oxidized with ease to a certain degree by O_2. As a result, the functional form of the enzyme must be the fully reduced or partially reduced form.

The level of copper reduction can be estimated by examining the Cu K-edge spectrum of these preparations. Edge analysis suggests that ~70% ± 10% of total copper

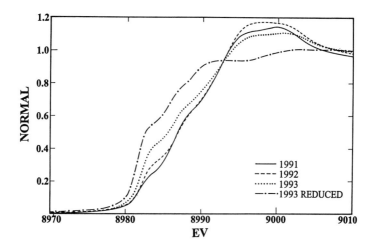

Cu-MMO SAMPLES
Figure 4. Cu K-edge X-ray absorption spectra of pMMO membranes as isolated.

ions exists as Cu(I) ions in a recent preparation of which high and unusually stable activity was observed (Figure 5). Assuming that all clusters are involved in hydroxylation chemistry and undergo synchronized turnover with dioxygen, then ~1/3 of total membrane-bound copper ions should remain as Cu(I) after dioxygen-turnover. Upon exposing a completely-reduced preparation to air, a significant fraction of the copper ions is rapidly oxidized as indicated by the fast recovery of the sample EPR intensity to the level prior to reduction within <10 min. (Figure 6). Edge analysis of this preparation revealed that ~70% of total copper ions in this synchronized sample still exists as Cu(I) ions. This higher-than-expected level of Cu(I) ions suggests that these pMMO-associated copper ions do not exhibit the same reactivity toward dioxygen. As mentioned above, if these membrane-bound copper ions are pMMO site(s) of dioxygen activation, increasing dioxygen tension should yield a more oxidized sample. Treating as-isolated or fully reduced pMMO membranes with pure dioxygen for 2 hrs yields a much more oxidized preparations as indicated by the Cu K-edge absorption spectrum. Edge analysis suggests that ~40%-50% of the total membrane-bound copper ions exists as Cu(I) ions. In parallel EPR measurements, the dioxygen-saturated sample exhibits features typical of magnetically-isolated or weakly-coupled copper ions (Figure 6). Thus the level of reduced copper ions in these synchronized preparations is in reasonable agreement with the picture of the copper clusters as the dioxygen-activation site(s) of the enzyme. The remaining Cu(I) ions appears to resist further oxidation by dioxygen as demonstrated by the failure to obtained a highly or fully oxidized preparation upon prolonged O_2-exposure.

The copper ions in the as-isolated pMMO membranes have been found to be heterogeneous. This observation has led us to attempt to prepare the enzyme with a more well-defined redox state than the as-isolated form. We have discovered that these membrane-bound copper ions can be reduced by a variety of compounds, ranging from dithionite, organic dyes (thionine), and others (NADH, quinol analogs, etc...). Dithionite yields the best result, capable of reducing most if not all of the copper ions as observed by EPR and X-ray absorption spectroscopy (Figure 5 & 6). The reduced form reacts readily with dioxygen as expected if the copper clusters are indeed the monooxygenase-site(s) of the enzyme. The fully-oxidized form of the enzyme is much more difficult to prepare. In all oxidants tested to date (quinone analogs, $Co(phen)_3^{3+}$, etc...), ferricyanide is found to be the best, capable of oxidizing most if not all of the membrane-bound copper ions. (Residual Cu(I) ions have been observed occasionally, however). The bulk of the copper ions are seen to be oxidized in this ferricyanide-oxidized preparation as indicated by the complete loss of edge absorption at ~8984 eV (Figure 5). This result allows us to ascertain that most if not all of the membrane-bound copper ions is Cu(II) in this sample. The EPR spectrum of the fully-oxidized preparation is similar to the copper cluster signal reported earlier. The EPR spectrum exhibits the broad and quasi-isotropic features of the previously proposed copper clusters. The origin of this EPR absorption has been attributed to the |-1/2> -> |+1/2> transition in the ground state quartet manifold of a spin-coupled trinuclear copper cluster. Since there is no intense feature at high g values, the EPR spectrum is consistent with a quartet with very small zero-field splitting parameters (D&E). The appearance of this broad and isotropic signal indicates that the EPR signature of the copper cluster can be directly observed upon fully oxidizing the copper ions in the membranes. The overwhelming presence of this isotropic signal suggests that the bulk of the membrane-bound copper ions are organized into cluster units with well-defined magnetic and chemical properties.

Other oxidation states of these copper clusters, i.e. the two-electron reduced and one-electron-reduced species must also exist, even though they are not necessarily thermodynamically stable or catalytically competent. The two-electron reduced form of the copper cluster can be prepared by treating the highly-oxidized membranes with sub-stoichiometric amount of reductants (dithionite or NADH). This two-electron reduced copper cluster exhibits an isotropic 10-line EPR signal which has been characterized earlier [1]. The one-electron reduced species appears to be the resting form of the enzyme.

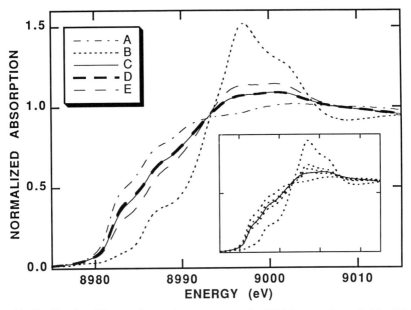

Figure 5. Cu K-edge X-ray absorption spectra of pMMO samples of A) dithionite-reduced; B) ferricyanide-oxidized; C) "as-isolated"; D) re-oxidized turnover; E) dioxygen-treated.

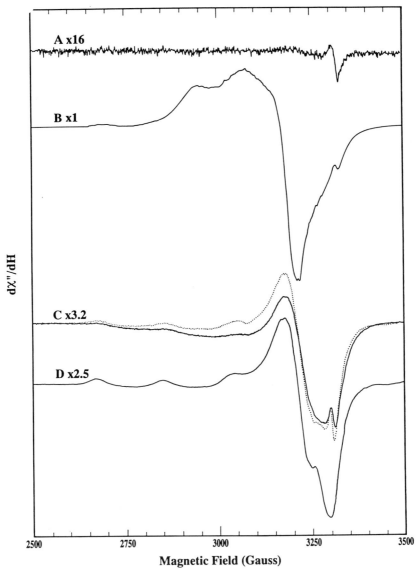

Figure 6. EPR spectra of pMMO samples of A) dithionite-reduced; B) ferricyanide-oxidized; C) (dotted-line) "as-isolated" prior to reduction, (solid line) re-oxidized turnover; D) dioxygen-treated.

This form can be obtained in a rather uniform state by treating the fully reduced membranes with pure dioxygen for an extended period of time (1-2 hrs). As mentioned above, X-ray absorption edge analysis indicates that ~40% - 50% of total copper ions exists as Cu(I) ions in these preparations.

4 HYDROXYLATION CHEMISTRY OF THE pMMO

In contrast to the sMMO which is a fairly catholic enzyme, known to oxidize a vast variety of hydrocarbons, the pMMO has a very narrow substrate range, capable of oxidizing C1-C5 hydrocarbons only. Products obtained from pMMO-catalyzed hydroxylation of alkanes and alkenes are tabulated in Table 1.

Table 1

Comparison of sMMO and pMMO hydroxylation regioselectivity

Substrate	Hydroxylation Products	
	pMMO	sMMO
Methane	Methanol	Methanol
Ethane	Ethanol	Ethanol
Propane	2-Propanol (~100%)	1-Propanol, 2-Propanol
n-Butane	2-Butanol (~100%)	1-Butanol (54%), 2-Butanol (46%)
1-Butene	1,2-Epoxybutane (58%) 3-Buten-2-ol (42%)	1,2-Epoxybutane (~100%)
n-Pentane	2-Pentanol (~100%)	1-Pentanol (27%), 2-Pentanol (70%), 3-Pentanol (3%)
2-Methylpropane	No product detected	NA
2-Methylbutane	No product detected	NA

Comparing to the sMMO, the pMMO exhibits unusual regio-selectivity in the case of C3 or higher alkanes. Hydroxylation catalyzed by pMMO at C-2 position is overwhelmingly preferred. Virtually no primary alcohol product has been detected. Stereochemical analysis of these 2-alcohol products also reveals highly unusual stereoselectivity. The chiral distribution of 2-alcohols formed by hydroxylation of n-butane, 1-butene, and n-pentane, deviates significantly from a racemic distribution (Table 2).

Table 2

Chiral Hydroxylation Product Distribution for pMMO

Substrate	Product	Temperature	R/(R+S) (ee)
n-Butane	2-Butanol	43 °C	73% (46%)
		38 °C	76% (52%)
		25 °C	79% (58%)
1-Butene	3-Buten-2-ol	43 °C	75% (50%)
		38 °C	81% (62%)
n-Pentane	2-Pentanol	43 °C	90% (80%)

(R)-2-alcohols clearly are the predominant species, thus the hydroxylation of these hydrocarbons catalyzed by the pMMO is biased toward the formation of (R)-products. This unusual chiral selectivity is enhanced by temperature such that low temperatures improve enantiomeric excess. If hydroxylation of alkanes catalyzed by pMMO follows a rebound-radical mechanism as commonly observed in parallel systems such as cytochrome P-450 [4], or sMMO isolated from _M. trichosporium_ OB3b [5], i.e. the radical intermediate has a sufficient life-time to undergo inversion, a racemic product distribution should be observed. Clearly this is not the case here for the pMMO. Even more dramatic result was obtained when chiral ethane ((R) and (S)-[1-^{3}H$_1$,^{2}H$_1$]ethane) was used as mechanistic probes (Table 3).

TABLE 3

Percentage Distribution of ^{3}H Label in the Methylene Group of the (2R)-2-Acetoxy-2-Phenylethanoate Derivatives of Ethanol Generated from Chiral Ethane by _M. capsulatus_ (Bath) pMMO membranes

Product	OR Me ""D \T #4	OR Me ""T \H #1	OR Me ""T \D #2	OR Me ""H \T #3
Substrate				
T CH$_3$ ""D \H (S)	8b	2b	75b	15b
		Retention = 97% a,c		
T CH$_3$ ""H \D (R)	77b	17b	4b	2b
		Retention = 100% a,c		
T CD$_3$ ""D \H (S)	2b	3b	75b	20b
		Retention = 100% a,c		
T CD$_3$ ""H \D (R)	80b	14b	2b	4b
		Retention = 101% a,c		

aOverall retention data have been corrected for the enantiomeric purity of substrates. bThese data obtained directly from spectra. cThe deviation from 100% is well within the compounded error range of NMR integration for product and substrate spectra.

Using these substrates, one should be able to follow the stereochemical course of the reaction upon analyzing the chiral distribution of alcohol products via tritium-NMR. After being corrected for the enantiomeric purity of starting materials, tritiated product

158

distribution data clearly indicates the hydroxylation of chiral ethane catalyzed by pMMO proceeds with 100% retention of configuration [6]. This startling result implies that the hydroxylation of methane and ethane catalyzed by the pMMO may follow a concerted mechanism. In another word, the activation of primary C-H bond and the oxygen insertion step must be synchronous. This finding strongly suggests that the mechanism employed by pMMO to oxidize alkanes may be drastically different from other monooxygenase mechanisms known to date.

Acknowledgment. This work is supported by the NIH (Grants GM 22432 to S.I.C., GM 32333 to H.G.F., RR 01209 to K.O.H., and RR 01237 to the National Tritium Labeling Facility) and NSF (CHE 9423181 to K.O.H.). The X-ray absorption data were collected at SSRL, which is supported by U.S. Department of Energy, Office of Basic Energy Sciences, Divisions of Chemical and Materials Sciences with further support provided by the NIH, National Center for Research Resources, Biomedical Resource Technology Program, and the Department of Energy, Office of Health and Environmental Research.

REFERENCES

1. Nguyen, H.-H. T.; Shiemke, A.K.; Jacobs, S.J.; Hales, B.J.; Lidstrom, M.E.; and Chan, S.I. *J. Biol. Chem.* **1994**, *269*, 14995-15005.
2. Nguyen, H.-H. T.; Nakagawa, K.H.; Hedman, B.; Lidstrom, M.E.; Hodgson, K.O.; Chan, S.I. *Manuscript in preparation.*
3. Kau, L.-S.; Spira-Solomon, D.J.; Penner-Hahn, J.E.; Hodgson, K.O.; and Solomon, E.I. *J. Am. Chem. Soc.* **1987**, *109*, 6433-6442.
4. Ortiz de Montellano, P.R. in *Cytochrome P-450 : Structure, Mechanism, and Biochemistry*; Ortiz de Montellano, P.R., Ed., Plenum Press : New York, 1986; pp. 217-271.
5. Priestley, N.D.; Floss, H.G.; Froland, W.A.; Lipscomb, J.D.; Williams, P.G.; Morimoto, H. *J. Am. Chem. Soc.* **1992**, *114*, 7561-7562.
6. Wilkinson, B.; Zhu, M.; Priestley, N.D.; Nguyen, H.-H. T.; Morimoto, H.; Williams, P.G.; Chan, S.I.; Floss, H.G. *Submitted to JACS.*

AMMONIA MONOOXYGENASE FROM *Nitrosomonas europaea*

D. J. Arp, N. G. Hommes, M. R. Hyman. L. Y. Juliette, W. K. Keener,
S. A. Russell and L. A. Sayavedra-Soto

Department of Botany and Plant Pathology, Oregon State University,
Corvallis, Oregon, USA

Introduction

Nitrosomonas europaea is an obligate chemolithoautotroph which obtains energy for
growth from the oxidation of ammonia (NH_3) to nitrite (NO_2^-). This bacterium, along
with other bacteria which obtain their energy from the oxidation of NH_3, contribute to
the biogeochemical cycling of inorganic nitrogen. The product of NH_3 oxidation, NO_2^-,
becomes the growth substrate for bacteria which oxidize NO_2^- to nitrate (NO_3^-).
Together, these two groups of bacteria carry out the process known as nitrification. In
croplands where NH_3 is applied as a nitrogen fertilizer, nitrification can lead to loss of
nitrogen because the product, NO_3^-, is readily leached into ground and surface waters
and serves as a substrate for denitrification. Gaseous products of nitrification include
the greenhouse gas nitrous oxide. In the treatment of sewage, nitrification plays a beneficial role in the conversion of NH_3 to NO_3^- which is readily denitrified to N_2. More recently, a potential role for NH_3 oxidizers in the degradation of some environmental
pollutants (e.g. chlorinated aliphatics) has been recognized.

Nitrification is initiated by the oxidation of NH_3 to hydroxylamine (NH_2OH). This
reaction is catalyzed by ammonia monooxygenase (AMO). NH_2OH is then oxidized to
NO_2^- in a reaction catalyzed by hydroxylamine oxidoreductase (HAO). The electrons
released in this oxidation are partitioned among the reductant requirement of AMO, the
biosynthetic reactions of the cell, and oxidative phosphorylation (Fig. 1). AMO has not
yet been purified with activity so our knowledge of
this enzyme is limited to observations with intact
cells and inactive proteins. A 27,000 MW
polypeptide is labeled by the mechanism-based inactivator $^{14}C_2H_2$ (Hyman and Wood, 1985). This
labeled (and therefore inactive) polypeptide copuri-
fies with a 40,000 MW polypeptide (McTavish et
al., 1993a). The genes which code for each of these
proteins are adjacent (McTavish et al., 1993a) and

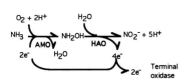

Fig. 1 The reactions catalyzed by
AMO and HAO

are cotranscribed (Sayavedra-Soto and Arp, unpublished). Therefore, a reasonable
model for the minimum subunit composition of AMO would include polypeptides of
27,000 and 40,000 MW. No direct knowledge of the prosthetic group composition of
AMO is available; however, copper is a likely constituent based on the inhibitor profile

M. E. Lidstrom and F. R. Tabita (eds.), Microbial Growth on C₁ Compounds, 159–166.
© *1996 Kluwer Academic Publishers. Printed in the Netherlands.*

of the enzyme (e.g. allylthiourea; Bedard and Knowles, 1989) and a stimulation of activity *in vitro* by the addition of copper ions (Ensign et al., 1993).

The substrate range of AMO is remarkably broad and includes the oxidation of alkanes (to alcohols), alkenes (to epoxides), aromatics (to phenols), ethers (to alcohols and aldehydes), thioethers (to sulfoxides) and chlorinated aliphatics (often to dehalogenated products) (Hyman et al., 1988; Rasche et al., 1990; Vannelli et al., 1990; Juliette et al., 1993a; Hyman et al., 1994; Keener and Arp, 1994). The bacterium does not appear to derive any benefit from the oxidation of these alternative substrates. Furthermore, sustained oxidation of these alternative substrates requires the presence of NH_3 as a source of reductant. Therefore, these alternative substrate oxidations are referred to as "cooxidations".

AMO is similar to the particulate methane monooxygenase of methanotrophs. This similarity extends to the putative subunit compositions, the inhibitor profiles, the broad (though not identical) substrate ranges and, most recently, to the DNA sequences of the genes coding for the proteins (Keeney and Nelson, 1982; Burrows et al., 1984; Bedard and Knowles, 1989; Semrau et al., 1995). Therefore, one expects that knowledge of AMO will continue to provide insight to the study of particulate methane monooxygenase, and vice versa.

Results

Towards a suitable in vitro assay for ammonia monooxygenase. Two problems which have plagued efforts to characterize AMO at the molecular level are 1) the development of a suitable *in vitro* assay and 2) stabilization of AMO activity in cell-free extracts. Because a direct electron donor (either natural or artificial) to AMO has not been identified, the *in vitro* assay relies on a coupling between AMO and HAO. In cell-free extracts of *N. europaea*, low levels of NH_3 oxidation activity (less than 10% of the intact cell rates) were measured for short periods of time (usually less than two hours) following cell breakage (Suzuki et al., 1981). The observed instability could be due to a loss of either AMO activity or to loss of any of the components required in this undefined, coupled assay.

Ammonia-dependent O_2 uptake in cell extracts was stimulated by the addition of Cu ions to the reaction mixture (Ensign et al., 1993). Bovine serum albumin (BSA) had previously been shown to be necessary to observe activity (Suzuki et al., 1981). The amount of Cu required for maximal stimulation varied with the amount of BSA added to the reaction mixture. BSA binds Cu, thereby competing with the beneficial role of Cu in the assay. A fraction (10 -20%) of the AMO retained activity in the cell-free extracts and could, therefore, be inactivated by C_2H_2. The remaining fraction of AMO was only susceptible to C_2H_2 inactivation following the addition of Cu. We proposed that Cu could dissociate from AMO resulting in loss of activity and that Cu could reassociate with AMO and restore activity when sufficient concentrations of Cu were supplied in the extracts. Upon treatment with Cu, all of the active AMO in intact cells was active in the cell-free extracts because labeling experiments with $^{14}C_2H_2$ revealed that the same amount of the 27,000 MW polypeptide was labeled in extracts treated with Cu and in intact cells. Nonetheless, the maximal specific activity of NH_3 oxidation in extracts was only about 10-20% of the specific rate of NH_3 oxidation in intact cells.

The role of BSA in the assay remained an enigma until Juliette et al. (1995) carried out a series of experiments which indicated that the role of BSA was to release the inhibition caused by fatty acids which accumulated in the extracts. Extracts exhibited three stages with regard to the requirement for BSA to observe NH_3 oxidation activity. BSA was not required in freshly prepared extracts, but was required for extracts aged as little as 30 min. With additional aging (150 min after cell breakage), activity was not detected even with the addition of BSA. Another fatty acid binding protein, lactoglobulin, replaced BSA while other proteins, such as ovalbumin, did not. Addition of palmitoleic acid (0.24 mM) inhibited activity in freshly prepared extracts and the inhibition was reversed by BSA or lactoglobulin. Furthermore, the concentration of palmitoleic acid in cell free extracts increased from 0.2-0.4 mM in freshly prepared extracts to 1.6-2.8 mM in extracts aged 105 min. Therefore, the experimental evidence supports the notion that the role of BSA is to bind fatty acids which would otherwise inhibit the oxidation of NH_3.

However, the mechanism by which fatty acids inhibit NH_3 oxidation is not apparent. Fatty acids did not inhibit the coupling enzyme, HAO, because oxidation of NH_2OH was not inhibited by fatty acids. This result suggested that the inhibition was specific for components associated with AMO activity. Given the broad substrate range of AMO, we considered the possibility that fatty acids were acting as substrates and outcompeting the binding of NH_3. However, depletion of fatty acids was not observed and predicted products were not observed in HPLC profiles of derivatized fatty acids (Juliette, 1995). We also considered the possibility that fatty acids were acting as protonophores and disrupting membrane gradients. The inhibition by fatty acids was not mimicked by protonophores such as dinitrophenol (100 µM) or carbonylcyanide m-chlorophenylhydrazone (10 µM). However, the electron transport inhibitor n-heptylhydroxyquinoline N-oxide (HQNO, 200 µg/ml) inhibited AMO activity and the inhibition was relieved by either BSA or reductant sources for AMO, such as NH_2OH or hydrazine (Fig. 2). Similar to the inhibition of NH_3 oxidizing activity by HQNO, the inhibition by fatty acids was relieved by either BSA or reductant sources for AMO. These preliminary results suggested that the inhibition of activity by fatty acids may involve disruption of electron transport. Although fatty acids completely inhibited NH_3 oxidation (Juliette et al., 1995), lower concentrations of fatty acids caused a lag in the onset of activity (Fig. 3). The lag was relieved by BSA or reductant sources for AMO. However, the cause of the lag and the mechanism for the inhibition of activity by fatty acids remains unknown.

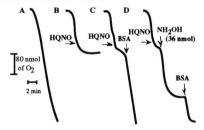

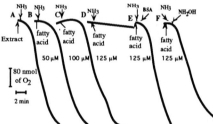

Fig. 2 Inhibition of NH_3-dependent O_2 uptake by n-heptylhydroxyquinoline N-oxide (HQNO) and relief of the inhibition by either BSA or NH_2OH. The O_2 electrode chamber contained cell extracts (200 µl, 2.2 mg of protein), $CuCl_2$ (13 µM) and $(NH_4)_2SO_4$ (2.8 mM). HQNO (200 µg/ml), BSA (5 mg/ml) (trace D) or NH_2OH (20 µM) were added at the indicated times.

Fig. 3 Effect of palmitoleic acid concentration on the initiation of NH_3-dependent O_2 uptake. Sequentially, $CuCl_2$, extract (200 µl, 2.2 mg of protein), palmitoleic acid, and $(NH_4)_2SO_4$ (2.8 mM) were added to the O_2 electrode chamber buffer. BSA (5 mg/ml) or NH_2OH (20 µM) was added at the indicated times.

Given the role of fatty acids in the inhibition of NH_3 oxidation activity and the accumulation of fatty acids in extracts during storage, we considered the possibility that production of fatty acids was also associated with the loss of NH_3 oxidation activity upon aging of extracts. Inclusion of BSA in the extracts extended the lifetime of activity from about 2 hrs to greater than 60 hrs (Juliette et al., 1995). This extended lifetime was not due to binding of fatty acids, but rather was due to the inhibition of lipolysis because 1) the amount of fatty acids produced was greatly decreased in extracts with BSA and 2) addition of fatty acids did not lead to destabilization of activity. In our survey of additional lipolysis inhibitors, we discovered that Cu ions effectively stabilized AMO activity in extracts and also inhibited lipolysis. Thus, two factors which are of primary importance in the assay of NH_3 oxidation activity, BSA and Cu, are also capable of stabilizing AMO activity in extracts. However, the mechanism of stabilization by both Cu and BSA, which is to inhibit lipolysis, is distinct from the two mechanisms by which these factors stimulate activity in the assay of NH_3 oxidation activity (Table 1). Another lipolysis inhibitor, phenylmethylsulfonylfluoride (which is more commonly known as a protease inhibitor), also stabilized activity for over 70 hrs.

TABLE 1: Proposed effects of copper and BSA on the activity and stability of ammonia oxidation

Additive	Effect on coupled AMO assay	Effect on stability of ammonia oxidation activity
Copper ions	Replaces Cu lost from the active site upon cell breakage	Inhibits phospholipase activity which leads to destabilization of activity
BSA	Removes fatty acids which inhibit the coupled ammonia oxidation assay	Inhibits phospholipase activity which leads to destabilization of activity

Use of inactivators to study the catalytic mechanism of ammonia monooxygenase.
Hyman and Wood (1985) determined that C_2H_2 is a mechanism-based inactivator of AMO in *N. europaea*. This work has led directly or indirectly to most of our molecular knowledge of AMO including the intracellular location, polypeptide composition, aggregation properties and amino acid sequence. The precise mechanism by which C_2H_2 inactivates AMO has not been determined. One possibility involves the formation of an unstable oxirene, analogous to the formation of a stable epoxide from an alkene, which would then react with amino acid residues at or near the active site of AMO. C_2H_2 also inactivates NH_3 oxidation in other nitrifiers. In a study of six phylogenetically diverse NH_3 oxidizing bacteria, inactivation with

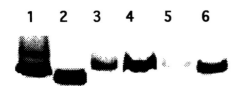

1   2   3   4   5   6

Fig. 4 Covalent incorporation of $^{14}C_2H_2$ into membrane-bound polypetides in nitrifying bacteria. 1, *N. europaea*; 2, *Nitrosococcus oceanus*; 3, *Nitrosolobus multiformis*; 4, *Nitrosomonas #56*; 5, *Nitrosospira briensis* and 6, *Nitrosovibrio tenuis*.

$^{14}C_2H_2$ led to the labeling in each species of a single polypeptide of about 27,000 MW (Fig. 4). To learn more about the catalytic mechanism of AMO and to support the development of nitrification inhibitors for field use, we have investigated a number of mechanism-based inactivators including additional alkynes, allylsulfide, phenylenediamine and dimethylcyclopropane.

The 1-alkynes (C_2 through C_{10}) are effective inactivators of AMO (Hyman et al., 1988; Table 2). 1-Hexyne is the most effective inactivator with regard to the rate of inactivation. 2-Alkynes were not inactivators of AMO, though 2-butyne and 2-hexyne were readily consumed by *N. europaea* in AMO dependent reactions. However, 3-alkynes were effective inactivators, although higher concentrations were required compared to their 1-alkyne isomers. 1,5-Hexadiyne was also an effective inactivator, but the presence of two triple bonds in the molecule did not increase the rate of inactivation. As with C_2H_2, the inactivations caused by the other alkynes required the catalytic turnover of AMO.

Keener (1995) investigated inactivation of AMO in intact cells by several compounds which are known as inactivators of other monooxygenases. 1-Aminobenzotriazole, cis/trans-1,2-dimethylcyclopropane, 1,3-phenylenediamine, p-anisidine and cyclopropyl bromide inactivated NH_3 oxidation by *N. europaea*. As expected for mechanism-based inactivators, catalytic turnover of AMO was required. The effects of the inactivators appeared to be specific for AMO because NH_2OH oxidation (catalyzed by HAO) was not affected. The inactivations by 1,3-phenylenediamine and 1,2-dimethylcyclopropane are best rationalized by single electron transfers to produce reactive radical intermediates. Cyclopropyl compounds inactivate dopamine β-monooxygenase via ring opening reactions resulting from one-electron oxidations (Fitzpatrick and Villafranca, 1987). Aniline and aniline derivatives inactivate dopamine β-monooxygenase. Inactivation by aniline hypothetically involved a one-electron oxidation of the N atom, with delocalization of the resultant radical in the ring leading to covalent attachment of the para carbon to the enzyme (Wimalasena and May, 1987).

TABLE 2: Alkyne inactivation of ammonia oxidation activity in intact cells of *Nitrosomonas europaea*[a]

Alkyne	Conc. (μM)	Time (sec) to 50% act.
Acetylene	20	84 ± 8
1-Propyne	102	89 ± 16
1-Butyne	113	85 ± 22
2-Butyne	80	Slight Inh.
1-Pentyne	20	44 ± 10
2-Pentyne	80	Slight Inh.
1-Hexyne	20	9 ± 1
2-Hexyne	80	Slight Inh.
3-Hexyne	20	35 ± 8
1-Heptyne	20	14 ± 4
3-Heptyne	80	53 ± 13
1-Octyne	20	22 ± 5
2-Octyne	200	Inh.
4-Octyne	200	103 ± 15
1-Decyne	20	20 ± 12
1,5-Hexadiyne	20	13 ± 1

[a] Cells were exposed to each alkyne while NH_3-dependent O_2 uptake was recorded. "Inh." indicates a reversible inhibition was observed, but no inactivation.

An interesting dichotomy is revealed when one considers the effects of NH_3 on the time required for these compounds to inactivate AMO. For a mechanism-based inactivator, one expects that the rate of inactivation will decrease in the presence of a substrate because the substrate binds to the active site and prevents the binding of the inactivator. Indeed, NH_3 increases the time required for inactivation by C_2H_2 (Fig. 5) and cyclopropylbromide (Keener, 1995). However, for allylsulfide (Juliette et al., 1993b), 1-hexyne (Fig. 5), 1,2-dimethylcyclopropane (Fig. 5), and several additional inactivators (Keener, 1995), increasing concentrations of NH_3 increased the rate of inactivation. This result is inconsistent with a competition between NH_3 and the inactivator for binding to the active site. However, the situation is complicated by the need for reductant to support the monooxygenase reaction. In intact cells, a low flux of reductant can be provided by the oxidation of endogenous storage compounds. The flux of reductant is increased in the presence of NH$_3$ as the NH$_3$ is oxidized to NH$_2$OH which is then further oxidized to NO$_2^-$ thereby releasing additional reductant which can be partitioned to AMO. Therefore, we considered the possibility that increasing rates of inactivation in the presence of increased NH$_3$ could be a result of an increased flux of reductant to AMO. However, when the flux of reductant was increased by providing cells with NH$_2$OH rather than NH$_3$, the rates of inactivation did not increase and NH$_3$ still enhanced the rates of inactivation (Keener, 1995). Therefore, we sought an alternative explanation for the enhancement of the rate of inactivation by NH$_3$. We currently favor the model that NH$_3$ enhances the inactivations by serving as an electron mediator. NH$_3$ would bind normally to the enzyme and transfer an electron to the activated oxygen species to form an aminium radical and a hydroxyl radical. In the absence of the inactivator, rebound between the aminium and the hydroxyl radical would result in the formation of the product, NH$_2$OH. However, with the inactivator bound to the enzyme, either the aminium radical or the hydroxyl radical might be quenched by reaction with the inactivator, thereby transforming the inactivator to its reactive state. This model would explain the dependency of the rates of inactivation on the NH$_3$ concentration and would also explain our difficulties in saturating the rate of inactivation by increasing the concentrations of inactivators. This model also requires that NH$_3$ and the inactivator can bind simultaneously to AMO. Noncompetitive interaction between NH$_3$ and several inhibitors of NH$_3$ oxidation indicates that simultaneous binding of NH$_3$ and other compounds to AMO can occur (Keener and Arp, 1993).

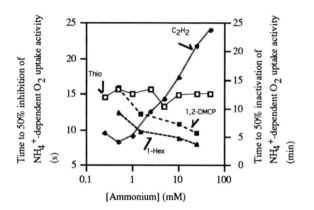

Fig. 5 The effects of NH$_4^+$ concentration on the rate of inhibition or inactivation of NH$_4^+$-dependent O$_2$ uptake activity in *N. europaea* in the presence of fixed concentrations of various compounds. The values for the reversible inhibitor, thiourea (Thio, 200 μM) and the irreversible inactivators C$_2$H$_2$ (200 μM) are in seconds (s). The values for the irreversible inactivators, 1-hexyne (1-Hex, 0.063 μM) and 1,2-dimethylcyclopropane (1,2-DMCP, 0.063 μM) are in minutes (min).

Gene expression and protein synthesis in N. europaea. We have recently investigated the effects of NH$_3$ on both protein synthesis and gene transcription in *N. europaea*. The 27 kDa component of AMO is the major polypeptide synthesized when cells are incubated simultaneously with (NH$_4$)$_2$SO$_4$ (as a potential inducer of AMO synthesis and nitrogen source), C$_2$H$_2$ (so NH$_3$ cannot serve as an energy source), NH$_2$OH (as an energy source) and ^{14}CO$_2$ (as a radiotracer to detect *de novo* protein synthesis in this autotrophic bacterium). The synthesis of the 27 kDa polypeptide under these conditions is inhibited by both chloramphenicol and rifampicin and is strongly affected by changes in the free NH$_3$ concentration (Hyman and Arp, 1995). Together these results suggest that the regulation of expression of AMO is influenced by the concentration of NH₃ available to cells. In addition to the 27 kDa component of AMO, the synthesis of a few other proteins was selectively induced by NH₃.

In recent years a number of *N. europaea* genes have been identified and their nucleotide sequences determined (e.g. *amoA* (McTavish et al., 1993a), *amoB* (Bergmann and Hooper, 1994), *hao* (Sayavedra-Soto et al., 1994), *cycA* (Bergmann et al., 1994) or *hcy* (Hommes et al., 1994), ORF2 (which likely codes for a tetraheme cytochrome; (Bergmann et al., 1994)) and the amino acid sequence of cytochrome *c*-552 (Fujiwara et al., 1995). We have investigated the regulation of expression of *amo* by NH₃. Using the sequence of *amo* we designed probes to detect the specific *amo* mRNAs induced during NH₃ incubations. In Northern hybridizations we detected the induction of specific *amo* mRNAs in cells treated as described above for *de novo* synthesis of AMO protein. The *amo* specific transcript was induced with NH₃ and correlated with the increase in AMO activity. There are two copies of the gene for AMO in the genome of *N. europaea* (McTavish et al., 1993a). Thus the *amo* specific mRNA detected might be the product of either one or both copies of *amo*.

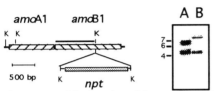

Fig. 6 Map of the insertion of the *npt* cassette into *amo*B 1 (left) and corroboration of the insertion (right). Hybridization of the probe (depicted as a line on *amo*B 1) to genomic DNA of wild type (lane A) and the mutant strain (lane B) is shown. The DNA was digested with EcoRI.

We have developed a transformation method which will allow us to further study the NH₃ response by selective mutagenesis . There are three copies of the gene for HAO, *hao*, (McTavish et al., 1993b) which were chosen as targets for mutagenesis. An antibiotic-resistance marker (*npt*, km^r) was inserted into a fragment of *hao* on a plasmid. The plasmid was transferred into *N. europaea* by electroporation and *npt* was mobilized into genomic copies of *hao* by homologous recombination, resulting in the insertion into (and likely inactivation of) the recipient gene. The transformed cultures were selected from liquid cultures and isolated in a newly developed protocol for growing the cells on solid media. The recombination events were corroborated by Southern hybridizations and DNA amplification. A collection of mutants with each one of the three copies of *hao* affected were produced and showed the feasibilty of the technique. No single copy of *hao* was essential to the cell. Similarly, the kanamycin cassette was inserted into one copy of *amo*B (Fig. 6). The insertion of the cassette conferred kanamycin resistance to *N. europaea*. We did not detect any differences between the phenotype of the mutant and the wild type which suggests that the affected copy of *amo*B also was not essential.

Conclusions

The major limitation to advances in our molecular knowledge of AMO continues to be the lack of purified and active preparations of this enzyme. Our work has revealed that the losses of NH₃ oxidizing activity which have plagued AMO studies in the past are not necessarily a result of an irreversible loss of AMO activity. Thus, AMO would appear to be more robust than previously thought. AMO has a remarkably broad substrate range and is inactivated by a variety of compounds with no structural similarity. Given this broad reactivity, generalizations regarding the chemical mechanism have not been forthcoming. Evidence for a radical-based mechanism is presented, but does not rule out other mechanisms for other substrates or inactivators. With the development of a transformation system for *N. europaea* and the characterization of a number of genes critical to NH₃ oxidation, one expects that progress in the area of gene expression and protein synthesis will continue to be rapid. NH₃ is required for the synthesis of AMOa and for the expression of *amo*. However, the mechanism of this induction by NH₃ remains to be determined. Our observation that each of the three copies of *hao* can be disrupted without affecting the phenotype of the cells provides an intriguing first look at the role of multiple gene copies.

References

Bedard C and R Knowles (1989). Microbiol. Rev. 53: 68-84.
Bergmann DJ et al., (1994). J. Bacteriol. 1767(11): 3148-3153.
Bergmann DJ and AB Hooper (1994). Biochem. Biophys. Res. Commun. 204(2): 759-762.
Burrows KJ et al., (1984). J. Gen. Microbiol. 130: 3327-3333.
Ensign SA et al., (1993). J. Bacteriol. 175(7): 1971-1980.
Fitzpatrick PF and JJ Villafranca (1987). Arch. Biochem. Biophys. 257: 231-250.
Fujiwara T et al., (1995). Current Microbiol. 31: 1-4.
Hommes NG et al., (1994). Gene 146: 87-89.
Hyman MR and DJ Arp (1995). J. Bacteriol. in press.
Hyman MR et al., (1988). Appl. Environ. Microbiol. 54(12): 3187-3190.
Hyman MR et al., (1994). App. Environ. Microbiol. 60(8): 3033.
Hyman MR and PM Wood (1985). Biochem. J. 227: 719-725.
Juliette LY et al., (1993a). Appl. Environ. Microbiol. 59(11): 3718-3727.
Juliette LY et al., (1993b). Appl. . Environ. Microbiol. 59(11): 3728-3735.
Juliette LY et al., (1995). J. Bacteriol. 177: in press.
Juliette LY (1995). Ph. D. Thesis: Oregon State University.
Keener WK (1995). Ph. D. Thesis: Oregon State University.
Keener WK and DJ Arp (1993). Appl. Environ. Microbiol. 59(8): 2501-2510.
Keener WK and DJ Arp (1994). Appl. Environ. Microbiol. 60(6): 1914-1920.
Keeney DR and DW Nelson (1982). Nitrogen-inorganic forms. Madison, WI, Amer. Soc. Agron.
McTavish H et al., (1993a). J. Bacteriol. 175(8): 2436-2444.
McTavish H et al., (1993b). J. Bacteriol. 175(8): 2445-2447.
Rasche ME et al., (1990). Appl. Environ. Microbiol. 56(8): 2568-2571.
Sayavedra-Soto LA et al., (1994). J. Bacteriol. 176(2): 504-510.
Semrau JD et al., (1995). J. Bacteriol. 177(11): 3071-3079.
Suzuki I et al., (1981). Can. J. Biochem. 59: 477-483.
Vannelli T et al., (1990). Appl. Environ. Microbiol. 56(4): 1169-1171.
Wimalasena K and SW May (1987). J. Am. Chem. Soc. 109: 4036-4046.

Quinocofactors in Copper Amine Oxidases and Lysyl Oxidase

Judith P. Klinman, Danying Cai, and Sophie Xuefei Wang
Departments of Chemistry and Molecular and Cell Biology, University of California,
Berkeley, CA 94720-1460

Introduction

Although quinones have been recognized as electron transfer reagents in biology, they had not been considered as important redox cofactors. However, this view has changed over the past five years with the emergence of the field of quinoproteins (J.P. Klinman, D. Mu, 1994). The first quinone to be described was pyrroloquinoline quinone (PQQ), which has been carefully documented to function as a dissociable cofactor in gram negative bacteria (cf. S.A. Salisbury et al., 1979). The PQQ-utilizing proteins are alcohol dehydrogenases, which are found in the periplasmic space of gram negative bacteria and often observed in greater quantities than the cofactor itself. This latter property has led to the observation that PQQ can function as a stimulant/vitamin when present in the growth medium of certain bacteria (M. Shimao et al., 1984; M. Ameyama et al., 1984).

The copper amine oxidases are a ubiquitous class of proteins which catalyze the oxidative deamination of primary amines, Eq. (1):

$$RCH_2NH_2 + O_2 + H_2O \longrightarrow RCHO + H_2O_2 + NH_3 \qquad (1)$$

In bacteria and fungi, these enzymes serve to generate free ammonium ion from precursor aliphatic and aromatic amines present in the growth medium. The mammalian enzymes have been proposed to play important physiologic roles in the regulation of blood plasma biogenic amine and intracellular polyamine levels. Despite intensive scrutiny over several decades, the active site cofactor in the copper amine oxidases has only recently been shown to be a quino-structure. Although originally attributed to PQQ, the structure was subsequently demonstrated to be a trihydroxyphenylalanine (topa quinone or TPQ, S.M. Janes et al., 1990) derived via a post-translational modification of a peptide-bound precursor tyrosine (D. Mu et al., 1992; D. Mu et al., 1994).

Lysyl oxidase is an extracellular, matrix-embedded enzyme catalyzing the oxidative cross-linking of elastin and collagen chains. The similarity of its catalytic reaction to the copper amine oxidases, together with the presence of protein-bound copper, has led to the classification of lysyl oxidase as a member of the copper amine

167

M. E. Lidstrom and F. R. Tabita (eds.), Microbial Growth on C₁ Compounds, 167–174.
© 1996 Kluwer Academic Publishers. Printed in the Netherlands.

oxidase family. A number of features support the idea that lysyl oxidase is a quinoprotein; these include: (i) its ability to support the redox cycling reaction characteristic of quinoproteins (M. Paz et al., 1991); (ii) its sensitivity to inhibitors selective for quinones (S.N. Gacheru et al., 1989); and (iii) its spectral properties (P.R. Williamson et al., 1986). There are, however, several fundamental features that distinguish lysyl oxidase from the remaining copper amine oxidases. First and foremost is the much reduced size of the lysyl oxidase: ca 32 kDa compared to ca. 80 kDa for the subunit of the copper amine oxidases (H.M. Kagan, 1986). Second, when the cDNA sequence for lysyl oxidase is compared to sequences for other copper amine oxidases, there is no obvious homology; in particular, it is not possible to find the Asn-Tyr-Asp/Glu consensus sequence found in all known TPQ-containing proteins (P.C. Trackman et al., 1990).

This paper addresses several features of quinoprotein structure and biogenesis. When the precursor-product relationship between tyrosine and topa quinone was first established, a novel mechanism involving "self-processing" was proposed, Scheme 1 (D. Mu et al., 1992; D. Cai, J.P. Klinman, 1994). According to Scheme 1, the active site copper lies in close proximity to the precursor tyrosine and is essential for the oxidative conversion of tyrosine to dopa. Subsequent to the oxidation of dopa to dopa quinone, a nucleophilic attack of a metal-bound water completes the process. Using a yeast expression system and site-directed mutagenesis, we have provided evidence in support of the mechanism shown in Scheme 1 for a methylamine oxidase cloned from *Hansenula polymorpha*. One outstanding question in the quinoprotein field has been the relationship of lysyl oxidase to other copper amine oxidases. As discussed in this paper, lysyl oxidase contains a quinostructure which is different from that seen with other members of the copper amine oxidase class. Despite these differences, it is possible to propose a biosynthetic mechanism which invokes a common precursor.

Scheme 1. Postulated self-processing pathway for topa quinone production from tyrosine involving the active site copper (D. Mu et al., 1992; D. Cai, J.P. Klinman, 1994).

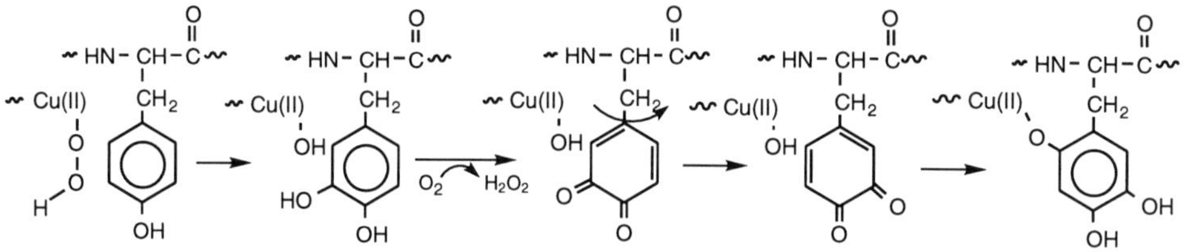

Results

Unlike many other species of yeast, *Saccharomyces cerevisiae* cannot use monoamines as a nitrogen source to support growth. This inability to utilize amines most likely correlates with a lack of amine oxidase production in *S. cerevisiae*. We have screened both the soluble and membranous compartments of *S. cerevisiae* to test for detectable levels of quinoprotein production. Using a gel electrophoresis redox staining method developed by Gallop and co-workers (M.A. Paz et al., 1991), no positive stain was observed with samples containing up to 370 mg of protein. We have concluded either that *S. cerevisiae* does not produce any quinoproteins or that these proteins are present at a level less than 0.1% of total protein (D. Cai, J.P. Klinman, 1994). For this reason *S. cerevisae* appears to be an excellent system in which to express a range of copper amine oxidases and to study their biogenesis.

Following the original report of expression of the gene for an amine oxidase from *H. polymorpha* in *S. cerevisiae* (M.J. de Hoop et al., 1992), we have shown generation of recombinant protein at a level of approximately 3% of total soluble protein. When *S. cerevisiae* cells transformed with the *H. polymorpha* gene were grown in a medium in which ammonium sulfate was substituted by 4 to 40 mM ethylamine, the cell density of cultures at saturation was typically 20 to 40% higher than that of control cells lacking ammonium sulfate or ethylamine. Although the final level of cell density was less than that reached with ammonium sulfate, this result indicates a functional activity of the recombinant amine oxidase *in vivo* to release ammonium ion from ethylamine and to support *S. cerevisiae* growth (D. Cai, J.P. Klinman, 1994).

Subsequent studies of the *S. cerevisiae* expression system have focused on the isolation of the recombinant *H. polymorpha* amine oxidase and its purification to homogeneity. Assay of purified enzyme with benzylamine and aliphatic amines indicated much higher activity with aliphatic amines, Table 1.

Table 1.   Oxidation of Amines by Yeast Amine Oxidases.

Enzyme	Benzylamine		Aliphatic amines[a]	
	Vmax (U/mg)	Km (mM)	Methylamine	Ethylamine
Recombinant enzyme	0.069	0.68	5.4	7.9
Benzylamine oxidase[b]	1.7	0.017	0.14	1.5

[a]Determined with 3 mM methylamine and 5mM ethylamine in 100 mM potassium phosphate, pH 7.2, at 37°C. Expressed as units/mg.
[b]Purified from *H. polymorpha* according to the method of Mu et al. (D. Mu et al., 1992).

These results indicate a very different substrate specificity for the recombinant amine oxidase, designated a methylamine oxidase, than for the amine oxidase previously isolated from *H. polymorpha* by Mu et al. (D. Mu et al., 1992), now designated a benzylamine oxidase. The presence of two isozymes of amine oxidases in *H. polymorpha* explains the previous failure to see a perfect alignment between the sequence of an active site derived topa quinone containing peptide (from *H. polymorpha* benzylamine oxidase) and the cDNA derived protein sequence (for *H. polymorpha* methylamine oxidase) (D. Mu et al., 1992). When the active site peptide from the recombinant protein was isolated and sequenced, it was found to give a perfect alignment with the cDNA derived protein sequence, Table 2.

Table 2. Topa Quinone Containing Peptide Sequences and DNA Derived Yeast
Methylamine Oxidase Protein Sequence.

Peptide from recombinant protein[a]	DNA derived sequence[b]	Peptide from yeast benzylamine oxidase[c]
Thr	Thr[401]	
Ala	Ala	Val
Ala	Ala	Ala
Asn	Asn	Asn
Blank	Tyr	Topa
Glu	Glu	Glu
Tyr	Tyr	Tyr
CmCys	Cys	Val
Leu	Leu	
Tyr	Tyr	

[a]Blank was an unidentified phenylthiohydantoin derivative and CmCys is
  carboxymethylated Cys.
[b]From P.G. Bruinenberg et al., 1989.
[c]From D. Mu et al., 1992.

The verification of a topa quinone moiety at position 405 in the recombinant
protein was carried out by a variety of techniques, which included redox staining of
purified protein, phenylhydrazine titrations which showed an end point of 1.5 mol of
phenylhydrazone per mol of enzyme dimer, and spectroscopic characterization of a p-
nitrophenylhydrazine derivative (D. Cai, J.P. Klinman, 1994). The later derivative
showed a λ_{max} of 472 nm at neutral pH and of 588 nm in 2.8M KOH. This 116 nm shift
as a function of pH is characteristic of topa quinone containing proteins and peptides
(Janes et al., 1992). These studies indicate that topa quinone can be generated from a
tyrosine precursor by expression of the *H. polymorpha* methylamine oxidase gene in a
strain of yeast that does not appear capable of quinoprotein biosynthesis itself. These
observations provide considerable support for the self-processing mechanism shown in
Scheme 1, and argue against a post-translational modification mechanism which requires
unique and specialized cellular processes.

Site-directed mutagensis has proven to be a powerful means of investigating
further the proposed self-processing mechanism. Once again, using the *H. polymorpha*
methylamine oxidase gene expressed in *S. cerevisiae,* mutations have been generated at
the active site consensus sequence and at the copper binding site. Three histidines have
been implicated as ligands to copper in the copper amine oxidases from spectroscopic
investigations (R.A. Scott, D.M. Dooley, 1985). Alignment of cDNA sequences for
copper amine oxidases shows a His-X-His sequence at the C-terminal domain of all
proteins (D. Mu et al., 1994; A.J. Tipping, M.J. McPherson, 1995). Given that His-X-His
is a well-known motif in copper-containing proteins (E.T. Adman, 1991), this region is

strongly implicated as participating in copper binding in the copper amine oxidases. In the yeast methylamine oxidase sequence, an extended motif, His-X-His-X-His, is observed. Reasoning that the central His456 was certain to be one of the His residues involved in metal binding, His456 was converted to Asp. Mutant protein was found to express and purify in a manner similar to the wild type recombinant protein, indicating a stable and properly folded mutant enzyme form (D. Cai, J.P. Klinman, 1994). When this protein was purified to homogeneity, it was found to contain copper at a very low level (*ca.* 4% of wild type protein). Importantly, we found that the H456D mutant neither exhibits amine oxidase activity nor contains a quinone moiety. These results indicate that disruption of the copper-binding site yields a protein incapable of cofactor production (D. Cai, J.P. Klinman, 1994).

We next turned to the conserved consensus site sequence surrounding the mature topa quinone. If the active site consensus sequence were the recognition and/or signaling for a post-translational processing of tyrosine, we reasoned that topa production would be curtailed or altered in a mutant protein. However, a mutant in which the acidic residue toward the C-terminus of topa was converted to Asn, E406N, showed the absorption spectrum characteristic of topa quinone, stained positive for quinoprotein using the redox cycling staining method and could be titrated with phenylhydrazine to a level of 1.9 mol reagent per mol of dimer (D. Cai, J.P. Klinman, 1994). Clearly, disruption of the conserved consensus sequence does not interfere with final titres of topa quinone in mature protein. One very unusual property of the E406N mutant is that it undergoes mechanism based inhibition in the course of methylamine oxidation. This indicates that while the consensus sequence is not required for cofactor synthesis, it plays an important role in maintaining the structural integrity of the active site. Overall, these studies provide strong support for a self-catalytic mechanism for TPQ biogenesis in a eukaryotic system. Similar conclusions have been reached by Tanizawa and co-workers regarding the expression of prokaryotic amine oxidases in *E. coli* (Y.-H. Choi et al., 1995; R.T. Matsuzaki et al., 1994).

In the course of ongoing studies of the TPQ-containing copper amine oxidases, a comparative in-depth study of the cofactor structure in lysyl oxidase was initiated. A major limitation in the structural characterization of lysyl oxidase has been the small amount of protein that can normally be obtained from mammalian sources, with a typical yield of 5 mg of protein per enzyme preparation. Although a CHO expression system is available for lysyl oxidase, this system only provides low (µg) levels of protein (H.M Kagan et al., 1995). Working with bovine aorta as a source of protein, we have recently succeeded in obtaining an active site peptide from lysyl oxidase in approximately 30% yield (S.X. Wang et al., 1995). Identification of the active site peptide was based on both absorbance and radioactivity arising from the [14C]-phenylhydrazine that had been used to label the active site cofactor. Sequencing of the purified peptide has indicated two amino acids in comparable yield at each cycle of Edman sequencing. Although this result could have indicated the presence of two peptides which had co-eluted at the same retention time from HPLC, subdigestion of the peptide and reanalysis indicated the same result of two amino acids at each round of sequencing. Confirmation of the cross-linked nature of the active site derived peptide has come from electrospray mass spectrometric analyses. Alignment of the two peptide sequences obtained from bovine aorta with the published cDNA sequence (P.C. Trackman et al., 1990) for rat aorta lysyl oxidase indicates the following result (S.X. Wang et al., 1995):

peptide 1	Asp	Thr	Blk	Asn	Ala	(Asp)
cDNA	Asp	Thr	Tyr^{349}	Ala	Ala	Asp

peptide 2	Val	Ala	Glu	Gly	His	(Lys)
cDNA	Val	Ala	Glu	Gly	His	Lys^{314}

We note the blank at position 3 of peptide 1, which aligns with tyrosine in the cDNA sequence and the absence of lysine at the end of peptide 2. We had expected to sequence through a lysine in the peptide, given the specificity of proteases employed in this study (thermolysin for the initial digest and Asp-N for the subdigest). We, therefore, have proposed that the lysyl oxidase cofactor is a cross-linked structure arising from a tyrosine and lysine designated lysine tyrosylquinone (LTQ):

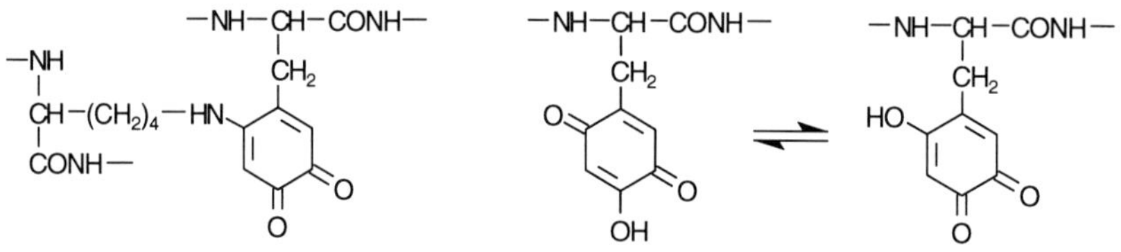

LTQ   TPQ

A number of experimental findings support the structure shown for LTQ. These include the known quino-structure of the cofactor in lysyl oxidase, the reactivity of the cofactor with phenylhydrazines, the mass of the cofactor structure deduced from mass spectrometric analysis, and the co-incidence of spectral properties of the lysyl oxidase derived peptide with a synthetic model compound (S.X. Wang et al., 1995).

In light of the new results with lysyl oxidase, the mechanism of biogenesis of LTQ becomes a particularly interesting question. One outstanding feature of the mechanism given in Scheme 1 is the generation of a dopa quinone intermediate. It is well-known that this type of o-quinone structure is highly reactive, undergoing facile nucleophilic attack at the C-2 position. We, therefore, propose that generation of LTQ may proceed in a manner highly analogous to TPQ. This is shown in Scheme 2, where the common intermediate dopa quinone is drawn as undergoing attack by a metal-bound hydroxide ion (to generate TPQ in the copper amine oxidases) or by an active site lysine (to generate LTQ in lysyl oxidase) (S.X. Wang et al., 1995). We note that an artifactual side reaction in which topa quinone reacts with a lysine side chain in the course of peptide isolation from lysyl oxidase can be ruled out from extensive model studies, which indicate that topa quinone reacts exclusively at its C-5 carbonyl in the presence of amines (M. Mure, J.P. Klinman, 1994).

Scheme 2. Postulated biogenetic pathways which invoke a common precursor, dopa quinone, in the generation of TPQ and LTQ.

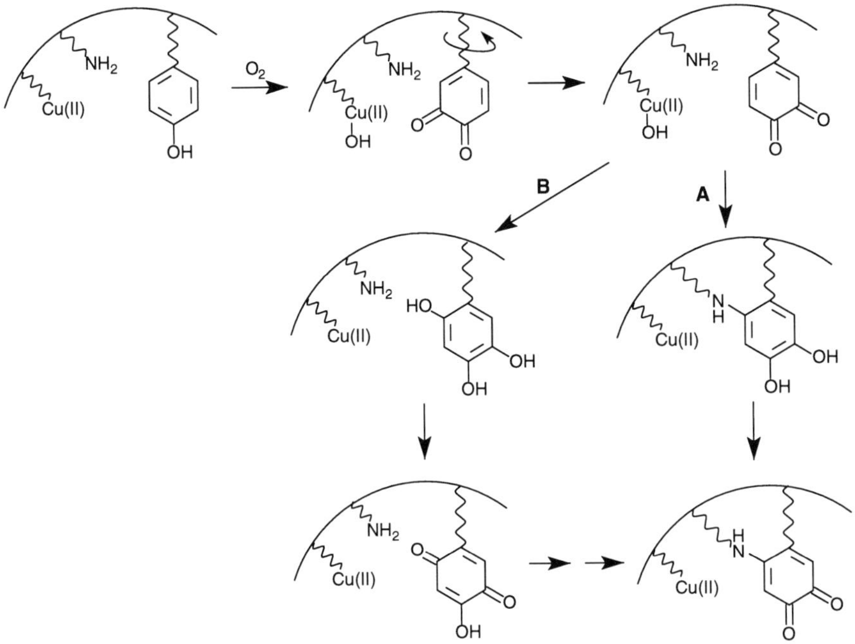

Conclusions

Using a recombinant system in which the gene for *H. polymorpha* methylamine oxidase is expressed in *S. cerevisiae*, we present extensive evidence in support of a self-processing mechanism for topa quinone biogenesis, Scheme 1. Recent studies with lysyl oxidase from bovine aorta indicate that this protein contains a cross-linked cofactor in the form of an aminoquinone. The biogenesis of this new species may be similar to that for TPQ, involving an initial formation of dopa quinone, which then undergoes attack by the ε-amino group of a lysine side chain, Scheme 2. The finding of a variant of TPQ in a mammalian system raises a number of provocative questions. For example, is it possible that additional variants of topa quinone also exist, in which the C-2 carbonyl of topa is replaced by polar side chains derived from amino acid residues such as Glu, Asp, Ser, or His? Additionally, is LTQ restricted to lysyl oxidase or will this new quino-structure appear in additional protein active sites?

References

Adman ET (1991) Adv. Protein Chem. 42, 144-197.
Ameyama M et al (1984) Agric. Biol. 48, 2909-2911
Bruinenberg PG et al (1989) Biochimica et Biophysica Acta 1008, 157-167.
Cai D, Klinman JP (1994) Biochem. 33, 7647-7653.
Cai D, Klinman JP (1994) J. Biol. Chem. 269, 32039-32042.

Choi Y.-H. et al (1995) J. Biol. Chem. 270, 4712-4720.

de Hoop MJ et al (1992) Yeast 8, 243-252.

Gacheru SN et al (1989) J. Biol. Chem. 264, 12963-12969.

Janes SM et al (1992) Biochem. 31, 12147-12154.

Janes SM et al (1990) Science 248, 981-987.

Kagan HM (1986) In Biology of Extracellular Matrix: A Series Regualtion of Matrix Accumulation, pp. 321-389, Academic Press, Orlando, Florida.

Kagan HM et al (1995) J. Cell. Biochem. 58, 1-10.

Klinman JP, Mu D (1994) Science 248, 299-344.

Matsuzaki R et al (1994) FEBS Lett. 251, 360-364.

Mu D et al (1992) J. Biol. Chem. 267, 7979-7982.

Mu D et al (1994) J. Biol. Chem. 269, 9926-9932.

Mure M, Klinman JP (1991) J. Am. Chem. Soc. 115, 7117-7127.

Paz MA et al (1991) J. Biol. Chem. 266, 689-692.

Salisbury SA et al (1979) Nature 280, 843-844.

Scott RA, Dooley DM (1985) J. Am. Chem. Soc. 107, 4348-4350.

Shimao M et al (1984) Agric. Biol. Chem. 48, 2873-2976.

Tipping AJ, McPherson MJ (1995) J. Biol. Chem. 270, 16939-16946.

Trackman PC et al (1990) Biochem. 29, 4863-4870.

Wang SX et al (1995) in preparation.

Williamson PR et al (1986) J. Biol. Chem. 261, 16302-16305.

CHAPERONING RUBISCO IN PURPLE BACTERIA

Hebe Dionisi, Susana Checa, Raúl Ferreyra and Alejandro Viale.
Programa Multidisciplinario de Biología Experimental (CONICET), Departamento de Microbiología, Facultad de Ciencias Bioquímicas y Farmacéuticas, Universidad Nacional de Rosario, 2000 Rosario, Argentina.

INTRODUCTION

Rubisco (ribulose 1,5-bisphosphate carboxylase/oxygenase) has played a pivotal role in the identification of the so-called molecular chaperones (Ellis 1994; Hartman, Harpel 1994), families of highly conserved proteins present in all organisms which play fundamental roles in the assembly, transport and degradation of cellular polypeptides as well as in the protection against cell injuries from many origins (Georgopoulos, Welch 1993; Ellis 1994). The chaperonins represent a subset of molecular chaperones, ubiquitously distributed among eubacteria, chloroplasts and mitochondria (Georgopoulos, Welch 1993; Ellis 1994; Viale et al. 1994). Two distinct members (GroEL and GroES) compose this family, which are encoded by a common operon (*groESL*) in most bacteria (Georgopoulos, Welch 1993; Viale et al. 1994). At present, *Escherichia coli* (*E. coli*) chaperonins constitute the best characterized members of the family (Georgopoulos, Welch 1993; Ellis 1994; Hartl et al. 1994), although studies on these proteins isolated from other bacteria suggest, in principle, a common mechanism of assisted folding (Terlesky, Tabita 1991; Soncini et al. 1992; Torres-Ruiz, McFadden, 1992; Saibil et al. 1993; Ellis 1994).

We described previously the cloning and characterization of Rubisco (Viale et al. 1989; Kobayashi et al. 1991; Viale et al. 1991) and *groESL* operons (Soncini et al. 1992; Ferreyra et al. 1993) from the purple bacterium *Chromatium vinosum* (*C. vinosum*), a phototrophic member of the γ-Proteobacteria (Woese 1987; Viale et al. 1994). We have characterized different activities of these proteins *in vitro*, and studied the role of chaperones in the assembly of bacterial Rubisco expressed in recombinant *E. coli* cells. Here, we present the result of these investigations.

RESULTS

In vitro characterization of Chromatium vinosum GroE molecular chaperones. C. vinosum chaperonins were purified from *E. coli* RF101 cells (see below) carrying plasmid pRF51 (Soncini et al. 1992; Ferreyra et al. 1993). Native molecular masses, as measured by size-exclusion chromatography, were *ca.* 900,000 (GroEL) and 90,000

M. E. Lidstrom and F. R. Tabita (eds.), Microbial Growth on C₁ Compounds, 175–182.

(GroES), respectively. These values, in addition to those deduced from the gene sequences (Ferreyra et al. 1993), are consistent with GroEL being composed by two stacked heptamers (GroEL$_{14}$) of identical 60-kDa subunits each, and GroES being a single homoheptameric toroid (GroES$_7$) of 10-kDa subunits (Ellis 1994).

Both biochemical and genetic analyses indicate that *in vivo* and *in vitro* GroEL and GroES physically and functionally interact (Georgopoulos, Welch 1993; Ellis 1994; Hartl et al. 1994). As for other chaperonins, *C. vinosum* GroEL shows a weak, K$^+$-stimulated ATPase which is specifically inhibited by GroES (Terlesky, Tabita 1991; Soncini et al. 1992; Torres-Ruiz, McFadden 1992; Georgopoulos, Welch 1993). Maximum inhibition (*ca.* 50%) in 10 mM KCl was obtained at a ratio GroES$_7$/GroEL$_{14}$ of approximately 1, thus suggesting that these two proteins are capable of forming assymetric complexes as shown for other chaperonins (Saibil et al. 1993; Ellis 1994).

A reaction thought to constitute a pivotal step in the physiological action of chaperonins consists in the ability of GroEL$_{14}$ to form binary, stable complexes with early folding intermediates of many proteins which possess no structural relationships when native (Ellis 1994). By slowly releasing the bound polypeptide, chaperonins are thought to effectively modulate the partition between productive folding and off-pathways (Ellis 1994; Hartl et al. 1994; Todd et al. 1994; Jaenicke 1995). We thus characterized the effects of *C. vinosum* chaperonins on the folding of rhodanese, a monomeric, two-domain mitochondrial protein widely employed as a model system for *in vitro* renaturation studies (Hartl et al. 1994; Mendoza et al. 1994 and references therein). As shown in Figs. 1 and 2, *C. vinosum* GroEL was able to arrest the spontaneous refolding of rhodanese. More than 95% folding arrest was obtained when the ratio GroEL$_{14}$/rhodanese reached around 0.5, either in the presence or absence of GroES (H.M. Dionisi et al., unpublished). Thus, *C. vinosum* GroEL$_{14}$ is capable of binding one or two molecules of unfolded polypeptides, as observed for *E. coli* GroEL (Ellis 1994; Hartl et al. 1994; Mendoza et al. 1994). The recovery of the highest yields of correctly folded rhodanese required the complete chaperonin system (*i.e.*, GroEL, GroES, ATP). The spontaneous dissociation of the binary complex to yield correctly folded rhodanese in the absence of GroES and ATP was found to be too slow to be of physiological significance (Fig. 1). It is still controversial whether any extent of folding occurs in association with the chaperonin, or after release into bulk solution (Ellis 1994; Hartl et al. 1994; Todd et al. 1994; Jaenicke 1995). In general, higher temperatures reduce productive folding by enhancing the aggregation of intermediates (Ellis 1994; Jaenicke 1995). This is consistent with the observations that the yields of sucessfully refolded rhodanese under chaperonin-assisted conditions were higher than those of spontaneous refolding at 37°C, and that these differences decreased when the temperature was lowered (*e.g.*, the yields of spontaneous rhodanese refolding reached almost 70% of those of chaperonin-assisted at 25°C, H.M. Dionisi et al., unpublished). These results indicate that the role of the complete chaperonin system becomes more important as the environment results less favorable for spontaneous folding and suggests, in principle, that there is no need for the unfolded protein to achieve a committed state prior to its release from GroEL. In fact,

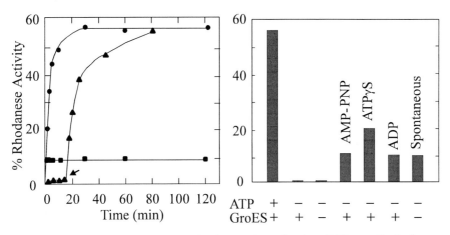

Figure 1 (left). Spontaneous and chaperonin-assisted refolding of rhodanese. Rhodanese (Sigma R1756) was unfolded at 4.5 µM in 50 mM Tris-HCl (pH 7.5), 1 mM β-mercaptoethanol, 6 M guanidinium-HCl at 25°C for 5 min. Refolding was initiated by a 50-fold dilution into a buffer maintained at 37°C containing 50 mM Tris-HCl (pH 7.5), 50 mM $Na_2S_2O_3$, 0.2 M β-mercaptoethanol, 10 mM $MgCl_2$, 10 mM KCl, 2 mM ATP, and 2.52 µM each of *C. vinosum* GroEL protomer and GroES protomer (circles). Refolding was conducted also in the same conditions in the absence of both chaperonins (squares), or in the absence of GroES and ATP (triangles). In the latter, GroES and ATP were added after 15 min (arrow). Rhodanese activity was assayed at 50°C (the temperature optimum for this enzyme, which blocks spontaneous folding during the assay) by a discontinuous method (Mendoza et al. 1994).

Figure 2 (right). Effect of adenine nucleotides on the assisted folding of rhodanese. Procedures were as above, except that ATP was replaced by AMP-PNP, ATPγS or ADP (2 mM final concentration). After 60 min at 37°C, rhodanese activity was assayed as described above.

recent data indicate that polypeptides (including Rubisco monomers) are released from GroEL in nonnative structure (Todd et al. 1994; Weissman et al. 1994).

The exact roles of ATP hydrolysis and GroES in the cycle of chaperonin-assisted folding remain a matter of speculation (Hartl et al. 1994; Schmidt et al. 1994; Todd et al. 1994; Jaenicke 1995). We found that the higher yields of assisted refolding were obtained for ATP (Fig. 2), in the presence of either non-hydrolyzable analogs of ATP or ADP the refolding yields were similar to those obtained spontaneously. Interestingly, these results also suggest that adenine nucleotide binding (rather than hydrolysis) is sufficient for the discharge of bound rhodanese from GroEL, an observation also reported for other proteins (Kawata et al. 1994; Schmidt et al. 1994 and references therein). In fact, it has been shown that the binding of ATP and other nucleotides elicit large conformational changes in GroEL, and in doing so weaken the affinity for nonnative proteins (Saibil et al. 1993). In such a case nucleotide binding to the GroEL-polypeptide complex would be sufficient to promote releasing of the polypeptide, which would then undergo (according to its intrinsic properties) either folding or off-pathway reactions, or would alternatively rebind to other GroEL complex. The hydrolysis of ATP by GroEL would then be employed to promote re-exposure of the high-affinity polypeptide binding sites. It is worth mentioning in this context that the major rearrangements produced in the GroEL tetradecamer after ATP binding are reversed upon hydrolysis (Saibil et al. 1993).

The ubiquitous distribution, high sequence conservation and lack of selectivity strongly suggest that the function of the GroE-type chaperonins has been strictly preserved during evolution (Ellis 1994; Viale et al. 1994). This conclusion is also supported by the interchangeability observed between GroEL and GroES components obtained from different sources: homologous or heterologous GroEL and GroES complexes from *C. vinosum* and *E. coli* chaperonins were equally effective in the assisted refolding of rhodanese (H.M. Dionisi et al., unpublished). Further evidence supporting a conservation of the promiscuity of chaperonins in recognizing unfolded polypeptides is provided by the successful replacement of the *E. coli groESL* operon by its *C. vinosum* homolog (R.G. Ferreyra et al., unpublished).

Participation of molecular chaperones in the in vivo assembly of Rubisco.

Role of GroE chaperonins. Two architecturally distinct forms of Rubisco are known (Hartman, Harpel 1994): one composed only by large (L) subunits (in L_2 to L_x oligomeric structures), and a hexadecamer of each eight large and small subunits (L_8S_8). These forms are evolutionary related, being the L_2 dimer the structural and functional motif of the L_8 core to which the small subunits are polarly attached (Hartman, Harpel 1994).

GroE chaperonins have been shown to interact and assist in the folding of Rubisco from bacteria and plastids both *in vivo* and *in vitro* (Gatenby, Ellis 1990; Ferreyra et al. 1993; Hubbs, Roy 1993; Hartman, Harpel 1994). A good correlation seems to exist between the *in vivo* levels and functional status of GroE chaperonins and those of assembled, dimeric Rubiscos (Fig. 3). In *E. coli*, the levels of GroE chaperonins are higher as the growth temperature increases (Georgopoulos, Welch 1993) and, accordingly, the levels of assembled Rubisco are also higher (Fig. 3, see also Gatenby, Ellis 1990). On the other hand, these levels were reduced in cells bearing mutations that impair the function in the *groEL* gene (*groEL140*, RF102), and this effect was aggravated when the growth temperature was increased (thus favoring aggregation over folding).

The role of chaperonins in facilitating the productive assembly of Rubisco *in vivo* was also confirmed by restricting GroE levels in the cells. This was attained by replacing the *E. coli groESL* operon by its *C. vinosum* homolog: the resulting *E. coli* cells (RF101) were found to possess around 50% of GroE chaperonin levels than those of wild-type *E. coli* at 30°C, and no substantial increases occurred at higher growth temperatures (R.G. Ferreyra et al., unpublished). As shown in Fig. 3, RF101 cells show severe impairments in the assembly of Rubisco, most probably as a consequence of the competition between the different folding intermediates in the cell for the scarce levels of free chaperonins available. In all cases mentioned above, the enhanced expression of *C. vinosum* GroE chaperonins from a multicopy plasmid increased the levels of assembled Rubisco. The

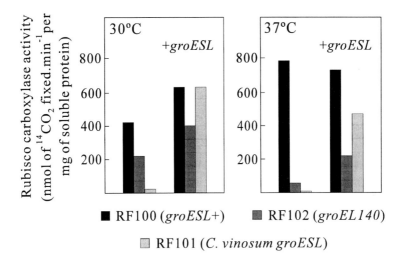

■ RF100 (*groESL*+)   ■ RF102 (*groEL140*)

▧ RF101 (*C. vinosum groESL*)

Fig. 3. Effect of growth temperature and GroE chaperonins levels on the expression of *R. rubrum* Rubisco in *E. coli*. RF100 (*groEL*+), RF102 (*groEL140*) or RF101 (in which the wild-type *groESL* operon was replaced by its *C. vinosum* homolog, R.G. Ferreyra et al., unpublished), transformed with plasmids pRR2119 (bearing *R. rubrum cbbM*) or pRRF2 (pRR2119 bearing, in addition, *C. vinosum groESL*) were grown at the indicated temperatures until the absorbance at 600 nm of the cultures reached 0.2. IPTG was added to 1 mM (final concentration), and further growth was allowed until the absorbance in all cases reached 1.0. Cells were collected, subjected to sonic disruption, and Rubisco carboxylase activity measured in cell extracts. Procedures and plasmids used here have been described previously (Ferreyra et al. 1993).

lower recoveries obtained at 37°C for RF102 (*groEL140*) may be related to the presence of defective GroEL particles derived from the *E. coli* host.

As for cyanobacterial Rubisco (Andrews 1988; Lee, Tabita 1990), the stable formation of the *C. vinosum* Rubisco L_8 core (which possesses about 0.5% of the carboxylase activity of the L_8S_8 enzyme) was obtained in *E. coli* in the absence of the simultaneous expression of small subunits. The successful *in vitro* reconstitution of functional hexadecameric enzymes when small subunits are added to preformed L_8 cores has been observed in Rubiscos from *C. vinosum* and cyanobacteria (Incharoensakdi et al. 1985; Andrews 1988; Lee, Tabita 1990), suggesting that no additional factors are required in this step.

The overall results support the notion that GroE chaperonins act at the level of formation of structured large (and probably small) subunit monomers, capable of self-association to generate the functional oligomeric structure.

Role of DnaK molecular chaperones. Members of the DnaK class of molecular chaperones appear to function by preventing aggregation at very early stages of folding,

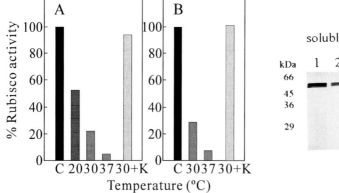

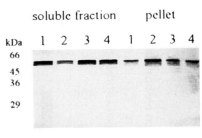

Figure 4 (left). Expression of assembled Rubisco in *dnaK* mutants. A, pRR2119 (*R. rubrum cbbM*); B, pCV17 (*C. vinosum cbbLS*). The cells were grown at the indicated temperatures and Rubisco carboxylase activity measured in cell extracts as described in the legend to Fig. 3. The values represent the percentages of activity in *dnaK* null mutants compared to those of the parental strain (C) at a given temperature. The bar at the right in both panels (30+K) indicates the percentage of activity obtained in the *dnaK* mutants expressing the E. coli *dnaK* genes from the multicopy plasmid pSH52.

Figure 5 (right). Immunoblot analysis of Rubisco expressed in *E. coli dnaK* mutants. Extracts were prepared from cells containing pCV17 (*C. vinosum cbbLS*) grown at 30°C (Ferreyra et al. 1993), and separated in soluble and pellet fractions by centrifugation at 105,000xg for 30 min. Pellets were resuspended in half of the original volume of SDS-loading buffer, and 50 µg of soluble protein (soluble fraction) or the equivalent pellet fraction were resolved by SDS-PAGE (10-18% polyacrylamide linear gradient). Immunoblotting using anti-Rubisco was done by transferring polypeptides from stained gels (Dionisi et al. 1995). Lane 1, parental strain; lane 2, *dnaK* null mutant; lane 3 and 4, *dnaK* null mutant containing pSH52.

e.g., when the polypeptide chain is emerging from the ribosome or from the import/export machinery (Ellis 1994; Hartl et al. 1994). Since both *in vitro* and *in vivo* studies suggest a functional and sequential cooperation between the DnaK and GroE classes of chaperones (Hartl et al 1994; Gaitanaris et al. 1994), we analyzed the possible participation of the DnaK chaperone family in the *in vivo* assembly of Rubisco.

We measured the levels of assembled L_2 (*R. rubrum*) or L_8S_8 (*C. vinosum*) Rubiscos in *E. coli* cells bearing *dnaK* null mutations (Bukau, Walker 1990). As shown in Fig. 4, in both cases Rubiscos carboxylase activities were significantly reduced in the mutants, an effect that was aggraviated at higher growth temperatures. Expression of the *dnaK* gene from a multicopy plasmid restored the levels of both dimeric and hexadecameric assembled Rubiscos to those observed in the parental strains (Fig. 4).

The total amounts of Rubisco expressed in the parental strains and *dnaK* mutants (as measured in the soluble and insoluble cell fractions) were similar (Fig. 5).

Nevertheless, the levels of soluble functional Rubiscos appeared substantially reduced in the mutants (*e.g.*, compare lanes 2). Therefore, the absence *in vivo* of DnaK chaperones resulted in a greater portion of Rubisco molecules forming insolubles aggregates, an effect which was almost completely reversed by restoring DnaK levels in these cells (lanes 3 and 4).

Mutations in genes coding for cohort members of the DnaK protein (*e.g.,* DnaJ or GrpE), also conduced to reduced levels of assembled Rubiscos (S.K. Checa et al., unpublished). Interestingly, the enhanced expression of GroE chaperones and other heat-shock proteins in the *dnaK* null mutants seemed not to alleviate the impairment in Rubisco assembly in these cells, supporting a specific role of DnaK chaperones in this process.

CONCLUSIONS

The above results indicate that ubiquitous molecular chaperones of the GroE and DnaK classes constitute necessary steps in the *in vivo* assembly of Rubisco. A large number of studies point now to a participation of these proteins in the acquisition of the functional structures at early stages of monomer folding, by blocking off-pathway reactions from the very moment the polypeptide emerges from the ribosome. The formation of Rubisco native dimeric or hexadecameric structures most probably occurs in the bulk solution, from structured monomers which associate spontaneously. Nevertheless, many aspects of chaperonin action remain to be clarified, including the exact roles of GroES and ATP in the binding/hydrolysis cycles and whether any extent of folding occurs in association with the chaperonins (Jaenicke 1995).

While the above chaperones show no selectivity in their action, the evolution of other proteins participating in the folding process of specific proteins appears a likely possibility. Rubisco may again provide clues in the identification of these putative specific assistants: a recent report suggests such a role for plant rubisco activase (Jimenez et al. 1995). It remains to be clarified whether a similar function is provided by the cyanobacterial homologs of these proteins (Li et al. 1993).

REFERENCES
Andrews TJ (1988) J. Biol. Chem. 263, 12213-12219.
Bukau B and Walker G (1990) EMBO J. 9, 4027-4036
Dionisi HG et al. (1995) BioTechniques 19, in press.
Ellis RJ (1994) Curr. Opin. Struct. Biol. 4, 117-122.
Ferreyra RG et al. (1993) J. Bacteriol. 175, 1514-1523.
Gaitanaris et al. (1994) Mol. Microbiol. 14, 861-869.
Gatenby AA and Ellis RJ (1990) Annu. Rev. Cell Biol. 6, 125-149.
Georgopoulos C and Welch WJ (1993) Annu. Rev. Cell Biol. 9, 601-634.
Hartl FU et al. (1994) Trends Biochem. Sci. 19, 20-25.
Hartman FC and Harpel MR (1994) Annu. Rev. Biochem. 63, 197-234
Hubbs AE and Roy H (1993) J. Biol. Chem. 268, 13519-13525.

Incharoensakdi A et al. (1985) Biochim. Biophys. Res. Comm. 126, 698-704.

Jaenicke R (1995) Phil. Trans. R. Soc. Lond. B 348, 97-105.

Jimenez ES et al. (1995) Biochemistry 34, 2826-2831.

Kawata Y et al. (1994) FEBS Lett. 345, 229-232.

Kobayashi H et al. (1991) Gene 97, 55-62.

Lee B and Tabita RF (1990) Biochemistry 29, 9352-9357.

Li LAL et al. (1993) Plant Mol. Biol. 21, 753-764.

Mendoza JA et al. (1994) J. Biol. Chem. 269, 2447-2451.

Saibil HR et al. (1993) Curr. Biol. 3, 265-273.

Schmidt M et al. (1994) J. Biol. Chem. 269, 10304-10311.

Soncini FC et al. (1992) In Murata N, ed, Research in Photosynthesis, Vol. III, pp 637-640, Kluwer Academic Publishers, The Netherlands

Terlesky KC and Tabita FR (1991) Biochemistry 30, 8181-8186.

Todd MJ et al. (1994) Science 265, 659-666.

Torres-Ruiz JA and McFadden BA (1992) Arch. Biochem. Biophys. 295, 172-179.

Viale AM et al. (1989) J. Bacteriol. 171, 2391-2400.

Viale AM et al. (1991) J. Bacteriol. 173, 5224-5229.

Viale AM et al. (1994) Int. J. Syst. Bacteriol. 44, 527-533

Weissman JS et al. (1994) Cell 78, 693-702.

Woese CR (1987) Microbiol. Rev. 51, 221-271.

The Biochemistry of CO Dehydrogenase in *Rhodospirillum rubrum*

P.W. Ludden[1], G.P. Roberts[2], R.L. Kerby[1,2], N. Spangler[1], J. Fox[1], D. Shelver[2], Y. He[2], and R. Watt[1]. Departments of Biochemistry[1] and Bacteriology[2], University of Wisconsin, Madison, 53706 USA

INTRODUCTION

Carbon monoxide metabolism by the photosynthetic bacteria was first noted by Uffen, who isolated *Rubivivox gelatinosus* (formerly *Rhodopseudomonas gelatinosus*) for its ability to tolerate CO (Uffen 1976). *R. gelatinosus* was shown to grow with CO as the sole energy and primary carbon source, anaerobically in the dark. *R. rubrum* was also shown to possess the ability to oxidize CO (Uffen 1981) and its ability to utilize CO as a carbon and energy source has been established (Kerby et al 1995). The growth of *R. rubrum* on CO depends on the enzyme carbon monoxide dehydrogenase (CODH) which catalyzes the reversible reaction shown in Eqn. 1:

Eqn. 1

$$CO + H_2O + Ferredoxin_{ox} \leftrightarrow CO_2 + 2H^+ + Ferredoxin_{red}$$

The CO oxidizing enzymes were named by Yagi (Yagi 1958; Yagi 1959), who first isolated the enzyme from *Desulfovibrio desulfuricans;* the name, which persists, is an unfortunate misnomer, as no dehydrogenation reaction occurs. Unlike the CODHs from acetogens and methanogens, the *R. rubrum* CODH does not catalyze the synthesis of acetylCoA and the CO-dependent autotrophy of *R. rubrum* is thought to occur by refixation of the CO product of the reaction.

CODH from *R. rubrum* is encoded by the *cooS* gene (Kerby et al 1992) and is an oxygen-labile 65 kDa monomer that contains one Ni, 8 Fe, and 8 acid-labile S atoms per molecule (Bonam, Ludden 1987). As described in the Results section of this manuscript, the metals are arranged in two clusters, an Fe_4S_4 B-cluster and a $NiFe_4S_4$ C-cluster (Hu et al 1995). A stable, catalytically inactive apo-form of the enzyme (lacking Ni) can be isolated from cells that have been starved for Ni (Bonam et al 1988) and this apoCODH can be activated in vitro by the addition of $NiCl_2$ to the enzyme solution (Stephens et al 1989; Ensign et al 1990). Co^{2+}, Fe^{2+} and Zn^{2+} atoms can replace Ni^{2+} in the Ni site of the enzyme, but of these, only the cobalt CODH exhibits activity (Ensign et al 1990). In vitro activation requires the reduced form of the apoCODH.

CODH exhibits a UV-visible spectrum characteristic of FeS clusters and the Ni atom does not appear to contribute to the spectrum. The oxidation state of the Fe_4S_4

183

M. E. Lidstrom and F. R. Tabita (eds.), Microbial Growth on C₁ Compounds, 183–190.
© 1996 Kluwer Academic Publishers. Printed in the Netherlands.

184

B-cluster can be monitored by its absorbance at 418 nm (Ensign et al 1989). The EPR spectrum of *R. rubrum* CODH shows features typical of Fe_4S_4 clusters in the dithionite-reduced state. In the thionine-oxidized state an unusual signal, similar to those of the more complex acetogen CODHs, is observed (see Results section) (Stephens et al 1989; Lindahl et al 1990).

A model for the CO oxidation pathway of *R. rubrum* is shown in Fig. 1.

Figure 1. Model for the CO oxidation system of *Rhodospirillum rubrum*

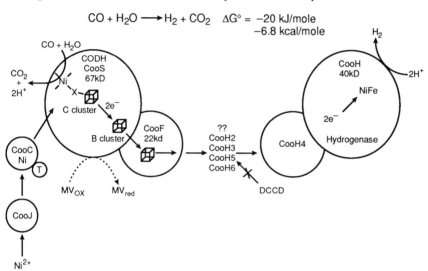

Synthesis of the enzyme is specifically induced by CO (Bonam et al 1989; Kerby et al 1992). Other compounds, including the CODH inhibitors CN^- and azide, do not induce the synthesis of the enzyme. Stoichiometric levels of Ni are not required for the synthesis of CODH, as CODH protein accumulates in CO-treated cells growing on medium that has been treated to remove Ni (Bonam et al 1988). A CO-insensitive hydrogenase activity is co-induced with CODH by CO and this enzyme is thought to be a Ni-enzyme as well. A regulatory protein involved in expression of the *coo* genes is described in the Results section of this manuscript (Shelver et al 1995). The redox state of the medium is crucial for the expression, and this finding is also described in the Results section.

The *coo* genes involved in CO-dependent growth of *R. rubrum* have been identified and most have been sequenced and mutagenized (Kerby et al 1992; Shelver et al 1995). A number of ancillary activities including the electron acceptor/donor for CODH (Ensign, Ludden 1991), Ni processing for CODH, and the CO-insensitive hydrogenase activity, are encoded by *coo* genes. Table 1 lists the *coo* genes and their products; the proposed roles of some of the gene products (e.g. CooH4, CooH2, CooH5, CooH6) are hypothetical and based on sequence similarities to other gene products of known function. Figure 2 shows the arrangement of the *coo* genes on the *R. rubrum* chromosome.

Table 1. The coo genes and the proposed properties/functions of their products.

Gene	Protein (kD)	Function or Proposed Function	Similar to:
cooF	22	Electron Acceptor and membrane anchor for CODH	dmsB, fdnH, narH
cooS	67	CODH, NiFeS protein	cmbB, cdHA
cooC	28	Ni metabolism/insertion; NTP binding site	ureG, hypB
cooT	7	Ni metabolism/insertion	thdF, norA
cooJ	14	Ni metabolism/insertion; His-rich C terminus	ureE, hypB
cooA	27	CO responsive transcriptional activator	crp, fnr
cooH	40	hydrogenase large subunit	hycE, nqo4, nuoD
cooH4	16	probably hydrogenase small subunit	hycG
cooH2	19	no E. coli hydrogenase equivalent gene	nqo5, nuoC
cooH3	18	Ferredoxin-like protein	hycF, frxB
cooH5	36	membrane bound, DCCD sensitive subunit	hycD
cooH6	?	? very large protein of unclear function	hycC, nuoL

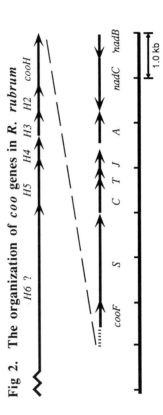

Fig 2. The organization of coo genes in R. rubrum

RESULTS

The metal clusters of R. rubrum CODH. R. rubrum CODH contains a B-cluster that appears to be a normal Fe_4S_4 cluster and is probably liganded to the protein via cysteine ligands. The B-cluster has a midpoint potential of - 418 mV as determined by direct electrochemical titration (Smith et al 1992) and is responsible for the $g_{av} = 1.94$ EPR signal of the dithionite-reduced enzyme. The Mössbauer spectrum of the B-cluster is observed as a large doublet that again is similar to the Mössbauer spectra of Fe_4S_4 clusters observed in ferredoxins (Hu et al 1995). The role of the B-cluster appears to be to accept/donate electrons from the C-cluster. The B-cluster of the Ni-deficient apoCODH is normal by spectroscopic criteria and is able to be chemically oxidized and reduced . The B-cluster in the apoCODH is unaffected by CO, suggesting that it is not a site of substrate binding. The EPR and Mössbauer spectra of the B-cluster are not affected by substitution of Ni isotopes (i.e., ^{61}Ni) and the EPR spectrum is not affected by substitution of other metals (Co^{2+}, Fe^{2+}); thus it does not appear that the B-cluster is electronically linked to Ni.

The remainder of the metal and acid-labile S atoms of CODH are found in the cluster called the C-cluster. The C-cluster is thought to be the catalytic site of CO oxidation in the *R. rubrum* enzyme (Ensign et al 1989; Ensign et al 1989) and in the more complex acetylCoA-synthesizing CODHs (Seravalli et al 1995). The C-cluster is defined by its EPR signal at $g_{av} = 1.89$ (Fig. 3) (in the acetogens, the signal has a g_{av} of 1.82) and by the presence of Ni in this cluster. The C-cluster is EPR silent in the dithionite-reduced state and is observed in the thionine-oxidized state of the enzyme. Fig. 3 shows the intensity of the $g_{av} = 1.89$ EPR signal at various points during the redox titration of CODH. As seen in Fig. 3, the $g_{av} = 1.89$ signal has a midpoint potential (E°') of -90 mV. It is interesting that apoCODH cannot be activated by Ni^{2+} at a potential of -300 mV (Ensign et al 1990), a potential that should allow the assembly of the EPR-detectable C-cluster. Perhaps the Fe_4S_4 component of the C-cluster must be reduced before Ni^{2+} can be accepted.

Figure 3. Redox titration of the *R. rubrum* CODH cluster as observed by EPR.

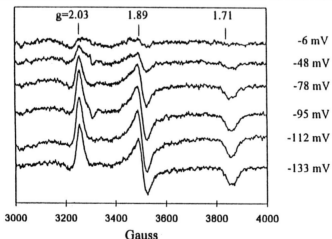

Until recently, it was not established whether CODH contained 2 Fe_4S_4 clusters and an additional Ni cluster that might contain one additional Fe atom **OR** whether CODH contained one conventional Fe_4S_4 cluster (the B-cluster) and that the Ni was associated with the other Fe_4S_4 unit to form the C-cluster. This ambiguity arose because of the imprecision of chemical assays for Fe and protein. Recent Mössbauer results convince us that the latter case (one B-cluster and one C-cluster) occurs in the *R. rubrum* CODH. All of the Fe in *R. rubrum* CODH can be accounted for in the Mössbauer spectra and, more importantly, there is no signal that is not attributable to one of the Fe_4S_4 clusters. Mössbauer has the advantage of detecting all Fe atoms present in a sample, and a Ni-Fe cluster distinct from the Fe_4S_4 clusters would have been detected. The Ni atom is associated with an Fe_4S_4 cluster, because the EPR signal that is attributed to the C-cluster is broadened by both [61]Ni and [57]Fe, demonstrating that the unpaired electron responsible for that signal resides on each of the atoms part of the time. The broadening due to [61]Ni is minimal, suggesting that the electronic link of Ni to the cluster is weak.

The Fe_4S_4 portion of the C-cluster is present in the Ni-deficient apoCODH, but the appearance of the $g_{av} = 1.89$ signal depends on the addition of Ni; addition of the non-paramagnet Zn to the enzyme does not result in generation of a $g_{av} = 1.89$ signal, so merely the presence of a metal atom at the Ni site is not sufficient.

CN^- is a tight-binding inhibitor of CODH and it is competitive vs. CO, suggesting that both CO and CN^- bind at the same site (Ensign et al 1989). [14]CN^- binds tightly to holoCODH but not to apoCODH, demonstrating that the Ni atom must be present for binding and that the Fe_4S_4 cluster precursor to the C-cluster is not sufficient for binding.

One of the Fe atoms of the C-cluster is distinct from the others as determined by Mössbauer spectroscopy. The Fe atom, referred to as ferrous component II (FCII), is thought to have an unusual ligation arrangement, perhaps like that found for one of the Fe atoms in the FeS cluster of aconitase. It is possible that this iron is involved in substrate and inhibitor binding in a manner analogous to that proposed for the A-cluster of the acetylCoA synthesizing CODHs (Qiu et al 1994).

Electron acceptor for CODH. The electron acceptor for CODH is encoded by the *cooF* gene (Kerby et al 1992). It is a 22 kD protein that contains a single Fe_4S_4 cluster and serves as both the electron acceptor and membrane anchor for CODH (Ensign, Ludden 1991). A complex of CODH and the 22 kD protein can be isolated by ethanol or deoxycholate treatment of the membrane fraction of the cell extract. The isolated 22 kD protein can bind to purified CODH and bind to the chromatophore membrane. The 22 kD protein is far more effective in accepting electrons than other, lower molecular weight ferredoxins from *R. rubrum* and other sources. Mutations in the *cooF* gene result in strains that are unable to grow on CO.

Gene products involved in Ni insertion. The *cooCTJ* gene products are thought to be involved in processing and insertion of Ni into CODH. Mutations in this region result in accumulation of apoCODH even when normal levels (10 uM) of Ni are added to the medium and the Coo$^-$ phenotype is cured only by high levels of Ni (500 uM). The sequence of *cooC* is similar to those of *ureG* (Lee et al 1992) and *hypB* (Maier et al 1993; Maier et al 1995) and the products of these genes are known to be involved in processing of Ni for urease and hydrogenase, respectively. The *cooJ* gene sequence is very similar to the *ureE* sequence (Lee et al 1992); *UreE* is known to be capable of binding multiple Ni atoms to its histidine-rich tail (Lee et al 1993). The carboxy terminus of CooJ contains even more his residues (15 of the last 30 residues) than does UreG. CooJ is a 14 kD protein and has been purified by binding to a his-bind column (Novagen). Purified apoCODH will accept several metal ions other than Ni^{2+} and several of these work at lower effective concentrations than does Ni^{2+} in the activation of apoCODH. Furthermore, Ni-starved cells do not produce CODH with Fe, Co or Zn in the Ni site, even if one of these ions is supplied to the culture. Therefore, one role of the Ni-processing *cooCTJ* gene products

188

may be to insure that only Ni is allowed to be inserted into CODH. It has not been possible to detect a CO-induced system for the uptake of Ni^{2+} by the cells.

Figure 4. Accumulation of CODH activity in photosynthetically-grown, CO-induced cultures of *R. rubrum* is dependent upon the cell-culture reducing potential. **A,** Na$_2$S was not added to the medium; **B,** Na$_2$S was added during CO induction.

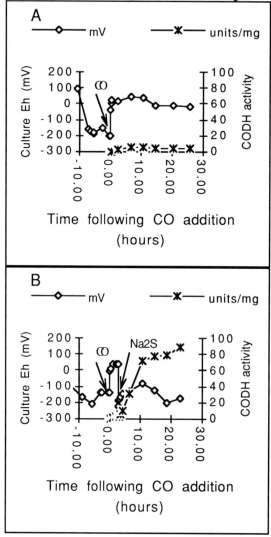

Regulation of expression of the coo genes. CODH in *R. rubrum* is expressed only in the presence of CO (Bonam et al 1989). This is in contrast to the CODHs of acetogens and methanogens, where CODHs are an obligate component of metabolism and thus always present (Kerby , Zeikus 1987; Thauer et al 1989). It appears that *R. rubrum* CODH is present for the purpose of extracting energy from CO, while the true role of CODH in acetogens and methanogens is in the synthesis and catabolism of acetate. Mutations in the *cooA* region render *R. rubrum* unable to accumulate CODH in response to CO and the CooA protein appears to be a transcriptional activator for the *coo* system (Shelver et al 1995). The sequence of *cooA* is very similar to that of *crp*, the cyclic AMP binding protein, and to that of *fnr*, the transcriptional activator involved in anaerobic response of *E. coli* and other organisms. The most notable difference between *cooA* and *crp* is at the region known to bind cAMP; in *cooA*, cysteine residues are found in this region in place of others involved in binding cAMP in *crp* and this suggests that a metal is bound at the effector binding site of this protein. A bound metal might be involved in binding CO, thus sensing the presence of the substrate for the *coo* system.

The accumulation of CODH is also affected by the redox of the medium as shown in Figure 4. Anaerobically growing, CO-treated cells at Emp > -100 mV fail to accumulate CODH, and it appears that CODH is not synthesized under these conditions. When Na_2S is added to lower the redox potential of the medium to < -200 mV, CODH accumulates rapidly in the cells. These results suggest a regulation of gene expression that responds to redox in the cell beyond the recognition of the aerobic/anaerobic switch. There is no O_2 in the medium of cells growing at > -100 mV, yet these cells fail to synthesize CODH.

The cooH genes. At least six gene products are involved in the production of an active, CO-induced hydrogenase. The *cooH* and *cooH4* products are the subunits of the hydrogenase, and several other genes have similarity to genes involved in energy transduction in other systems. It appears that the *cooH* transcript ends about 2 kb upstream of the *cooH6* gene and at this point, no Ni-processing genes analogous to *cooCTJ* have been detected in the *cooH* cluster. Perhaps a general set of Ni-processing gene products functions to incorporate Ni into the CO-induced hydrogenase; mutations in *cooCTJ* do not affect the production of functional hydrogenase in *R. rubrum*. While direct biochemical demonstration of Ni in the CO-induced hydrogenase has not been achieved, Ni-starved cells lack hydrogenase activity upon CO treatment and the deduced sequence for the large subunit (CooH) contains the conserved Ni-binding site found in other Ni-hydrogenases (Volbeda et al 1995). The CO-induced hydrogenase from *R. rubrum* appears to fall in the family of proteins that consists of Ni-containing, uptake hydrogenases (the *hycE* family), even though the role of this hydrogenase is the evolution of H_2. The CO-induced hydrogenase from *R. rubrum* lacks the C-terminal cleavage peptide found in all other members of the *hycE* family of hydrogenases (Rossman et al 1994). The cleavage peptide is thought to play a role in targeting the enzyme to the membrane.

CONCLUSIONS

R. rubrum is able to grow with CO as the sole C and energy source by virtue of a NiFeS-containing CODH and its associated hydrogenase. CODH contains an Fe_4S_4 B-Cluster and a $NiFe_4S_4$ C-cluster. The C-cluster is the site of CO binding and oxidation to CO_2. A precursor to the C-cluster consists of an Fe_4S_4 cluster in the enzyme and the C-cluster can be completed by the addition of Ni to the enzyme. The *coo* products are synthesized in response to CO when the E_{mp} of the medium is sufficiently negative. The *cooA* product appears to be a *crp*-like transcriptional activator. The *cooCTJ* products are involved in the specific processing of Ni for CODH and perhaps in specifying Ni as the only metal that is inserted into the Ni site of the enzyme. A CO-induced hydrogenase completes the pathway of electron transport from CO to proton, releasing H_2. The CO-induced hydrogenase is similar to the *hycE* family of hydrogenases, but is smaller and lacks the C-terminal cleavage peptide that is thought to be involved in targeting the hydrogenase protein to the membrane.

Acknowledgements.
Support for the authors' lab for this project has come from the Dept. of Energy (Grant DE-FG02-78ER13691 to PWL)

REFERENCES
Bonam D, et al. (1989) J. Bacteriol. 171, 3102-3107.
Bonam D, et al. (1987) J. Biol. Chem. 262, 2980-2987.
Bonam D, et al. (1988) Proc. Natl. Acad. Sci. USA 85, 31-35.
Ensign SA, et al. (1989) Biochemistry 28, 4968-4973.
Ensign SA, et al. (1990) Biochemistry 29, 2162-2168.
Ensign SA, et al. (1989) Biochemistry 28, 4973-4979.
Ensign SA, et al. (1991) J. Biol. Chem. 266, 18395-18403.
Hu Z, et al. (1995) submitted to J. Am. Chem. Soc.
Kerby RL, et al. (1987) J. Bacteriol. 169, 5605-5609.
Kerby RL, et al. (1992) J. Bacteriol. 174, 5284-5294.
Kerby RL, et al. (1995) J. Bacteriol. 177, 2241-2244.
Lee MH, et al. (1992) J. Bacteriol. 174, 4324-4330.
Lee MH, et al. (1993) Protein Sci. 2, 1042-1052.
Lindahl PA, et al. (1990) J. Biol. Chem. 265, 3873-3879.
Maier T, et al. (1993) J. Bacteriol. 175, 630-635.
Maier T, et al. (1995) Eur. J. Biochem. 230, 133-138.
Qiu D, et al. (1994) Science 264, 817-819.
Rossman R, et al. (1994) Eur. J. Biochem. 220, 377-384.
Seravalli J, et al. (1995) Biochemistry 34, 7879-7888.
Shelver D, et al. (1995) J. Bacteriol. 177, 2157-2163.
Smith ET, et al. (1992) Biochem. J. 285, 181-185.
Stephens PJ, et al. (1989) J. Biol. Chem. 264, 16347-16350.
Thauer RK, et al. (1989) Annu. Rev. Microbiol. 43, 43-67.
Uffen RL (1976) Proc. Natl. Acad. Sci. USA 73, 3298-3302.
Uffen RL (1981) Enzym. Microb. Technol. 3, 197-206.
Volbeda A, et al. (1995) Nature 373, 580-587.
Yagi T (1958) Biochim. Biophys. Acta 30, 194-195.
Yagi T (1959) J. Biochem. 46, 949-955.

Anaerobic Carbon Monoxide Dehydrogenase

S.W. Ragsdale[a], M. Kumar[a], J. Seravalli[a], D. Qiu[b], and T.G. Spiro[b].

[a]Department of Biochemistry, University of Nebraska, Lincoln, NE 68583-0718, USA
and [b]Department of Chemistry, Princeton University, Princeton, New Jesey 08544

Ten years ago, the scheme describing CO_2 fixation by the acetyl-CoA pathway was significantly changed (Ragsdale, Wood, 1985). The key enzyme in the pathway was proposed to be carbon monoxide dehydrogenase (CODH) and this enzyme, not a B_{12} enzyme, was proposed to bind, activate, and condense the methyl, carbonyl, and CoA groups to form acetyl-CoA. This was controversial and, at the time, none of the proposed intermediates had been identified. Most of the intermediates have now been detected and characterized. A significantly altered and more complete scheme for acetyl-CoA synthesis will be presented. A bimetallic mechanism for carbon monoxide oxidation in anaerobes will be proposed.

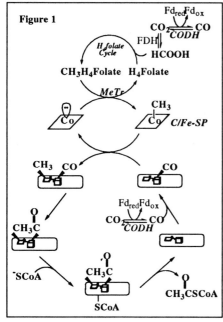

Figure 1

Three major points will be discussed.
(i) The synthesis of acetyl-CoA and the oxidation of CO to CO_2 occur at separate Ni-FeS clusters (Kumar et al., 1993). This is an unusual class of heterometallic center in which Ni is bridged to an FeS cluster.

(ii) A bimetallic mechanism of CO oxidation is postulated in which Ni and Fe both play catalytic roles (Qiu et al., 1995a). Proposed intermediates at center C are Fe-CO and Ni-OH, that react to form CO_2.

(iii) A bimetallic, bio-organometallic scheme describing acetyl-CoA synthesis has been proposed (Qiu et al., 1994; Kumar et al., 1995). This reaction sequence involves Fe-CO and methylnickel intermediates at center A of CODH.

191

M. E. Lidstrom and F. R. Tabita (eds.), Microbial Growth on C, Compounds, 191–196.

In the Wood-Ljungdahl pathway (Figure 1), two mol of CO or CO_2 are converted to acetyl-CoA (Ragsdale, 1991). For autotrophic growth, acetyl-CoA can serve as a source of all the macromolecules in the cell. It also is an energy source since cleavage of the thioester bond results in ATP synthesis. In this pathway, one mol of CO_2 is converted to methyltetrahydrofolate (CH_3-H_4folate) by formate dehydrogenase and tetrahydrofolate enzymes. The step that commits the methyl group to acetyl-CoA synthesis is catalyzed by methyltransferase, which catalyzes transfer of the methyl group from CH_3-H_4folate to a C/Fe-SP (Drake et al., 1981; Zhao et al., 1995). Methylcobalamin formation is an interesting reaction that will not be covered here.

The reactions performed by CODH are interesting because they involve organometallic intermediates. CO_2 is converted to CO at center C of CODH. CO is at the same oxidation state as the carbonyl group of acetyl-CoA. CO then combines with center A to form an intermediate metal-carbonyl species. Then the components of acetyl-CoA are successively assembled on CODH. The methyl group is transferred from the C/Fe-SP; then the methyl group and CO are combined to form an acetylmetal species. Finally, CoA adds and cleaves the acetyl group to form acetyl-CoA.

CODH is one of the four known nickel enzymes, containing 2 Ni, 11-14 Fe, ~14 inorganic sulfides, and zinc (Ragsdale et al., 1983). These metals are organized into three clusters. **Center A** contains 1 Ni and ~ 4 Fe in an unusual Ni-FeS cluster. This is the site of acetyl-CoA assembly (Gorst, Ragsdale, 1991; Shin, Lindahl, 1992; Kumar et al., 1993). **Center B** is a [4Fe-4S] cluster that is involved in electron transfer reactions. **Center C**, like center A, contains 1 Ni and an unknown amount of Fe also as a Ni-FeS cluster. Center C is the site of CO oxidation/CO_2 reduction (Kumar *et al.*, 1993).

MECHANISM OF CO OXIDATION

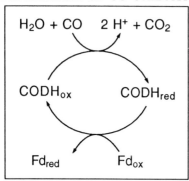

Figure 2.

Figure 2 summarizes the mechanism for CO oxidation by center C of CODH. Extensive steady state and presteady state kinetics of this reaction have been performed in both the CO oxidation and CO_2 reduction directions (Seravalli et al., 1995). It was demonstrated that CO oxidation follows a ping-pong mechanism in which the enzyme is reduced in the first half reaction, producing CO_2. The second half reaction involves transfer of the electrons through center B to external electron acceptors, like ferredoxin. A ping-pong mechanism was predicted earlier using cell extracts of *C. thermoaceticum* (Diekert, Thauer, 1978). In recent work, Javier Seravalli has used a variety of electron acceptors, that includes one- and two-electron mediators with redox potentials spanning 0.5 volts. These include ferredoxin, methyl viologen, flavins, cytochrome C, and thionin. All these were shown to give the same V/K for CO, which provides strong evidence for a ping-pong mechanism with two separable half reactions.

Complete sets of initial velocity patterns using viologen (Seravalli et al., 1995), ferredoxin, and cytochrome C also fit a ping-pong mechanism. The pH profiles for the V_{max} and V/K for CO overlay that of the EPR spectrum of center C, the site for CO oxidation (Seravalli et al., 1995). We interpret both of these profiles to indicate that enzyme-bound water is undergoing deprotonation to form the E-OH species that acts as a nucleophile to attack bound CO.

Insight into the mechanism of CO oxidation has been gained by using cyanide. Cyanide is a slow binding competitive inhibitor of the *C. thermoaceticum* ($K_i < 10$ mm) (Morton, 1991) and *Rhodospirillum rubrum* CODHs (Ensign et al., 1989). It is slow binding because the enzyme gets trapped in this complex from which escape is very slow (Morton, 1991). The CODH-CN adduct has been characterized by spectroscopy. Identification of a ^{13}C-CN ENDOR signal from center C proved that CN binds to center C (Anderson et al., 1993). By FTIR and Resonance Raman studies, Fe and Ni were shown to be bound simultaneously to cyanide (Qiu et al., 1995a). Three Raman bands were located at 719, 383 and 343 cm[-1], that were sensitive to [13]CN- or C[15]N- isotopic substitution. These bands were also sensitive to the incorporation of both [54]Fe and [64]Ni into the enzyme. Thus, both metals are implicated simultaneously in the

$$Fe-C\equiv N_{\diagdown Ni}$$
$$140°$$

vibrational modes. By modeling the frequencies with normal mode calculations for a variety of bridging geometries, a good fit was obtained with a Fe-CN-Ni complex having an essentially linear Fe[II]CN unit interacting with a Ni[II] ion at an angle of about 140°.

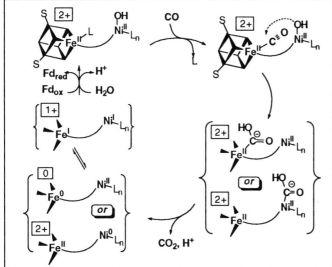

Figure 3

This model of the inhibited active site provides the basis for a newly proposed mechanism of CO oxidation (Qiu et al., 1995a) (Figure 3). The Fe[II] ion serves as the binding site for CO. The Ni[II] ion is then properly positioned to act as a Lewis acid, activating the bound CO. We posit that Ni(II) could also deliver a metal-bound hydroxide nucleophile to the bound CO. Evidence for a metal-assisted hydroxide attack comes from the pH profile of the CO oxidation activity (above) (Seravalli et al., 1995).

Hydroxide attack would produce bound $^-$COOH, a strong reductant. It could release CO_2 and H^+ by transferring an electron pair to the Fe[II]. Alternatively, the mechanism could be described as a CO insertion into a metal-hydroxide bond that would produce a Ni-bound $^-$COOH. This step might be energetically advantageous, since Ni[II] is generally easier to reduce than is Fe[II]. After two-electron delivery to the metal that is bound to the carboxyl group, inner sphere electron transfer within center C could redistribute the electrons to the most stable configuration. Finally, the reduced form of center C would pass electrons to an external electron acceptor, like ferredoxin.

BIMETALLIC MECHANISM OF ACETYL-COA SYNTHESIS

There is recent evidence for a bimetallic and bio-organometallic mechanism of acetyl-CoA synthesis (Figure 4). After CO is generated at center C from CO_2, CO binds to one of the Fe sites in center A (Qiu et al., 1994; Qiu et al., 1995b), the methyl group binds to Ni (Kumar et al., 1995), then these are joined to form an acetylmetal species that is then combined with CoA to form acetyl-CoA. It is not known if CO or the methyl group binds first. It may be a random mechanism since binary CODH-CO and CODH-CH_3 complexes have been demonstrated.

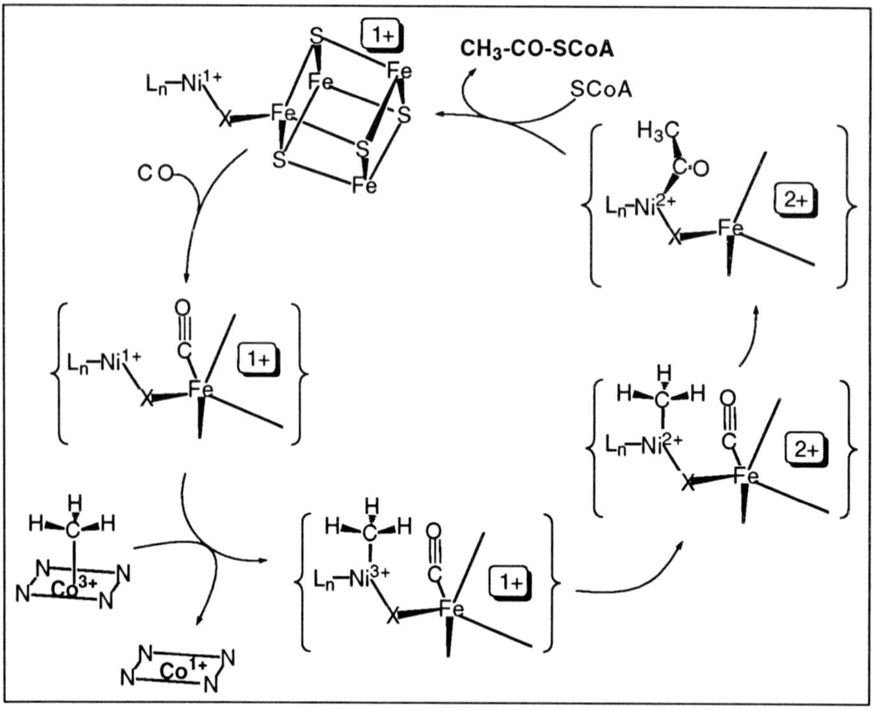

Figure 4

When CO binds to center A of CODH, a complex is formed that has been studied by EPR, IR, ENDOR, Mössbauer, and Raman spectroscopy. A variety of studies have been performed to demonstrate that this adduct is a catalytic intermediate in the acetyl-CoA pathway. (i) By freeze quench EPR, it was shown that the rate constant for formation of the NiFeC EPR signal is faster than k_{cat} for the overall synthesis of acetyl-CoA (Kumar et al., 1993). (ii) The CODH-CO complex reacts with methylated C/Fe-SP with a rate constant of 10 s^{-1} (five-fold faster than k_{cat}) (Kumar et al., 1995). (iii) The NiFeC EPR signal decays in the mixing time when reacted with CH_3I and the C/Fe-SP (Gorst, Ragsdale, 1991). (iv) The rate of formation of the NiFeC signal equals the rate of formation of the 360 cm^{-1} Fe-C vibrational band (Qiu et al., 1995b). (v) The NiFe^{13}C EPR signal converts to NiFe^{12}C by exchange of the bound carbonyl group with CH_3-^{12}CO-SCoA (Gorst, Ragsdale, 1991). (vi) The Fe-^{13}CO IR and Raman vibrational bands are replaced by Fe-^{12}CO bands as a result of exchange with CH_3-^{12}CO-SCoA (Kumar, Ragsdale, 1992; Qiu et al., 1994). (vii) The NiFeC EPR signal is formed from acetyl-CoA in the absence of CO (Gorst, Ragsdale, 1991). (viii) Substrates, CoA & acetyl-CoA (Ragsdale et al., 1985), and the inhibitor, N-bromosuccinimide (Shanmugasundaram et al., 1988), affect the NiFeC EPR spectrum. These combined results demonstrate conclusively that the Fe-CO species at center A is a catalytically relevant intermediate in the acetyl-CoA pathway.

The results just described proved that an Fe-carbonyl is the precursor of the carbonyl group of acetyl-CoA. Then what is the Ni doing in the enzyme? We proposed a bimetallic mechanism for acetyl-CoA synthesis in which Ni acted as the methyl group acceptor (Qiu et al., 1994). We recently confirmed this mechanism by detecting and characterizing the methylnickel intermediate (Kumar et al., 1995).

A vibrational band at 429 cm^{-1} was assigned to the methyl-Co stretching band from the methylated C/Fe-SP based on the magnitude of the ^{13}C isotope shift. When the methylated C/Fe-SP was reacted with CODH, the 429 cm^{-1} band decreased and a Raman band at 422 cm^{-1} appeared. This was assigned to a methyl-metal vibration by an isotope shift of the expected magnitude with both $^{13}CH_3$ and CD_3 substitution. Then, samples of CODH that had been substituted with ^{54}Fe, ^{58}Fe, and ^{64}Ni were methylated. The band remained at 422 cm^{-1} with either ^{54}Fe or ^{58}Fe. Therefore, a methyl-Fe intermediate was ruled out. However, the band shifted by the expected magnitude of 5 cm^{-1} with the ^{64}Ni isotope substitution. This proves that methylation of CODH generates a methyl-Ni bond.

Manoj Kumar has done further mechanistic work to determine whether CODH removes the methyl group of the methylated C/Fe-S protein by a nucleophilic substitution or a radical mechanism. This can be distinguished by determining whether the transmethylation product is Co^{1+} (for a nucleophilic mechanism) or Co^{2+} (for a radical mechanism). Our data clearly show that the product of the reaction is Co^{1+} which is formed at a rate 10-fold faster than k_{cat} for the overall synthesis of acetyl-CoA. Therefore, the reaction involves a Ni^{1+} nucleophile that attacks the methyl group forming the equivalent of methyl-Ni^{3+}. This is not expected to be very stable, therefore, the cluster would be able to rapidly transfer an electron by an inner sphere mechanism to make a more stable methyl-Ni^{2+} species.

CONCLUSION

The field of enzymology has been dominated by bio-organic chemists. However, it is clear that nature has evolved some interesting metal-centered catalysts. When we think about metabolic pathways and intermediates in these pathways, we think about organic compounds like glucose-6-phosphate. We may now also consider the interesting bio-organometallic chemistry occurring in nature, in fact, in our own intestines at this very minute (Lajoie et al., 1988). In this lecture, recent studies of the mechanism of acetyl-CoA formation have been described that demonstrate new biological roles for nickel and FeS clusters. In addition, novel bioinorganic enzymatic mechanisms of activating one-carbon units and forming C-C and C-S bonds have been described.

The work described here supported by the Department of Energy (DE-FG02-91ER20053) and NIH (GM39451) grants to SWR.

BIBLIOGRAPHY
Anderson, ME et al. (1993) J. Am. Chem. Soc. 115, 12204-12205.
Diekert, GB, Thauer, RK (1978) J. Bacteriol. 136, 597-606.
Drake, HL et al. (1981) J. Biol. Chem. 256, 11137-11144.
Ensign, SA et al. (1989) Biochemistry 28, 4973-4979.
Gorst, CM, Ragsdale, SW (1991) J. Biol. Chem. 266, 20687-20693.
Kumar, M et al. (1993) J. Am. Chem. Soc. 115, 11646-11647.
Kumar, M et al. (1995) Science, in press.
Kumar, M, Ragsdale, SW (1992) J. Am. Chem. Soc. 114, 8713-8715.
Lajoie, SF et al. (1988) Appl. Environ. Microbiol. 54, 2723-2727.
Morton, TA (1991) Ph.D., University of Georgia.
Qiu, D et al. (1994) Science 264, 817-819.
Qiu, D et al. (1995a) Science, submitted.
Qiu, D et al. (1995b) J. Am. Chem. Soc. 117, 2653-2654.
Ragsdale, SW (1991) CRC Crit. Rev. Biochem. Mol. Biol. 26, 261-300.
Ragsdale, SW et al. (1983) J. Biol. Chem. 258, 2364-2369.
Ragsdale, SW, Wood, HG (1985) J. Biol. Chem. 260, 3970-3977.
Ragsdale, SW et al. (1985) Proc. Natl. Acad. Sci. USA 82, 6811-6814.
Seravalli, J et al. (1995) Biochemistry 34, 7879-7888.
Shanmugasundaram, T et al. (1988) BioFactors 1, 147-152.
Shin, W, Lindahl, PA (1992) Biochemistry 31, 12870-12875.
Zhao, S et al. (1995) Biochemistry, submitted.

Enzymology of Methylamine Dehydrogenase

Antonius C.F. Gorren[1], Carol J.N.M. van der Palen[1], Rob J.M. van Spanning[2], and Johannis A. Duine[1].
[1]Department of Microbiology & Enzymology, Delft University of Technology, Julianalaan 67, 2628 BC Delft, The Netherlands.
[2]Department of Microbiology, BioCentrum Amsterdam, Free University, De Boelelaan 1087, 1081 HV Amsterdam, The Netherlands.

KINETIC STUDIES.

Introduction. Methylamine dehydrogenase (MADH) belongs to the class of quinoprotein enzymes because it has tryptophyl tryptophanquinone (TTQ) as cofactor. In most cases, it is clear that the small blue copper protein, amicyanin, functions as natural electron acceptor. Sofar, the kinetic behaviour of MADH has mainly been studied with artificial electron acceptors. Here we present an in depth study on the kinetics of MADH and amicyanin from *Thiobacillus versutus* (nowadays called *Paracoccus versutus* (Katayama et al., 1995)). Since H^+ and NH_4^+, being products of the reaction, could affect the kinetics of MADH (NH_4^+ could react with the quinone moiety of TTQ, forming an iminoquinone, certainly modifying the kinetic properties of the enzyme), special attention is paid to the effect of these cations on enzyme behaviour. Although the effect of NH_4^+ on the spectral and kinetic properties of MADH is already known for a long time, the question is whether it is unique in this or whether other cations show similar behaviour. This point will also be addressed here with respect to the fully oxidized, the semiquinone, and the fully reduced forms of the en zyme ($MADH_{ox}$, $MADH_{sem}$, $MADH_{red}$, respectively).

Binding of monovalent cations to MADHox. It has long been known that NH_4^+ changes the absorbance spectrum of $MADH_{ox}$ from *Methylophilus methylotrophus* W3A1 (Kenney, McIntire, 1983). Recently, Kuusk and McIntire demonstrated that a whole range of other monovalent cations affect the absorbance spectrum of $MADH_{ox}$ as well (Kuusk, McIntire, 1994). Some cations (($CH3)_3NH^+$, ($CH3)_4N^+$) induce a red-shift of the absorbance maximum (designated type I), whereas others (K^+, Na^+) cause changes, resembling those observed upon reduction of the enzyme (type II). Most cations display both effects at different concentrations and/or pH (Li^+, Rb^+, Cs^+, NH_4^+). Very similar observations were made with MADH from *T. versutus* (Gorren, Duine,

197

M. E. Lidstrom and F. R. Tabita (eds.), Microbial Growth on C₁ Compounds, 197–204.

1994). In both cases type-I and type-II absorbance changes occur; for both enzymes cation affinities tend to increase at higher pH. Both groups postulated that the effects were due to the random binding of cations to two binding sites on MADH, with (at least for some cations) strong negative cooperativity between the two sites. In both cases it was deduced that the spectrum of MADH with both sites occupied resembles the type-II spectrum. The same order of affinity of the cations for both sites was established for the two enzymes. The main difference lies in the observations with NH_4^+ and the interpretation thereof. Whereas the results for *T. versutus* MADH were explained by binding of NH_4^+ to the type-I and type-II sites, an additional NH_4^+-binding site was postulated for MADH from *M. methylotrophus* W3A1. Although the absorbance changes did not differ significantly from those of the other cations, the observations with NH_4^+ were indeed different in that only in titrations with NH_4^+ a red-shift could be observed <u>after</u> a first phase, that looked like a hybrid of a type-I and type-II spectral transition. The first phase was tentatively interpreted as the covalent binding of NH_4^+ to TTQ to form the iminoquinone, whereas the second phase was ascribed to type-I binding. By contrast, no absorbance or other spectral data suggest the formation of an iminoquinone from $MADH_{ox}$ and NH_4^+ in the case of *T. versutus*.

Both the research groups of McIntire and Duine suggested that type-I binding occurs at the substrate binding site and that, consequently, methylamine binds to $MADH_{ox}$ electrostatically, as a methylammonium cation (Kuusk, McIntire, 1994; Gorren, Duine, 1994). Evidence for this contention comes from a variety of observations: type-I binding is observed for the non-convertible substrate-analogues $(CH3)_2NH_2^+$, $(CH3)_3NH^+$, and $(CH3)_4N^+$, as well as for the simple alkali metal ions Cs^+, Rb^+, and Li^+, with the highest affinity displayed by the methylammonium mimics $(CH3)_2NH_2^+$ and $(CH3)_3NH^+$ (Kuusk, McIntire, 1994; Gorren, Duine, 1994). All type-I binding cations are competitive inhibitors of the reduction of $MADH_{ox}$ by methylamine with a strong correlation between K_d-values derived from the optical titrations on the one hand and the kinetic data on the other (Gorren, Duine, 1994). Moreover, type-I absorbance changes occur transiently in the reduction of $MADH_{ox}$ by methylamine (Gorren, Duine, 1994) or propylamine (McWhirter, Klapper, 1989). Resonance Raman spectra of $MADH_{ox}$ in the presence of $(CH3)_3NH^+$, Cs^+, and NH_4^+, demonstrated that the cations are not covalently bound to TTQ, but do interact directly with the C6-carbonyl oxygen (Gorren et al., 1995 b).

The identity of the type-II binding site is unknown. Type-II binding appears to be rather aspecific (Kuusk, McIntire, 1994; Gorren, Duine, 1994), and the affinities of cations for this site are relatively low. For several reasons it has so far been impossible to apply alternative spectroscopic techniques to study type-II binding. However, the large optical changes, as well as the strong negative cooperativity between type-I and type-II binding suggest that type-II binding occurs in the vicinity of the TTQ in the active site pocket.

Binding of monovalent cations to MADHsq. The binding of NH_4^+ to $MADH_{sq}$ was first demonstrated by the strong broadening of the radical EPR-signal of $MADH_{sq}$ from *M. methylophilus* W3A1 (Kenney, McIntire, 1983). Similar observations were recently made with $MADH_{sq}$ from *T. versutus*, not just for NH_4^+, but also for Cs^+, which demonstrates that no covalent interaction is involved (Gorren et al., 1995a). A small effect was observed with Na^+ as well. However, whereas in the case NH_4^+ and Cs^+ the broadening can be ascribed with some confidence to the coupling between the electron spin of $MADH_{sq}$ and the nuclear spin of the cation, alternative interpretations are possible in the case of Na^+. Binding of monovalent cations to $MADH_{sq}$ is accompanied by subtle changes in the optical (Gorren et al., 1995a) and resonance Raman spectra. By comparison of the EPR-intensities of $MADH_{sq}$ in the presence and absence of monovalent cations, it was shown that cations that bind to $MADH_{sq}$ (NH_4^+, Cs^+, Na^+) stabilize $MADH_{sq}$ relative to $MADH_{ox}$ and $MADH_{red}$. By contrast, $(CH3)_3NH^+$, which binds most efficiently to $MADH_{ox}$, does not bind to $MADH_{sq}$, nor does it stabilize the semiquinone state (Gorren et al., 1995a).

Although there is at present no spectroscopic evidence for it, the kinetic experiments described below imply that most monovalent cations, but not $(CH3)_3NH^+$, also bind with low affinity to $MADH_{red}$ (Gorren et al., 1995a).

Kinetic studies of MADH with methylamine and amicyanin. The first in depth kinetic study of MADH with both physiological substrates (Gorren, Duine, 1994) showed that, with amicyanin as the reoxidizing substrate, k_{cat} varies little between pH 7 and 10, and the sharp optimum at pH 7.5 that is found in assays with the artificial electron acceptors PMS or PES (phenazine (m)ethosulfate) (Eady, Large, 1968; Mehta, 1977; Matsumoto, 1978; Husain, Davidson, 1987), is not observed. Similarly, the maximal rate of reduction of $MADH_{ox}$ by methylamine is not very sensitive to pH-variation. On the other hand, the apparent MADH-methylamine association rate constant, as well as the specificity constant (k_{cat}/K_m) of MADH for amicyanin increase by several orders of magnitude when the pH increases from 5 to 10 (Gorren, Duine, 1994). The ionic strength dependence of the reactions is hard to establish as it tends to be drowned by the much stronger specific cationic effects. However, if it can be assumed that Mg^{2+} does not specifically bind to MADH, since it has no optical effects (Kuusk, McIntire, 1994; Gorren, Duine, 1994), it can be concluded that increasing the ionic strength lowers both the rate of reduction of $MADH_{ox}$ by methylamine and the rate of oxidation of $MADH_{red}$ by amicyanin (Gorren, Duine, 1994), probably by lowering the affinity of MADH for its substrates.

Monovalent cations competitively inhibit the reduction of $MADH_{ox}$ by methylamine (Gorren, Duine, 1994; Gorren et al., 1995b), with the order of effectivity perfectly agreeing with the order of affinity observed in optical titrations. For the type-I binding cations, inhibition is achieved by direct competition for the substrate binding site, whereas type-II binding cations might exert their effect by the negative cooperativity between type-I and type-II binding. It should be noted that inhibition by the low-affinity, type-II binding cations (Na^+, K^+) is (partly) due to a non-specific ionic strength effect. Interestingly, the type-I binding cation Cs^+ appears unable to completely inhibit the reduction of MADH at saturating concentrations (Gorren, Duine, 1994), suggesting an alternative pathway of methylamine oxidation, perhaps by binding to the type-II site.

Monovalent cations also strongly enhance the rate of oxidation of $MADH_{red}$ by amicyanin (as well as by other electron acceptors) (Gorren, Duine, 1994). The first oxidation step ($MADH_{red}$-to-$MADH_{sq}$), which is particularly slow in the absence of cations, is accelerated more than the second step ($MADH_{sq}$-to-$MADH_{ox}$), leading to the transient observation of $MADH_{sq}$ in the presence, but not in the absence of cations. The cations mainly affect the actual electron transfer, with minor effects at best on the affinity of MADH for amicyanin (Gorren, Duine, 1995a). A clear correlation exists between the rate enhancement and the ability of the cations to stabilize $MADH_{sq}$, but a quantitative assessment shows that the rate enhancement must be largely due to a change in the reorganization parameter I (Gorren, Duine, 1995a). A picture emerges in which monovalent cations strongly increase the rates of both oxidation steps by significantly lowering the reorganizational barrier, while additionally stimulating the first oxidation step by preferentially binding to $MADH_{sq}$, thereby increasing the driving force. The results suggest that during reoxidation the product ammonium ion remains bound in the MADH active site cavity, only to be expelled by the binding of $CH_3NH_3^+$ to $MADH_{ox}$ at the start of a new reaction cycle.

REACTION MECHANISM.

It has been proposed (McWhirter, Klapper, 1989; Backes et al., 1991; Davidson, 1993; Klinman, Mu, 1994) that the reductive half reaction follows a pathway that resembles the one deduced for the TPQ-containing amine oxidases (Klinman, Mu, 1994). According to this model the reaction starts by covalent binding of methylamine to the TTQ-C6 to form a carbinolamine, followed by dehydration to a Schiff base, tautomerization, proton abstraction via a carbanionic intermediate, and hydrolysis to the aminoquinol and formaldehyde. Some support for a carbanionic intermediate is derived from structure reactivity correlations for the reaction of $MADH_{ox}$ with p-substituted benzylamines (Davidson et al. 1992), and, recently, ESEEM data provided evidence for aminotransfer (Warncke et al., 1993). However, the structure correlation studies were too limited to furnish conclusive evidence and the central assumption in the ESEEM study, that the observed effects prove the methylamine-derived nitrogen to be covalently bound to the TTQ, requires corroboration in view of the large effect that the alkali metal cation Cs^+ has on the EPR-properties of $MADH_{sq}$ (Gorren, Duine, 1995a). All other previous observations of specific ammonia-adducts to TTQ in optical absorbance (Kenney, McIntire, 1983; Backes et al., 1991), resonance Raman (Backes et al., 1991) and EPR-spectroscopy (Kenney, McIntire, 1983) have now been demonstrated not to involve covalent interactions.

At any rate, the first reaction step definitely is not the formation of an carbinolamine, but the electrostatic binding of $CH_3NH_3^+$ in the active site pocket in direct interaction with the C6-carbonyl oxygen (Kuusk, McIntire, 1994; Gorren, Duine, 1994; Gorren et al., 1995b). It is this reaction that is accompanied by a red-shift in the absorbance spectrum (McWhirter, Klapper, 1989; Gorren et al., 1995b). The subsequent spectral phase, which already yields the spectrum of $MADH_{red}$, exhibits a very large deuterium kinetic isotope effect (McWhirter, Klapper, 1989; van Wielink et al., 1989; Brooks et al.; 1993; Gorren et al., 1995b), suggesting that it involves proton/hydrogen abstraction. Therefore, although it remains a possibility that, following the formation of the red-shifted intermediate, methylamine binds covalently to TTQ, thus starting the reaction sequence described above, the putative carbinolamine intermediate is not detected. A third and final spectral phase was demonstrated to involve the release of formaldehyde to the medium (McWhirter, Klapper, 1989). In view of the different affinities that $(CH3)_3NH^+$ and NH_4^+ have for MADH in its three redox states (Kuusk, McIntire, 1994; Gorren et al., 1995a; Gorren et al., 1995b), it seems likely that this reaction is triggered by protein conformational changes, that, following reduction, contract the active site pocket and expel formaldehyde, while retaining, covalently or otherwise, the ammonium ion.

Since some monovalent cation needs to be present in the active site pocket to sustain electron transfer from $MADH_{red}$ to an electron acceptor (Gorren, Duine, 1994; Gorren et al., 1995a), NH_4^+ probably remains in the active site pocket until it is expelled by methylamine at the start of a new cycle. Other cations (K^+ or Na^+) can, at sufficiently high concentrations, replace NH_4^+ in stimulating electron transfer from MADH to amicyanin.

It was recently found that apo-amicyanin, which is structurally very similar to native amicyanin, has a strongly decreased affinity for MADH (Davidson et al., 1993). Moreover, from studies with Zn^{2+}- and Ag^+-substituted amicyanin, as structural analogues for oxidized and reduced amicyanin (Ubbink et al., 1994), it was deduced that only oxidized amicyanin binds stron gly to MADH, suggesting that after reduction, amicyanin rapidly dissociates from MADH, thereby allowing the binding of a new oxidized molecule and preventing product inhibition by reduced amicyanin.

From recent studies of the temperature dependence of the electron transfer from $MADH_{red}$ (from *Paracoccus denitrificans*) to amicyanin, and as-suming E_m-values for amicyanin and the MADH red/sq-couple of +221 mV and +100 to +190 mV, respectively, it was deduced that the value for the reorganization parameter 1 is exceptionally high (225 kJ/mol) for a biological electron transfer reaction (Brooks, Davidson, 1994a). Since these studies were performed in a low-salt buffer (10 mM KP$_i$; pH 7.5), and in view of the effects that cations have on the oxidation rate, it seems likely that much lower values will be found when the cation concentration is sufficiently high. In a subsequent paper the other three redox reactions between MADH and amicyanin, i.e. from Amic$_{red}$ to MADH$_{sq}$, from MADHsq to Amic$_{ox}$, and from Amic$_{red}$ to MADH$_{ox}$, were studied as well (Brooks, Davidson, 1994b). However, the rate constants for the separate redox steps reported in this paper can not be taken at face value. The value for the rate constant for electron transfer from Amic$_{red}$ to MADH$_{ox}$, in particular, (0.05 s^{-1}) must be an artefact. This reaction, which was monitored at 428 nm in the presence of a large excess of amicyanin, was reported to be sufficiently slow for initial rates to be measured. However, as the rate constant for the reverse reaction (from MADH$_{sq}$ to Amic$_{ox}$) was found to be 240 s^{-1} and since k$_{obs}$ should equal the sum of forward and reverse reaction rate constants, a rate of at least 240 s^{-1} should have been observed, which is equivalent to a t$_{1/2}$ of 3 ms. Moreover, if the calculated rate constants of 240 s^{-1} and 0.05 s^{-1} were correct, a 4800-fold excess of amicyanin would be needed to obtain 50 % conversion of MADH$_{ox}$ to MADH$_{sq}$ (on [TTQ]-basis). The conversion of 1 mM MADH$_{ox}$ (2 mM TTQ) to MADH$_{sq}$ would then require the addition of 9.6 mM Amic$_{red}$. Also, since 428 nm is not isosbestic for the transition of MADH$_{sq}$ to MADH$_{red}$, and since from the estimated rate constants the equilibrium concentration of MADH$_{sq}$ can be calculated to be only 7.5*10^{-4} of total MADH, accumulation of MADH$_{sq}$ would be expected not to occur. Therefore the correlation between ln k$_{ET}$ and DG^0 reported by Brooks and Davidson must be fortuitous, particularly since there seems to be no reason to *a priori* assume that l is equal for both MADH-redox steps.

ELECTRON TRANSFER PATHWAYS.

It has become increasingly clear that all MADHs that use amicyanin as the electron acceptor, require a small soluble c-type cytochrome to tranfer electrons between amicyanin and a membrane bound *aa$_3$*-type oxidase. In *P. denitrificans* cytochrome c$_{550}$ seems to fulfill this role but cytochrome c$_{552}$ is able to replace it. Moreover, the mutational studies also revealed that cyto-chrome $_{bc1}$ is able to accept electrons from amicyanin and donate them to a cytochrome $_{cbb3}$ functioning as an alternative oxidase (de Gier et al., 1995). In methylamine grown *T. versutus*, which appears to contain only an *aa$_3$*-type oxidase, amicyanin is not able to reduce this oxidase directly (P. van Bastelaere, unpublished). The electron transfer pathway for both *T. versutus* and *P. denitrificans* therefore is best described by a linear scheme:
methylamine -> MADH -> amicyanin -> cytochrome c -> cytochrome *aa$_3$* -> O$_2$. Amicyanin is not present in *M. methylotrophus* W3A1, and the gene for amicyanin that is found in the cluster that contains the MADH genes of *P. denitrificans* (van Spanning et al. 1994; van der Palen et al, 1995), *T. versutus*

(Huitema et al, 1993) and *Methylobacterium extorquens* AM1, is missing in the case of *M. methylotrophus* W3A1 (Chistoserdov et al., 1994a; 1994b).

Electron transfer from amicyanin to cytochrome c_{551i} in *P. denitrificans* is reported to require the formation of a ternary complex between MADH, amicyanin and the cytochrome (Gray et al., 1988), since the midpoint potential of free amicyanin, which is unfavorable for electron transfer to cytochrome c_{551i} (+294 mV and +190 mv for amicyanin and cytochrome c_{551i}, respectively) is lowered (to +221 mV) upon binding to MADH. A high-resolution 3-D structure of the ternary complex has now been obtained (Chen et al., 1994). The interaction between MADH and amicyanin is very similar to that in the binary complex. The relevance of the determined structure remains to be established in view of the unprecedented distance (24.8 Å) across which electrons have to be transferred from the amicyanin copper to the heme iron in the complex and the doubts whether cytochrome c $_{551i}$ is the electron acceptor for amicyanin. The requisite formation of a ternary complex appears to be specific for the *Paracoccus* system. In *T. versutus* direct electron transfer from amicyanin to cytochrome c_{550} does not require the presence of MADH (van Wielink et al., 1989), and *M. methylophilus* W3A1 lacks amicyanin (Chistoserdov et al., 1994b).

BIOSYNTHESIS OF TTQ.

The results of cloning and sequencing of the structural genes for MADH (see above) suggest that in maturation of this enzyme, first a pro-enzyme is formed after which a post-translational modification process converts the two tryptophans concerned into TTQ. Apart from the chemical challenge in this (selective dimerization of two specific amino acids far apart in a protein chain, hydroxylation of trp at the 6 or 7 position, sites which are not preferred in chemical derivatizations), also the question where the modification occurs is interesting as the solutions which can be envisaged, either the cytoplasm, periplasm or both, are difficult to envisage in relation with current views on protein transport and folding. With respect to the latter, renaturation of the denatured α and β subunits to active enzyme has been achieved *in vitro* (only with the β subunit in the oxidized form, not in the reduced form (F. Huitema, unpublished)). Mutationional studies have revealed that a cytochrome c (Page, Ferguson, 1993) and a protein with sequence similarity to cytochrome c peroxidase of *Pseudomonas aeruginosa* (Chistoserdov et al., 1994a; van der Palen et al., 1995) are involved in TTQ biosynthesis. In the absence of the "peroxidase" (in a *mauG* mutant of *P. denitrificans*), we found production of a protein which reacts with antibodies raised against MADH, behaves chromatographically similar to MADH, is inactive in the assay, gives negative results in the quinone staining assay, and does not contain a chromophore (van der Palen et al., 1995; unpublished). Several of these properties were also exhibited by a protein produced in a cytochrome c-less mutant strain (Page, Ferguson, 1993). Since it is unknown whether dimerization of the tryptophans had already occurred, it is clear that the peroxidase/cytochrome c couple is involved in TTQ formation but the precise step in this cannot be indicated yet. In view of the periplasmic location of the couple, the step must occur after the protein has been transported across the cytoplasmic membrane. Characterization of the mutant protein and of the peroxidase is in progress.

REFERENCES

Backes G et al. (1991) Biochemistry 30, 9201-9210.
Brooks HB et al. (1993) Biochemistry 32, 2725-2729.
Brooks HB, Davidson VL (1994a) Biochemistry 33, 5696-5701.
Brooks HB, Davidson VL (1994b) J. Am. Chem. Soc. 116, 11201-11202.
Chen L et al. (1994) Science 264, 86-90.
Chistoserdov AY et al. (1994a) J. Bacteriol. 176, 4052-4065.
Chistoserdov AY et al. (1994b) J. Bacteriol. 176, 4073-4080.
Davidson VL et al. (1992) Biochemistry 31, 3385-3390.
Davidson VL (1993) In Davidson, ed, Principles and Applications of Quinoproteins, pp. 73-95 Marcel Dekker, New York, USA.
Davidson VL et al. (1993) Biochim. Biophys. Acta 1144, 39-45.
de Gier et al. (1995) Eur. J. Biochem. 229 148-154.
Eady RR, Large PJ (1968) Biochem. J. 106, 245-255.
Gorren ACF, Duine JA (1994) Biochemistry 33, 12202-12209.
Gorren ACF et al. (1995a) Biochemistry 34 (1995) 9748-9754.
Gorren ACF et al. (1995b) Biochemistry, in press.
Gray KA et al. (1988) J. Biol. Chem. 263, 13987-13990.
Huitema F et al. (1993) J. Bacteriol. 175 , 6254-6259.
Husain M, Davidson VL (1987) J. Bacteriol. 169, 1712-1717.
Katayama et al. (1995) Microbiol. 141, 1469-1477.
Kenney WC, McIntire W (1983) Biochemistry 22, 3858-3868.
Klinman JP, Mu D (1994) Annu. Rev. Biochem 63, 299-344.
Kuusk V, McIntire WS (1994) J. Biol. Chem. 269, 26136-26143.
Matsumoto T (1978) Biochim. Biophys. Acta 522, 291-302.
McWhirter RB, Klapper MH (1989) In Jongejan JA and Duine JA, eds, PQQ & Quinoproteins, pp. 259-268, Kluwer Academic Publishers, Dordrecht.
Mehta RJ (1977) Can. J. Microbiol. 23, 402-406.
Page MD, Ferguson SJ (1993) Eur. J. Biochem. 218, 711-717.
Ubbink M et al. (1994) Eur. J. Biochem. 222, 561-571.
van der Palen CJNM et al. (1995) Eur. J. Biochem. 230, 860-871.
van Spanning et al. (1994) Eur. J. Biochem. 226, 201-210.
van Wielink JE et al. (1989) In Jongejan JA and Duine JA, eds, PQQ & Quinoproteins, pp. 269-278, Kluwer Academic Publishers, Dordrecht.
Warncke K et al. (1993) J. Am. Chem. Soc. 115, 6464-6465.

Structural studies of methylamine dehydrogenase

F.S. Mathews, L. Chen, R.C.E. Durley, Z.-w. Chen and W. S. McIntire[1]
Washington University School of Medicine, Dept. of Biochemistry and Molecular Biophysics, St. Louis, MO, USA, [1]Molecular Biology Division, Department of Veterans Affairs Medical Center, San Francisco, CA 95121, and the Department of Biochemistry and Biophysics and the Department of Anesthesia, University of California, San Francisco CA 94143.

INTRODUCTION

Methylamine dehydrogenase (MADH), found in methylotrophic bacteria, is an inducible, periplasmic quinoenzyme which catalyzes the oxidation of methylamine to formaldehyde and ammonia (De Beer R et al. 1980). It contains the novel cofactor tryptophan tryptophylquinone (TTQ), which is derived from two tryptophan side chains (McIntire WS et al. 1991). Subsequently, electrons are transferred to the membrane bound terminal oxidase, cytochrome aa_3 via a series of soluble electron carrier proteins. In facultative autotrophs, such as *Paracoccus denitrificans*, the initial electron acceptor in this chain is amicyanin, a blue copper protein (Husain M, Davidson VL 1985), while in *Methylophilus methylotrophus* W3A1, which lacks the gene for amicyanin (Chistoserdov AY et al. 1994) the initial acceptor is a cytochrome c_{552} (Chandrasekar R Clapper MH 1986). In the case of *P. denitrificans*, *in vitro* studies suggest that the acceptor following amicyanin in the electron transfer chain is cytochrome c_{551i} (Husain M, Davidson VL 1986). All three proteins are induced when these bacteria are grown on methylamine as the sole carbon source.

MADH is a hetero-tetramer consisting of two identical pairs of heavy (H) and light (L) subunits of approximate molecular weights 47 kDa and 15.5 kDa, respectively. Amicyanin is a cupredoxin of 11.5 kDa molecular weight. It is most similar to plastocyanin, consisting of two antiparallel β-sheets joined along one edge like a clam shell but having one additional β-strand at the amino terminus (Durley RCE et al. 1993). The copper is coordinated by one Cys, one Met and by two His side chains, one of which is partially exposed to solvent. Complex formation between MADH and amicyanin *in vitro* causes perturbation of the absorption spectrum of TTQ and a shift in the redox potential of the copper center of amicyanin from 294 mV to 221 mV (Davidson VL 1993) Cytochrome c_{551i} is an acidic protein of molecular weight 17.5 kDa which contains approximately 40 additional residues at the amino terminus and 30 at the carboxyl terminus compared with most bacterial c-type cytochromes (Chen et al. 1994). The majority of the charged residues are located on the side and rear portions of the cytochrome surface leaving the front surface, containing the exposed CBC methyl of the heme group, surrounded by hydrophobic

M. E. Lidstrom and F. R. Tabita (eds.), Microbial Growth on C₁ Compounds, 205–212.
© 1996 *Kluwer Academic Publishers. Printed in the Netherlands.*

residues.

The structures of MADH from *Thiobacillus versutus* (TV-MADH) (Vellieux FMD et al. 1989) and *P. denitrificans* (PD-MADH) (Chen L et al. 1992a) were first solved at 2.2 Å and 2.8 Å resolution, respectively. Subsequently, the structures of a binary complex between PD-MADH and amicyanin (Chen L et al. 1992b) and of a ternary complex between PD-MADH, amicyanin and cytochrome c_{551i} (Chen L et al. 1994) were determined at 2.6 Å and 2.4 Å, respectively. Recently, the structure of PD-

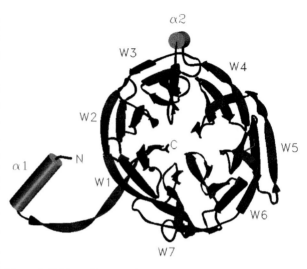

Figure 1. Ribbon diagram of the heavy subunit of PD-MADH, viewed along the pseudo 7-fold axis of symmetry. The W's represent the seven 4-stranded β-sheets.

MADH was refined at 1.75 Å resolution (Chen L, Mathews FS unpublished) and the structures of 4 forms of MADH from *M. methylophilus* W3A1 (WA-MADH) were determined at 2.5-3.0 Å resolution.

The present paper describes some of these structural results.

RESULTS

Structure of PD-MADH and its complexes. The H subunit of PD-MADH is a 7-fold β-propeller consisting of a disk-shaped domain of seven 4-stranded antiparallel β-sheets arranged about a pseudo 7-fold axis of symmetry (Fig. 1). The L subunit contains 5 β-strands in two antiparallel β-sheets of two and three strands each (Fig. 2). The TTQ prosthetic group is located in the L subunit, and is formed by fusion of Trp 57 and Trp 108 by a covalent bond linking their $C^{\epsilon 3}$ and $C^{\delta 2}$ atoms, respectively. Trp 57 also contains quinone oxygen atoms attached to $C^{\eta 2}$ (O6) and $C^{\zeta 2}$ (O7). The orthoquinone of Trp 57 is

Figure 2. Ribbon diagram of the light subunit of MADH. The 6 disulfide bridges and the TTQ are also shown.

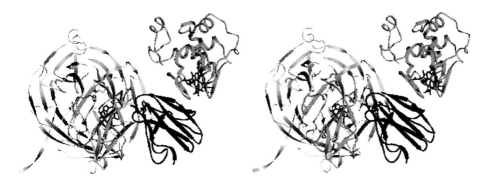

Figure 3. Stereo diagram of the Ternary complex. PD-MADH is on the left, amicyanin on the lower right and cytochrome c_{551i} on the upper right.

located in the active site pocket in the interior of the L subunit (positioned approximately on the pseudo 7-fold axis of the H subunit) while Trp 108 is packed among hydrophobic side chains except for one end of the indole ring which is exposed to solvent. The L subunit is highly crosslinked, containing six disulfide bridges in addition to the TTQ crosslink.

Since the interactions between amicyanin and MADH are nearly the same in the binary and ternary complexes only the latter structure is described here. The structure of ternary complex between MADH, amicyanin and cytochrome c551i is shown in Fig. 3. Amicyanin forms a bridge between MADH and the cytochrome which are otherwise are not in contact with each other. MADH and amicyanin are oriented so that the benzene ring of Trp 108 of MADH faces the exposed histidine ligand of

Table I

MADH-AMICYANIN INTERACTIONS: PD *VS* W3A1

Amicyanin	PD-MADH	Distance (Å)	WA-MADH
Van der Waals $\leq$ 3.5 Å			
Ala^{50}O	Leu^{71}CD2 (L)	3.5	Leu
Met^{71}CG	Thr^{54}OG1 (L)	3.5	*Pro*
Lys^{73}CG	Val^{127}O (L)	3.3	Val
His^{91}NE2	Pro^{145}O (H)	3.3	Pro
His^{95}NE2	Glu^{101}CB (L)	3.1	Glu
Phe^{97}CA	Asp^{167}OD2 (H)	3.3	Asp
Phe^{97}CE2	Pro^{100}CG (L)	3.5	Pro
Phe^{97}CZ	Arg^{184}NH2 (H)	3.4	Arg
Salt Bridge			
Arg^{99}NH2	Asp^{167} OD1 (H)	2.7	Asp
Arg^{99}NH2	Asp^{167} OD2 (H)	2.7	Asp
Potential Salt Bridge			
Lys^{68}NZ	Asp^{115}OD2 (L)	4.4	Asp

Additional Interface residues (> 3.5 Å separation)

Amicyanin	PD-MADH	W3A1-MADH
Met28	Phe43(H)	Phe
Met51	Ser144(H)	Ser
Pro52	Ala146(H)	*Ser*
Asn54	Tyr169(H)	Tyr
His56	Ala55(L)	*Ser*
Val58	Ser56(L)	Ser
Lys68	Val58(L)	Val
Gly69	Phe102(L)	Phe
Thr93	Trp108(L)	Trp
Pro94	Phe110(L)	Phe
Pro96	Gly111(L)	Gly
Arg99	Asp115(L)	Asp
	Gly128(L)	Gly
	Lys129(L)	Lys

Table II. Crystallographic data for four crystal forms of WA-MADH

Crystal	Native pH 4.0	Native pH 10.4	Br-φ-hydrazine Ph 7.0	Semicarbazide pH 7.4
Shape	Needle	Rec. plate	Rec. plate	Square plate
Size	2.2 x .32 x .1 mm	.4 x .3 x .08 mm	.7 x .6 x .08 mm	.4 x .4 x .3 mm
Color	Light yellow	Light yellow	Red	Red
Resolution	2.5 Å	2.8 Å	3.0 Å	2.8 Å
R_{merge}	11.9	15.3	14.6	9.2
Space Group	$P2_1$	$P2_1$	$P2_1$	$P2_1$
a (Å)	80.4	69.5	69.8	88.6
b (Å)	83.5	83.1	82.9	121.8
c (Å)	86.6	90.0	90.2	92.5
β (°)	114.8	91.6	91.5	97.0
Mol./a. u.	1	1	1	2
R-factor	0.174 (7-2.4 Å)	0.196 (10-2.8 Å)	0.173 (10-3.0 Å)	0.165 (10-2.8 Å)
rmsd (Å)	0.014	0.016	0.013	0.014
Waters	227	None	None	53
B- Refinement	Yes	Yes	Yes	Yes

copper. The distance from the O6 atom of TTQ, where substrate oxidation is believed to occur, to the copper is 16.8 Å. The closest distance from TTQ to the copper atom is 9.3 Å. Amicyanin and the cytochrome are oriented so that β-Strand 3 of amicyanin which forms the hinge of the β-clam shell makes contact with a peptide segment of the cytochrome. This segment is located between the histidine ligand to the heme iron and the start of a short helix located between the histidine and the methionine ligands and close to one of the heme propionates. The copper to iron distance is 24.7 Å and the distance from copper to the nearest atom of the heme is about 21 Å. Within the complex, the O6 of TTQ, the copper and the iron are nearly collinear, with the O6 atom separated from the iron atom by 41.2 Å.

The MADH-amicyanin interface is largely hydrophobic and covers an area of approximately 750 Å². Approximately two thirds of the residues in the MADH portion of the interface come from the L subunit (Table I). The amicyanin portion of the interface contains the normally exposed histidine ligand to the copper as well as seven surrounding hydrophobic residues (three methionines, three prolines and a phenylalanine) which form a hydrophobic patch. A similar hydrophobic patch centered on an exposed histidine ligand is found in the other cupredoxins and has been implicated in electron transfer (Farver O et al. 1982). About 40% of the residues in the interface are hydrophilic but only the aliphatic portions of their side chains lie within the interface leaving the polar portions of the residues

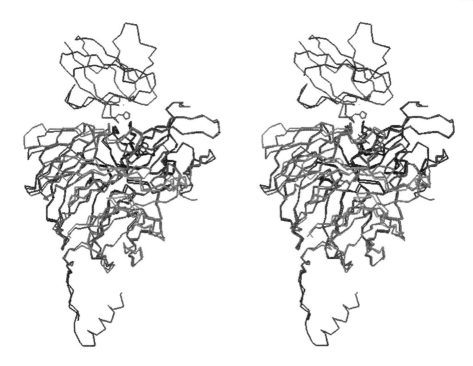

Figure 4. α-Carbon tracings of WA-MADH superimposed on the binary complex of PD-MADH with amicyanin. The helix at the bottom of the H subunits starting at residue 1 belongs to PD-MADH while the H subunit in WA-MADH begins at position 18.

pointing into solution. There are two salt bridges and 3 water molecules connecting the two proteins. The chemical makeup of the MADH-amicyanin interface suggests that this complex is stabilized by both hydrophobic and electrostatic forces. The electrostatic component of the interface may contribute to the 73 mV decrease in the redox potential of amicyanin upon complex formation.

The amicyanin-cytochrome interface is smaller than the quinoprotein/cupredoxin interface, covering approximately 425 Å^2. It is also more polar, with approximately 65% of its residues hydrophilic. The amicyanin and cytochrome are joined by 1 salt bridge, 4 hydrogen bonds and two solvent molecules. Based on its coordination, one of the solvent molecules may be a cation such as sodium or potassium.

W3A1 MADH. Data were collected from four different types of WA-MADH crystals (Table II). Two of these were native crystals, grown at pH 4.0 and pH 10.4. The remaining two were derivatized forms of the enzyme, one treated with bromophenylhydrazine and the other with semicarbazide. Both ligands are known to be potent inhibitors of MADH (Kenny WC, McIntire WS 1983) and were reacted with the enzyme in solution at neutral pH prior to crystallization. All four crystal types were grown under similar conditions, from 6-18 % polyethylene glycol solution in the presence of 100 mM NaCl. In three of the four cases the crystals were small thin plates, less than 0.1 mm in thickness while in the fourth type

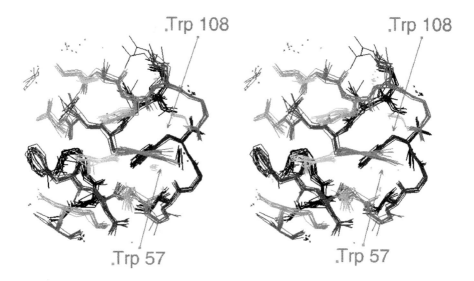

Figure 5. View of the active sites of ten L subunits from 4 crystalline forms of WA-MadH. Four of the subunits are from native uncomplexed MADH and six are from MADH inhibited by semicarbazide or bromophenylhydrazine. The six side chains of Trp 57 are oriented very similarly to one another and are tilted by about 10°-15° from the mean plane of Trp 57 in the four native structures. In contrast Trp 108 shows no concerted movements among the several structures.

(semicarbazide) somewhat larger crystals could be obtained. Three of the four structures

were solved by molecular replacement while the fourth (bromophenylhydrazine) was isomorphous to one of the remaining (pH10.4 native, Table II).

No significant changes in conformation can be perceived between the low and high pH forms of native WA-MADH at the current stage of refinement. The rms deviation between them is approximately 0.40 Å. Thus, although WA-MADH undergoes a pH-dependent change in its spectral properties (Kuusk V, McIntire WS 1994), no concerted change in structure appears to occur. WA-MADH contains no amicyanin, utilizing instead cytochrome c_{552} (Chandrasekar R, Clapper MH 1986). Comparison of WA-MADH with PD-MADH indicates that close structural homology is maintained (rms deviation of 0.75 Å between equivalent C_α atoms), despite limited sequence homology (51% between H-subunits, 72% between L-subunits). Except for one deletion, in the WA-MADH H subunit, the greatest difference is the presence of an additional disulfide bridge in the WA-MADH L subunit, linking Cys 39 to Cys 83 (based on the PD-MADH numbering). These residues are both Ser in PD-MADH; this additional cross link appears to have no effect on the active site of WA-MADH but might affect its thermal stability.

When WA-MADH is superimposed on the PD structure in the binary complex, no obvious deterrence to complex formation with amicyanin can be discerned (Fig. 4). Comparison of the amicyanin/MADH interface residues (Table I) indicates that there are no major replacements of the PD-MADH side chains which would inhibit complex formation. This result is in agreement with findings that oxidized TV-amicyanin reacts efficiently with reduced WA-MADH to oxidize the enzyme (McIntire WS. unpublished).

The two inhibited forms of WA-MADH have been compared with the native forms of the enzyme. In both cases additional electron density was found adjacent to the O6 atom of TTQ, indicating that this is the attachment site for the inhibitors. However, despite the size of the substituants (bromophenylhydrazine or semicarbazide) there was only enough additional density to account for a single additional atom bound to the TTQ. Similar results were obtained for TV-MADH (Huizinga et al. 1992) and for cyclopropylamine-substituted PD-MADH (Chen L, et al. unpublished). However, significant conformational changes were observed for the TTQ moiety (Fig. 5). The two derivatives undergo a 10°-15° increase in the dihedral angle of Trp 57 with respect to Trp 108, from about 45° to about 55°, the most substantial change in the MADH molecule. Similar changes were found for cyclopropylamine substituted MADH (Chen L, Mathews FS unpublished) and for the TV-MADH derivatives (Huizinga WE et al. 1992).

CONCLUSIONS

The structures of a binary complex between PD-MADH and amicyanin and of a ternary complex between them and cytochrome c_{551i} have been determined. MADH and amicyanin associate so that the exposed edges of Trp 108 of TTQ and of the His 95 ligand of copper are juxtaposed. Amicyanin and cytochrome c_{551i} associate so that one edge of the β-clam shell of amicyanin is in contact with a chain segment of the cytochrome close to the heme propionates. The distance from the catalytically active quinone oxygen of TTQ to the copper is 16.8 Å and from the copper to the iron is 24.8 Å, respectively. This study provides the first detailed analysis of a crystalline complex of 3 sequential components of an electron transport chain and the first description of an acidic c_L class cytochrome. The structures of four forms of WA-MADH are very similar to each other, even though two are of the native enzyme at two extremes of pH and the other two are inhibited forms of the enzyme. The amicyanin binding site of PD-MADH appears to be maintained in WA-MADH despite its absence from the organism and probable substitution by a cytochrome c_{552}. The relative tilts of the two tryptophan rings of TTQ appear to increase by about 10° to 15° from the native value of about 45° when inhibitor is bound to the enzyme. This conformational change in TTQ may be important for the electron transfer step of the reaction sequence.

Acknowledgements. This work has been supported by NSF Grant No. MCB-9119789 (F.S.M.) and by NSF Grant No. MCB-9206952 (W.S.M.) and by the Department of Veterans Affairs (W.S.M).

References

Chandrasekar R, Clapper MH (1986) J. Biol. Chem. 261, 3616-3619.
Chen L et al. (1992a) Proteins 14, 288-299.
Chen L et al.(1992b) Biochemistry 31, 4959-4964.
Chen L et al.(1994) Science 264, 86-90.
Chistoserdov AY et al. (1994) J. Bacteriol. 176, 4073-4080.
De Beer R et al. (1980) Biochim. Biophys. Acta 662, 370-374.

Durley RCE et al. (1993) Protein Science 2, 739-752.

Farver O et al. (1982) Biochemistry 21, 3356-3361.

Huizinga EG et al. (1992) Biochemistry 31, 9789-9795.

Husain M, Davidson VL (1985). J. Biol. Chem. 260, 14626-14629.

Husain M, Davidson VL (1986) J. Biol. Chem. 261, 8577-8580.

Kenny WC, McIntire WM (1983) Biochemistry 22, 3858-3868.

Kuusk V, McIntire WM (1994) J. Biol. Chem. 269, 2616-26143.

McIntire WS et al. (1991) Science 252, 817-824.

Vellieux FMD et al. (1989) EMBO J. 8, 2171-2178.

The biochemistry of methanol dehydrogenase

Christopher Anthony and Simon L. Dales
Biochemistry Department, University of Southampton, UK

INTRODUCTION

Methanol dehydrogenase (MDH) is responsible for oxidation of methanol to formaldehyde in almost all bacteria growing aerobically on methane or methanol; it is a periplasmic quinoprotein which passes its electrons to a specific c-type cytochrome which is usually called cytochrome c_L (Anthony, 1986, 1992, 1993, Goodwin and Anthony, 1995). Our understanding of this enzyme has increased dramatically in the last 3 years with publication of the 3-dimensional structures of the enzyme from *Methylotrophus* W3A1 (at 2.6 Å) (Xia *et al.*, 1992; White *et al.*, 1993) and from *Methylobacterium extorquens* (at 1.94Å) (Anthony *et al.*, 1994; Ghosh *et al.*, 1995). The system considered in this brief review will be the MDH and cytochrome c_L from *M. extorquens*; in particular, the following questions will be considered: what does the structure tell us about the mechanisms of methanol oxidation, and electron transfer between the two proteins, and the initial interaction ('docking') between MDH and cytochrome c_L.

RESULTS

The mechanism of methanol dehydrogenase. All the evidence is consistent with reduction by methanol of the prosthetic group (PQQ) to the quinol (PQQH$_2$), followed by 2 sequential single electron transfers to the cytochrome c_L; during this process the semiquinone is produced and it is this free radical form that is usually isolated. Before the structure of MDH was determined, it had been proposed that the very reactive C-5 carbonyl of PQQ, together with a basic group, is likely to feature in the mechanism (Frank *et al.*, 1989), as is the recently demonstrated Ca^{2+} ion, shown to be essential for activity, and located near the PQQ in the active site (Richardson and Anthony, 1992).

Figure 1 shows the active site of MDH from *M. extorquens* and Figure 2 illustrates a mechanism, involving Asp$_{303}$ acting as a base, initiating the reaction by abstraction of a proton from the alcohol substrate. In this mechanism the Ca^{2+} is given a role in addition to a structural role in maintaining PQQ in an active configuration; it is proposed that the Ca^{2+} acts as a Lewis acid by way of its coordination to the C-5 carbonyl oxygen of PQQ thus providing the electrophilic C-5 for attack by the oxyanion. It is also possible that the Ca^{2+} ion coordinates to the substrate oxygen atom. This hemiketal mechanism is based on that previously proposed by Frank and colleagues

M. E. Lidstrom and F. R. Tabita (eds.), Microbial Growth on C$_1$ Compounds, 213–219.
© 1996 *Kluwer Academic Publishers. Printed in the Netherlands.*

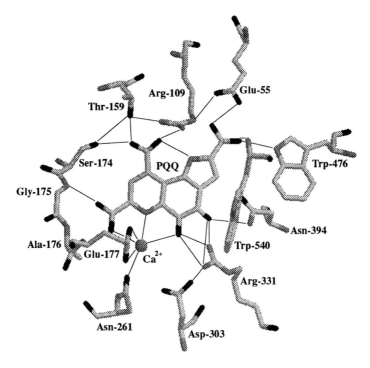

Figure 1 The coordination of the Ca^{2+} and the PQQ at the active site of methanol dehydrogenase. In addition to these equatorial interactions PQQ is 'sandwiched' between the indole ring of Trp-243 and the disulphide ring made up of Cys-103/104 (see Figure 4).

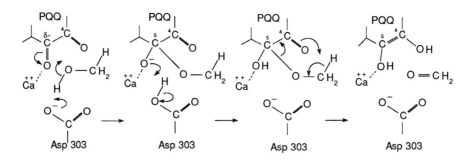

Figure 2 MDH mechanism involving a hemiketal intermediate. It is suggested that the Ca^{2+} acts as a Lewis acid, and the reaction is initiated by proton abstraction by a base (possibly Asp-303); this is followed by attack by the oxyanion on the electrophilic C-5, leading to formation of the hemiketal intermediate as proposed by Frank *et al.* (1989).

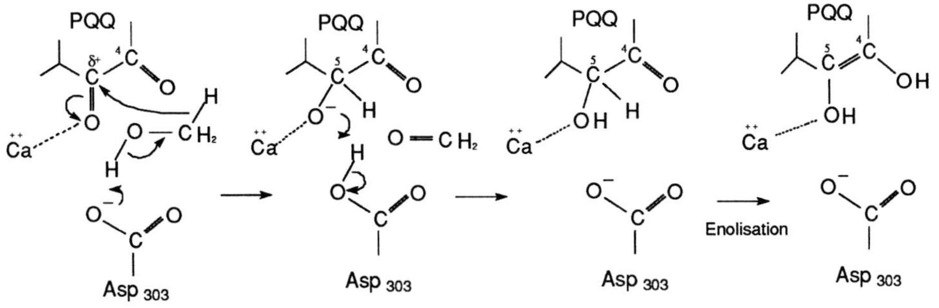

Figure 3 MDH mechanism involving an acid/base-catalysed hydride transfer. The initiation of the reaction is similar to that in Figure 4 but the electrophilic C-5 is attacked by a hydride from the substrate rather than by the oxyanion.

(Frank *et al.*, 1989); an alternative mechanism might involve an acid/base-catalysed hydride transfer mechanism which is similar except that the electrophilic C-5 would be attacked by a hydride and no hemiketal derivative would be formed (Figure 3) (Anthony, 1993; Anthony *et al.*, 1994).

Is the novel disulphide ring structure in the active site involved in electron transfer? The PQQ prosthetic group lies within a hydrophobic active site chamber, the floor being formed by Trp_{243}, whose indole group is parallel to, and in contact with, the planar ring system of PQQ. The ceiling of the chamber is formed by a novel structure arising from a disulphide bridge between adjacent cysteine residues ($Cys_{103-104}$). These are bonded by an unusual non-planar *trans* peptide bond within a strained 8-membered ring (Ghosh *et al.*, 1995). Figure 4 shows the PQQ 'sandwiched' between Trp-243 and this novel ring structure. It also shows the out-of-plane carbonyl at C-4, consistent with the predicted free radical semiquinone form of the prosthetic group. The rarity of disulphide bridges between adjacent cysteines in proteins suggests that this novel ring structure in MDH must serve an important biological function. This conclusion is supported by the observation that reduction of MDH with dithiothreitol led to loss of all activity with cytochrome c_L, whereas activity with an artificial electron acceptor (phenazine ethosulphate, PES) was retained, indicating that the ring structure may have particular importance in electron transport from $PQQH_2$ to cytochrome c_L (Blake *et al.*, 1994). However, we have now demonstrated that the reason reduced enzyme appears to remain active with PES is because the dye re-oxidises the adjacent thiols back to the original disulphide bridge (Avezoux *et al.*, 1995). It is not possible, therefore to determine whether or not the novel disulphide is necessary for activity when measured with PES. We have subsequently shown that no free thiols are ever detected during the reaction cycle with cytochrome c_L. Furthermore, carboxymethylation of the thiols produced by reduction of the disulphide ring led to formation of active enzyme (Table 1). We have concluded from these experiments that the disulphide ring in the active site

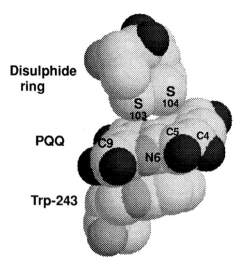

Disulphide ring

PQQ

Trp-243

S 103 S 104

C9 C5 C4

N6

Figure 4 The relationship between PQQ and the disulphide ring structure in the active site. The Ca^{2+} is coordinated between the C-9 carboxylate, the N-6 of the PQQ ring and the carbonyl oxygen at C-5.

of MDH does not function as a redox component of the reaction. We have confirmed that the inactivation that occurs on reduction is not a result of major structural change or to modification of the prosthetic group, and experiments with MDH isolated from a *mxaA* mutant have shown that it has no special function in the process of Ca^{2+} incorporation into the active site (Avezoux *et al.*, 1995). An alternative explanation is that the disulphide ring merely acts to hold the prosthetic group in place in the active site; cleavage of the ring by reduction might thus displace the PQQ into an inactive configuration, which is then reversed by carboxymethylation.

The disulphide ring is not an absolute requirement for PQQ-containing enzymes. Sequence analysis and molecular modelling have confirmed that the disulphide ring and the active site region of the quinohaemoprotein alcohol dehydrogenase of acetic acid bacteria is very similar to that of MDH (Cozier *et al.*, 1995). By contrast, the adjacent cysteines are not present in the membrane glucose dehydrogenase, whose electron acceptor is membrane ubiquinone. This would seem to imply that the disulphide does have a direct role in electron transfer, but there is an alternative explanation. In native MDH the PQQ is always found in the semiquinone form, consistent with its role in mediating between the two electron oxidation of methanol and the single electron reduction of cytochrome c_L (Frank *et al.*, 1988; Dijkstra *et al.*, 1989). By contrast, the process in GDH is a two electron transfer from glucose to ubiquinone (Matsushita *et al.*, 1989). Although the semiquinone might be involved as an intermediate, it has never been observed during the reaction. It is possible therefore that the disulphide ring in the active site of MDH functions in the stabilisation (or protection) of the free radical form of PQQ. In this respect the ring would be acting in a manner analogous to the

Table 1 The activity of methanol dehydrogenase after reduction, followed by carboxymethylation, or oxidation by air or PES

MDH was reduced with dithiothreitol (DTT). It was subsequently oxidised in air for 1h, or carboxymethylated with iodoacetate. A sample was incubated in 50 mM-Tris buffer (pH 8.0), PES (27.5 mM) was added and MDH immediately separated from the incubation mixture by rapid gel filtration. The total time between addition of PES and separation from PES was about 1 minute. Specific activity is rate of reduction of PES or cytochrome c_L (nmol min^{-1} mg^{-1}). This Table is taken from Avezoux *et al.* (1995).

Treatment of methanol dehydrogenase	Specific activity		Thiols per tetramer
	PES	Cytc_L	
None	356	32.8	0.02
Incubation with iodoacetate	342	32.3	0.05
Reduction with DTT (to MDH$_{red}$)	352	3.7	3.97
Carboxymethylation of MDH$_{red}$ (to MDH$_{cm}$)	339	30.1	0.05
Incubation of MDH$_{cm}$ with DTT	354	33.0	0.07
Oxidation of MDH$_{red}$ in air	350	30.7	0.39
Oxidation of MDH$_{red}$ with PES (to MDH$_{PES}$)	349	16.7	0.26
Incubation of MDH$_{PES}$ with iodoacetate	310	17.3	0.37

tryptophans in the active sites of galactose oxidase (Ito *et al.*, 1994) and methylamine dehydrogenase (Chen *et al.*, 1992) which stabilise a tyrosine free radical and a tryptophan quinone free radical respectively. Reduction of the disulphide ring in MDH clearly diminishes the protection it provides; subsequent carboxymethylation presumably restores the protection. It is worth noting that in these three enzymes the proposed protecting groups are clearly visible in surface views of the active site, limiting access to the prosthetic group from the surface.

The interaction of MDH and cytochrome c_L. MDH and cytochrome c_L interact in the periplasm, which during most methylotrophic growth is likely to be at more or less neutral pH and at relatively low ionic strength. The affinity of MDH for cytochrome c_L is very high, the K_m for the cytochrome being 1-7 µM when tested in normal (low ionic strength) conditions. There is considerable evidence from work with a number of methylotrophs that the interactions between the two proteins are ionic in nature; and chemical modification studies have indicated that the interaction is by way of a small number of lysyl residues on MDH and carboxylates on cytochrome c_L (Chan and Anthony, 1991; Cox *et al.*, 1992; Anthony, 1992). It has been proposed that the well-established inhibition of methanol oxidation by EDTA (Anthony and Zatman, 1964) is a result of its binding to these lysyl residues, thus preventing the specific electrostatic interaction of MDH with cytochrome c_L (Chan and Anthony, 1992). However, it has recently been suggested that, in *Paracoccus denitrificans*, hydrophobic interactions between the two proteins are predominant (Harris and Davidson, 1993), and this is supported perhaps by the recently reported structure of MDH from *M. extorquens* (at 1.94Å resolution). This shows that the PQQ is buried within an internal chamber which communicates with the exterior of the protein by way of a hydrophobic funnel in the surface which is perhaps the most likely place for interaction with cytochrome c_L (Anthony *et al.*, 1994; Ghosh *et al.*, 1995). Because the X-ray structures of MDHs from different bacteria are very similar (Xia *et al.*, 1992; Ghosh *et al.*, 1995), it seems unlikely that MDH and cytochrome c_L interact exclusively by electrostatic interactions in *M. extorquens* and exclusively by hydrophobic interactions in *P. denitrificans*.

Our previous methods of studying the interaction of MDH and cytochrome c_L have depended upon measurements of the overall process, including initial 'docking' and subsequent electron transfer. We have therefore developed an alternative method for investigating the initial interaction by a method based on the decrease in the tryptophan fluorescence of MDH that occurs on binding to cytochrome c_L due to quenching by its haem prosthetic group (Dales and Anthony, 1995). The K_d for binding of cytochrome c_L measured by this method was 1.3 μM and raising the ionic strength to 0.05 increased the K_d about 20-fold. That the change in fluorescence depended upon binding, and not on the activity of MDH, was demonstrated by repeating this experiment with inactive MDH that lacked Ca^{2+}, purified from a mutant (*mxaA*). That the mechanism of action of EDTA is not as straightforward as previously suggested, however, is indicated by the unexpected observation that relatively high concentrations of EDTA (1 mM) had no effect on the initial binding of MDH to cytochrome c_L as shown by the fluorescence assay described above, whereas this concentration of EDTA completely inhibited electron transfer between methanol and the terminal electron acceptor.

We suggest that the initial 'docking' of MDH and cytochrome c_L is by way of ionic interactions between lysyl residues on its surface and carboxylate groups on the surface of cytochrome c_L. This interaction is not inhibited by EDTA which we suggest acts by binding to nearby lysyl residues, thus preventing movement of the 'docked' cytochrome to its optimal position for electron transfer, which probably involves interaction with the hydrophobic funnel in the surface of MDH (Dales and Anthony, 1995). The conclusion of Harris and Davidson (1993) that predominantly hydrophobic interactions are involved in the interaction of MDH and cytochrome c_L was surprising, but inevitable because all their measurements were made at a high ionic strength (about 0.3), at which ionic interactions are likely to be negligible. These authors measured a K_d value of 375μM at this ionic strength compared with the rate of 0.8μM measured by us when we first described this interaction in *P. denitrificans* (Long and Anthony, 1991). A further investigation using *P. denitrificans* has now confirmed that ionic strength does have some effect on this system, and this has led to Davidson and his colleagues to an apparently similar conclusion to ourselves; that after a non-optimal collision, there is a rearrangement of the proteins to produce the most efficient orientation for electron transfer (Harris *et al.*, 1994).

CONCLUSION

In summary, the determination of the structure of methanol dehydrogenase has provided confirmation of some previous suggestions about its function, but has raised even more questions. These include questions about its mechanism, the role of the novel disulphide ring at the active site and the interaction with its cytochrome electron acceptor.

REFERENCES

Anthony, C. & Zatman, L.J. (1964) Biochem. J. **92**, 614-621

Anthony, C. (1986) Adv. Microbial. Physiol. **27**, 113-210

Anthony, C. (1992) Biochim. Biophys. Acta **1099**, 1-15

Anthony, C. (1993) in *Principles and applications of quinoproteins* (Davidson, V.L. ed.), Methanol dehydrogenase in Gram-negative bacteria. pp. 17-45, Marcel Dekker, New York

Anthony, C., Ghosh, M. & Blake, C.C.F. (1994) Biochem. J. **304**, 665-674

Avezoux, A., Goodwin, M.G. & Anthony, C. (1995) Biochem. J. **307**, 735-741

Blake, C.C.F., Ghosh, M., Harlos, K., Avezoux, A. & Anthony, C. (1994) Nature Struct. Biol. **1**, 102-105

Chan, H.T.C. & Anthony, C. (1991) Biochem. J. **280**, 139-146

Chan, H.T.C. & Anthony, C. (1992) FEMS Microbiol. Lett. **96**, 231-234

Chen, L.Y., Mathews, F.S., Davidson, V.L., Huizinga, E.G., Vellieux, F.M.D. & Hol, W.G.J. (1992) Protein-Struct. Funct. Genet. **14**, 288-299

Cox, J.M., Day, D.J. & Anthony, C. (1992) Biochim. Biophys. Acta **1119**, 7-106

Cozier, G.E., Giles, I.G. & Anthony, C. (1995) Biochem. J. **307**, 375-379

Dales, S.L. & Anthony, C. (1995) Biochem. J. In Press

Dijkstra, M., Frank, J. & Duine, J.A. (1989) Biochem. J. **257**, 87-94

Frank, J., Dijkstra, M., Duine, A.J. & Balny, C. (1988) Eur. J. Biochem. **174**, 331-338

Frank, J., van Krimpen, S.H., Verwiel, P.E.J., Jongejan, J.A., Mulder, A.C. & Duine, J.A. (1989) Eur. J. Biochem. **184**, 187-195

Ghosh, M., Anthony, C., Harlos, K., Goodwin, M.G. & Blake, C.C.F. (1995) Structure **3**, 177-187

Goodwin, P.M. & Anthony, C. (1995) Microbiology **141**, 1051-1064

Harris, T.K. & Davidson, V.L. (1993) Biochemistry **32**, 14145-14150

Harris, T.K., Davidson, V.L., Chen, L.Y., Mathews, F.S. & Xia, Z.X. (1994) Biochemistry **33**, 12600-12608

Ito, N., Phillips, S.E.V., Yadav, K.D.S. & Knowles, P.F. (1994) J. Mol. Biol. **238**, 794-814

Long, A.R. & Anthony, C. (1991) J. Gen. Microbiol. **137**, 415-425

Matsushita, K., Shinagawa, E., Adachi, O. & Ameyama, M. (1989) J. Biochem. **105**, 633-637

Richardson, I.W. & Anthony, C. (1992) Biochem. J. **287**, 709-715

White, S., Boyd, G., Mathews, F.S., Xia, Z.X., Dai, W.W., Zhang, Y.F. & Davidson, V.L. (1993) Biochemistry **32**, 12955-12958

Xia, Z.X., Dai, W.W., Xiong, J.P., Hao, Z.P., Davidson, V.L., White, S. & Mathews, F.S. (1992) J. Biol. Chem. **267**, 22289-22297

Methanol dehydrogenase structure.

Z.-X. Xia, W.-W. Dai, Y.N. He, S.A. White[1], G.D. Boyd[1], F.S. Mathews[1] and V.L. Davidson[2]
Shanghai Institute of Organic Chemistry, Chinese Academy of Sciences, Shanghai 200032, China, [1] Department of Biochemistry, Washington University School of Medicine, St. Louis, MO 63110, U.S.A., [2] Department of Biochemistry, University of Mississippi Medical Center, Jackson, MS 39216, U.S.A.

INTRODUCTION

Methanol dehydrogenase (MEDH, EC 1.1.99.8) is a soluble quinoprotein located in the periplasmic space of many methylotrophic bacteria (Anthony, 1986). The enzyme catalyzes the oxidation of methanol to formaldehyde, utilizing the single carbon compound as the sole source of carbon and energy. Other primary alcohols are also substrates of the enzymatic reaction $RCH_2OH ----- >RCHO +2H^+ + 2e^-$ (Duine, Jongejian, 1989). The natural electron acceptor is cytochrome c_L , an acidic c-type cytochrome with molecular mass of 17 Kda (Anthony, 1992).

MEDH is an H_2L_2 heterotetramer (Mr $\approx$140 KDa) with approximate subunit molecular masses of 62 and 8 KDa for heavy (H) and light (L) subunits, respectively (Nunn et al., 1989). Each tetramer contains two molecules of pyrroloquinoline quinone (PQQ) which are noncovalently bound to the polypeptide chains (Salisbury et al.,1989), and it was reported that there is approximately one Ca^{2+} in each tetramer which is essential for the enzymatic function (Richardson, Anthony, 1992).

We have determined the three-dimensional structures of MEDH from three different bacteria: *Methylophilus* W3A1, *Methylophilus methylotrophus* and *Paracoccus denitrificans*. The structures of MEDH from *M. methylotrophus* and *M.* W3A1 in crystal form A (Xia et al.,1989) were determined at 2.6 Å resolution (Xia et al.,1992) using multiple isomorphus replacement (MIR) method, based on the amino acid sequences from homologous proteins. The resulting structures represent the first examples of the "β-propeller' superbarrel with pseudo symmetry of 8-fold axis.

We then determined the amino acid sequence of MEDH from *M.* W3A1 by DNA sequencing method (G.D.Boyd, F.S.Mathews, unpublished), which contains 571 and 69 residues for H and L subunits respectively, giving sequence identity of 64% and 69% for H and L subunits, respectively, with those used for the MIR structure determination. On the basis of the DNA-derived amino acid sequence the structure of MEDH from *M.* W3A1 was refined at 2.4 Å resolution, which revealed the active site structure of the enzyme (White et al.,1993) and a conserved tryptophan packing motif in the 'β-propeller'

M. E. Lidstrom and F. R. Tabita (eds.), Microbial Growth on C$_1$ Compounds, 220–226.

superbarrel (Mathews et al.,1994).

Recently the x-ray diffraction data of MEDH from *M.* W3A1 in crystal form B (Z.-X. Xia et al. unpublished) have been collected at high resolution using synchrotron radiation. The structure has been solved by molecular replacement method using the refined structure of crystal form A as a search molecule, and the structure of the form B crystal has been refined at 1.9 Å resolution.

The crystals of MEDH from *M. methylotrophus* used for structure determination are isomorphus with the form A crystals of MEDH from *M.* W3A1; however, the amino acid sequence of its H subunit is unavailable at present, therefore, that of *M.* W3A1 MEDH was used instead, along with the L subunit sequence of MEDH from *M. methylotrophus* (Anthony et al., 1993), to refine the structure at 2.4 Å resolution.

We have also crystallized MEDH from *P. denitrificans* (W.-W.Dai et al., unpublished) and determined the structure by molecular replacement method using the structure of MEDH from *M.* W3A1 as a search molecule. The sequences of the two species are 64% identical (Harms, 1987; Spanning, 1991). While there is no deletion comparing the sequence of MEDH from *P. denitrificans* with that from *M.* W3A1, there are a total of 8 insertions as well as the C-terminal extensions of 20 and 14 residues for H and L subunits, respectively. In addition, the sequences of the 42 N-terminal residues of the H subunit are very different for the two species. The refinement at 2.5 Å resolution is in progress.

Here we present the three-dimensional structures of MEDH from the three methylotrophic bacteria. The emphasis is placed on the structural features of MEDH from *M.* W3A1, which are conserved in the two different crystal forms and in the structures from the other two species as well as in the recently reported structure of MEDH from *Methylobacterium extorquens* AM1 (Blake et al.,1994;Ghosh et al., 1995). The structures of MEDH from *Methylobacterium organophilum* XX and *Methylosinus trichosporium* OB3b (D.H. Ohlendorf, personal communication) have also been determined.

RESULTS

Crystallographic data The crystallographic data of MEDH from the three bacteria are listed in Table 1.

Table 1 Crystallographic data

Source	(Crystal form)	Crystal system	Space group	Cell parameters a(Å)	b(Å)	c(Å)	β(°)	resolution (Å)	R-factor (%)
M. W3A1	(A)	monoclinic	$P2_1$	124.70	62.87	85.02	93.63	2.4	15.3
	(B)	monoclinic	$P2_1$	98.12	69.74	109.84	110.29	1.9	18.1
M. methylotrophus		monoclinic	$P2_1$	124.87	62.68	85.01	93.40	2.4	21.0*
P. denitrificans		orthorhombic	$P2_12_12$	113.34	122.28	107.36		2.5	22.2**

* H subunit sequence of MEDH from *M.* W3A1 was used and no water is included in the model.
** C-terminal extension and refinement is in progress and no water is included in the model.

Overall structure and folding topology The dimensions of the H_2L_2 tetramer are about 110 x 60 x 60 Å with the non-crystallographic 2-fold axis of symmetry relating a pair of HL dimers. The two H subunits come into contact across the non-crystallographic 2-fold axis with the interface of about 1700 $Å^2$ in area. In contrast, the two L subunits make no

222

contact with each other and each L subunit interacts only with the H subunit within an HL dimer. (Fig. 1)

The L subunit is quite unusual since it does not form an independent globular domain but instead stretches across the surface of the H subunit, folding into the shape of the letter 'j'. The N-terminal segment (about 32 residues) of the L subunit is irregular, containing a few β-turns, and forms the hook part of the 'j', whereas the C-terminal segment contains an 8 turn α-helix forming the stem part of the 'j'. The 12 C-terminal residues show higher temperature factors

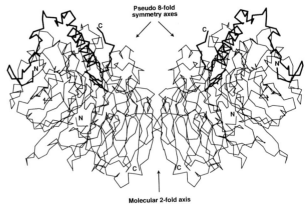

Fig.1 α-carbon diagram of the MEDH molecule.
(The H subunits are drawn in thin lines ; the L subunits and PQQ are drawn in thick lines. All the N- and C- termini are indicated).

than those of the rest of the structure in form A crystals of MEDH from *M.* W3A1, indicating that they are more flexible, and in form B crystals they are probably disordered.

The major structural feature of the H subunit is the 'β-propeller' superbarrel. The H subunit contains 8 antiparallel twisted β-sheets, or 'blades', each consisting of 4 β-strands arranged topologically like the letter 'W', and the 8 'W's are arranged circularly, forming the main disc-shaped body of the subunit (Fig. 2). The 8 'W's are very similar to one another in the folding topology, related by a pseudo 8-fold axis of symmetry, which is inclined by about 45° from the non-crystallographic 2-fold axis. In each W-like-β-sheet, the 4 β-strands running from N- to C- termini (βna, βnb, βnc and βnd, n=1-8) are arranged progressively from the central pseudo 8-fold axis. W8 is composed of 3 inner strands ($\beta 8a, \beta 8b$ and $\beta 8c$) from the C-terminal segment and one outer strand ($\beta 8d$) from the N-terminal segment. While one 6-fold and two 7-fold 'β-propeller' superbarrels

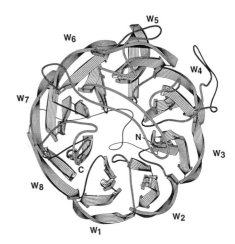

Fig.2 Ribbon diagram of the H subunit
(Only the secondary structures within the 8 'W's are shown. The N- and C- termini are indicated.)

were previously observed in three other protein structures (Varghese et al., 1983; Vellieux et al.,1989; Chen et al.,1992; Ito et al., 1991), MEDH structures are the first examples of the 8-fold 'β-propeller' superbarrel. The 8 'W' disc is located at the bottom of the H subunit, and above the disc there are some additional secondary structures.

Each HL dimer of MEDH from *M*.W3A1 contains 8 cysteines which form 4 intra-chain disulfide bridges, three in the H subunit (Cys103-Cys104, Cys144-Cys167 and Cys379-Cys408), and one in the L subunit (Cys6-Cys12). Six of the 8 cysteins are conserved in all of the MEDH with known sequences, however, Cys144 and Cys167 are replaced by serines in MEDH from *P. denitrificans* and two other bacteria.

In each HL dimer, four peptides are found to be in *cis*-configuration instead of normal *trans*-peptides, i.e. Pro72, Pro258, Trp270 and Pro380, all of which are conserved along with their preceding residues. *cis*-Trp270 is a non-proline residue, which is unusual, and is stabilized by two hydrogen bonds involving the main-chain oxygen and nitrogen atoms of the *cis*-peptide.

Conserved tryptophan packing motif In 7 of the 8 'W's, there is a tryptophan located in the middle of the outer strand of each β-sheet, i.e. strand βnd (n=1,2,3,4,6,7,8). The only exception is W5 which contains Val340 instead of tryptophan at the corresponding position of β5d. Six of the 7 tryptophans are conserved in 4 MEDHs, whereas Trp286 in W4 is replaced by phenylalanine in the other 3 MEDH sequences. Val340 in W5 is not conserved,and the corresponding residues in the other 3 MEDHs are serines. The positions of the 7 tryptophans and Val340 obey the pseudo symmetry of 8-fold axis.

The 7 tryptophans in βnd (n=1,2,3,4,6,7,8) are involved in three different types of interactions (Table 2) which help stabilize the 8-fold 'β-propeller' assembly:

Table 2   Three types of interactions involving 7 tryptophans

Trp	type (1) O of Trp to N of res.	type (2) NE1 of Trp to O of res.	type (3) indole ring of Trp to peptide
Trp 88 (W1, β1d)	**Ala** 77 (W1, β1c)	Ala 132 (W2, β2c)	Thr 134---**Gly** 135 (W2, β2d)
Trp 139 (W2, β2d)	**Ala** 129 (W2, β2c)	Leu 183 (W3, β3c)	Thr 185---**Gly** 186 (W3, β3d)
Trp 190 (W3, β3d)	**Ala** 180 (W3, β3c)	Asp 280 (W4, β4c)	Thr 281---**Gly** 282 (W4, β4d)
Trp 286 (W4, β4d)	Gly 276 (W4, β4c)	Arg 333 (W5, β5c)	Asn 335---**Gly** 336 (W5, β5d)
Trp 458 (W6, β6d)	**Ala** 448 (W6, β6c)	Asn 492 (W7, β7c)	Asp 494---**Gly** 495 (W7, β7d)
Trp 499 (W7, β7d)	**Ala** 489 (W7, β7c)	Val 40 ---	--- --- --- ---
Trp 44 (W8, β8d)	Val 568 (W8, β8c)	--- --- ---	--- --- --- ---

(1) The carbonyl oxygen of each tryptophan in βnd hydrogen bonds to the backbone nitrogen of a residue in the adjacent strand βnc in the same 'W' to form one of the main-chain hydrogen bonds of the β-sheet.. The latter residues in βnc are limited to small ones by steric hindrance, i.e. 5 alanines, one glycine and one valine, all of which are conserved in 4 MEDHs with known H subunit sequences.

(2) In 6 of the 7 tryptophans in βnd (n=1,2,3,4,6,7), atom NE1 of each side chain forms a hydrogen bond to the main-chain oxygen of a residue in the adjacent β-sheet Wn+1, and for 5 of the 6 (n=1,2,3,4,6) the latter residue is located on the β(n+1)c side of the[β(n+1)c --- β(n+1)d] turn.

(3) Each indole ring in 5 of the 7 tryptophans in βnd (n=1,2,3,4,6) is parallel to and in van der Waals contacts with a peptide plane in Wn+1,which is located on the β(n+1)d side of the [β (n+1)c --- β(n+1)d] turn . The second residues linked by the peptides are always glycines, also limited by steric hindrence. These glycines are conserved except Gly186 in W3 which is replaced by phenylalanine in MEDH from *M.organophilus* XX.

Trp499 in W7 and Trp 44 in W8 are two exceptions to the second and third types of interactions. For these two tryptophans, the Wn+1 are W8 and W1 respectively, and neither (β8c-β8d) or (β1c-β1d) turn exists; β8c and β8d are from C- and N-terminal segments respectively while the segment connecting β1c and β1d forms a very large β-buldge instead of a regular turn. In addition, the orientation of the indole ring of Trp499 is completely different from those of the other 6 tryptophans.

Active site structure   Each HL dimer contains one PQQ and one Ca⁺², which are located in the funnel-shaped central channel at the top of the 8'W'disc, with the normal of the PQQ ring plane inclined form the pseudo 8-fold axis by about 20°. The PQQ ring is sandwiched between the indole ring of Trp237 and the disulfide bridge Cys 103 - Cys 104 , making van der Waals contacts with both (Fig.3). The structure refined at high resolution shows that O4 is out of PQQ ring plane by about 0.9Å, toward the Trp237 side, suggesting the reduced form of PQQ.

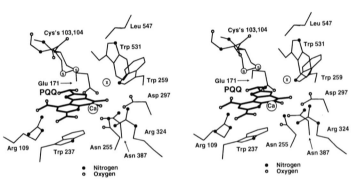

Fig. 3 Stereoscopic view of the active site of MEDH
(The circle labeled X represents the solvent molecule bound to O5 of PQQ; this site in the hydrophobic pocket may serve to bind substrate or product.)

The disulfide bridge Cys103-Cys104 is very unusual, formed by two cysteines which are adjacent   in the amino acid sequence. Although such a disulfide bridge was predicted from the energy calculation to possess a cis-configuration for the linking peptide (Chandrasekaran, Balasubramanian,1969), the refined structures of MEDH from *M.*W3A1 in both crystal forms show that the linking peptide is in a *trans*- and non-planar configuration indicated by the dihedral angle ω of -152°. The main-chain dihedral angles φ and ψ of the two cysteines are normal, but Asp105, the residue adjacent to the disulfide bridge, is forced to be located in an unfavorable region of the Ramachandran plot. The dihedral angle χ1 of Cys103 is 70° which is very different from the preferred value -60° (Richardson, 1981) while the other side-chain dihedral angles are close to the normal values. The unusaual values of $\chi 1$ of Cys103 as well as of ϕ and ψ of Asp105 help stabilize the disulfide bridge Cys103-Cys104 in the distorted *trans-* configuration for the linking peptide.

The six-coordinate Ca^{2+} is bound to O5, N6 and O7A of PQQ as well as the 3 side-chain oxygen atoms of Glu171 and Asn255, at an average distance of 2.66Å. Thirteen amino acid residues are involved in the interactions with O4, O5 and the 3 carboxyl groups of PQQ through hydrogen bonds and salt bridges (Fig.4).

Two mechanisms have been proposed for the reaction catalyzed by MEDH (Anthony, 1994), both requiring a base to initiate the reaction by abstracting a proton from the substrate. Ca^{2+} acts as a Lewis acid by its coordination to O5, providing the electrophilic C5.The substrate binding site was proposed to be

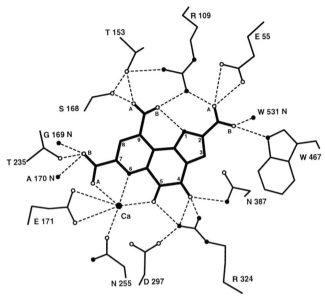

Fig. 4   Schematic diagram of the active site of MEDH.
(The hydrogen bonds and the coordination of the Ca^{2+} are indicated.
Solid and open circles denote nitrogen and oxygen atoms, respectively.)

close to O5 (Duine et al.,1987). In the structure of MEDH from *M.*W3A1 (crystal form A) refined at 2.4 Å resolution, a piece of electron density was identified to be a water molecule (White et al. 1993), which hydrogen bonds to O5 and is located above the plane of PQQ . This site is surrounded by the hydrophobic residues Trp259 and Trp531 as well as the disulfide bridge; since it is accessible by the bulk solvent along the axis of the superbarrel, it might be the best candidate for substrate binding. It also hydrogrn bonds to both side-chains of Glu171 and Asp297; the latter is 4Å from PQQ, but its two carboxyl oxygen atoms interact with both side-chains of Arg324 and Asn255 respectively. Either Glu171 or Asp297 might play the role of the catalytic base. When the resolution was extended to 1.9Å for the form B crystals, which were grown in the presence of 3-5mM methanol, the "omit"map, calculated from a model with this water molecule omitted, showed a large piece of electron density at the omitted water site in each H subunit. This piece of electron density is elongated, stretching from O5 toward the disulfide bridge, approximately along the normal of the PQQ ring plane. It seems reasonable for this site to be interpreted as methanol or formaldehyde, and it is more likely to be the latter, implied by the reduced form of PQQ.

CONCLUSIONS

The three-dimensional structures of MEDH from bacteria *Methylophilus* W3A1, *Methylophilus methylotrophus* and *Paracoccus denitrificans* have been determined and that from *M.* W3A1 has been refined at high resolution. The two HL dimers are related to each other by the non-crystallographic 2-fold axis of symmetry to form the H_2L_2 heterotetamer. The structures represent the first examples of 'β-propeller' superbarrel with pseudo symmetry of 8-fold axis. There is a conserved tryptophan packing motif in the superbarrel, and 7 tryptophans are involved in three different types of interactions. PQQ is bound non-covalently in the funnel-shaped central channel, and it is sandwiched between a tryptophan indole ring and a novel disulfide bridge which is formed by two adjacent cysteines and linked by the peptide in a distorted *trans-* configuration. The six-coordinate Ca^{2+} is bound to O5, N6 and O7A of PQQ as well as three side-chain oxygen atoms provided by the protein. A substrate binding site is proposed.

REFERENCES

Anthony C (1992) Biochem. Biophys. Acta 1099, 1-15.

Anthony C (1986) Adv. Microb. Physiol. 27, 113-210.

Anthony C et al. (1993) In Murrell, Kelly Eds., Microbial Grouwth on C1 Compounds, 221-233, Intercept Ltd., Andover, Hampsher, U.K.

Anthony C et al. (1994) Biochem. J. 304, 665-674.

Blake C C F (1994) Nat. Struct. Biol. 1, 102-106.

Chandrasekaran R, Balasubramanian R (1969) Biochim. Biophys. Acta 188, 1-9.

Chen L et al. (1992) Proteins 14, 288-299.

Duine J A et al. (1987) Adv. Enzym. 59, 169-212.

Duine J A , Jongejian J A (1989) Annu. Rev. Biochem. 58, 403-426.

Ghosh M et al. (1995) Struct. 3, 177-187.

Harms N et al. (1987) J. Bacterial. 169, 3969-3975.

Ito N P et al.(1991) Nature 350, 87-90.

Mathews F S et al. (1994) The 3rd Symposium on PQQ and Quinoproteins, Capri.

Muzin A G (1992) Proteins,14, 191-201.

Nunn D N et al. (1989) Biochem. J. 260, 857-862.

Richardson J S (1981) In Antinsen C B Eds., Adv. Protein Chem. 34, 167-339, Acaemic Press Inc. New York.

Richardson I W, Anthony C (1992) Biochem. J. 287,709-715.

Salisbury S A et al. (1979 Nature (London) 280, 843-844.

Van Spanning R J M et al. (1991) J. Bacterial. 173, 6948-6961.

Varghese J N et al. (1983) Nature 303, 35-40.

Vellieux et al. (1989) EMBO J. 8, 2171-2178.

White S et al. (1993) Biochem. 32, 12955-12958.

Xia Z.-X. et al. (1989) FEBS Lett. 258, 175-176.

Xia Z.-X. et al. (1992) J. Biol. Chem. 267, 22289-22297.

Production of L-Lysine and Some Other Amino Acids by Mutants of *B. methanolicus*

R.S. Hanson, R. Dillingham and P. Olson, Department of Microbiology, University of Minnesota Medical School, Minneapolis, MN
G.H. Lee, D. Cue, F.J Schendel, C. Bremmon, and M.C. Flickinger Department of Biochemistry and Biological Process Technology Institute, College of Biological Sciences, University of Minnesota, St. Paul, MN.

INTRODUCTION:
Bacillus methanolicus is is a Gram-positive, endosporeforming, facultative methylotroph with an optimal growth temperature of 50-53 º C . This bacterium employs the efficient ribulose-monophosphate pathway for formaldehyde assimilation and possesses a novel nicotinamide adenine nucleotide-dependent methanol dehydrogenase (Dijkhuisen *et al*, 1988) rather than the cytochrome linked MDH found in gram-negative methylotrophs. Several isolates similar to those isolated by us (Schendel *et al*, 1990) have been described and have been shown to be related to *Bacillus frimus* and *B. azotoformans* (Arfman, Dijhuizen, 1993). . We have previously shown that mutants of *B. methanolicus* strains MGA3 and NOA2 which lacked homoserine dehydrogenase (methionine, threonine auxotrophs) and were also resistant to the lysine analogue, S-(2-aminoethyl)-L-cysteine (AEC) produced significant amounts of lysine when grown in a mineral salts medium under threonine limited conditions in fed-batch fermentations (Schendel *et al*, 1990).

Lysine is used to supplement feeds for poultry, swine and other livestock that contain cereal grains which contain limited quantities of this amino acid. Fermentation processes that employ strains of *Corynebacterium glutamicum* or *Brevibacterium lactofermentum* with starch hydrolysate or molasses as feedstocks have been used as the major methods for L-lysine production (Tosaka *et al*, 1983) Methanol is available in pure form, it is highly soluble in water, methanol solutions are not explosive and residual methanol can be readily removed from products after fermentations are completed. Methanol is also relatively inexpensive, the price is stable and it is easily stored and transported. Therefore, methanol has potential as a good substrate for the production of fermentation products (Lentin, Neikus, 1987).

Aspartokinase isoenzymes catalyze the first reaction required for the common pathway involved in the biosynthesis of lysine, threonine, methionine and isoleucine. (Figure 1) Previous studies have indicated that *B. methanolicus* MGA3 possesses three isoenzymes (Schendel, Flickinger, 1992).. One is inhibited by lysine alone (aspartokinase II), another is subject to concerted feedback inhibition by threonine and lysine (aspartokinase III) and a third is believed to be inhibited by diaminopimelic acid. The *lys*E gene which encodes aspartokinase II has been cloned and sequenced, and purified from *E. coli* clones that express the gene (Schendel, Flickinger, 1992). The gene product was shown to be similar in structure and sensitivity to lysine inhibition to the aspartokinase II from *B. subtilis*. Aspartokinase III accounts for the major portion of the activity of aspartokinases (approximately 50%) while aspartokinase II accounts for about 40% of the total activity.

The *lys* A gene which encodes diaminopimelate (DAP) decarboxylase, the enzyme that catalyzes the final reaction in lysine biosynthesis has also been cloned from *B. methanolicus* MGA3 (Mills, Flickinger, 1993). The gene was found to be 57% similar in nucleotide sequence to the *lys* A gene of *B. subtilis*. DAP decarboxylase was shown to be competetively inhibited by lysine ($(K_i=0.93mM)$.

M. E. Lidstrom and F. R. Tabita (eds.), Microbial Growth on C₁ Compounds, 227–236.

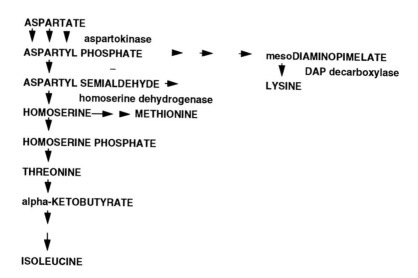

Figure 1. Biosynthesis of the amino acids of the aspartate family. Intermediates and endpreoducts of the pathway are shown in capital letters while the enzymes relevant to this manuscript are shown in lower case letters.

RESULTS

Several new mutants of *B. methanolicus* strains MGA3 and NOA2 have been isolated in attempts to increase lysine production since the first report of lysine excretion by mutants of these bacteria (Schendel *et al*, 1990).

It was observed that some lysine producing strains also produced significant amounts of diaminopimelic acid (DAP), alanine and glutamate (Figures 2 and 3). Alanine has been shown to inhibit growth of some mutants of *B. methanolicus* at levels of 1 g/liter or more. Growth of wild type strains was also inhibited by lysine at levels exceeding 10 g/liter. Lysine production in all strains was inhibited by excess threonine in the growth medium. Wild type strains and mutants derived from them lysed and rapidly lost viability during the stationary phase of growth in fermenters and lysine production ceased due to the presence of growth inhibitors or the lack of nutrients in high cell density fermentations. Therefore, selection strategies were designed to select for mutants with more desirable properties. These properties included resitance to high levels of lysine, resistance to 10 g/liter of alanine and resistance to diaminobutyrate (DAB) an analogue of diaminopimelate that inhibits diaminopimelate decarboxylase. It was reasoned that mutants resistant to AEC in the presence of DAB may have elevated levels of DAP decarboxylase. A mutant of strain NOA2 which lacked homoserine dehydrogenase and produced significant lysine (NA29#PO3) was mutagenized with diepoxyoctane. Cultures resistant to increasingly higher levels of lysine, alanine and DAB were selected. One strain 13A52 was shown to produce more lysine in shaken flasks and fermenters than previous isolates. After mutagenesis of this strain, cells which grew in MVtm broth supplemented 40 g/liter of lysine, 2.5 g/liter of alanine, 10 g/liter of threonine and 1.0 g/liter of diaminobutyrate were selected. The MVtm medium is the minimal vitamins medium described by Schendel et al (1990) suplemented with 0.15 mM of each threonine and methionine. All media contained also 8 g/liter methanol as a carbon and energy source in batch cultures or in agar plates.

Samples of cultures which grew in this medium after mutagenesis were spread onto the surface of plates containing MVtm medium supplemented with 10 g/liter of threonine and 0.5 g/liter of diaminobutyrate. Filter paper discs (Whatmann #3, 1 cm. diameter) that were soaked with a solution of AEC (250 g/liter) were placed in the center of the of the plates. After incubation for 1-2 days at 50°C, colonies that grew closest to the discs were inoculated into MYtm broth and samples of the culture were spread onto MVtm agar supplemented with 10 g/liter of threonine 0.5 g/liter of diaminobutyrate. Filter paper discs saturated with 250 g/liter of dehydrolysine were placed in the center of the plates and colonies that grew nearest the discs were selected. Dehydrolysine has been shown to be effective for selection of lysine producing strains of *Methylobacillus glycogenes* (Motoyama *et al*, 1993). Several colonies were isolated by streaking from the colonies growing nearest the DHL discs onto MVtm agar supplemented with 10 g/liter of threonine and 0.1 g/liter of AEC. These isolates were grown in the MVtm medium and those that produced the highest concentrations of lysine were stored at -80°C in MV media plus 10% glycerol. Strain 13A52-8 was selected for further studies.

This strain, like wild type strains and others derived from it , lysed after 2 days growth in MVtm broth. A culture was treated with mitomycin C (0.25 mg/liter) for 7 hours at 50°C in MYtm media (MVtm media containing 0.25% yeast extract). The surviving cells were plated onto MYtm agar and clones were grown in MYtm broth. Cultures that survived for 6 days were plated onto MYtm agar and were stored at -80°C in 10% glycerol. One strain with atypical colony morphology was selected (strain 13A52-8A6). Unlike parent strains that formed colonies with indented dimples in the center, colonies of strain 13A52-8A6 were smooth over the entire surface and did not lyse or die during the stationary phase of growth.

After mutagenesis of strain 13A52-8A6 with NTG, new mutants that grew in MVtm broth with increasing levels of lysine, threonine, DAB and alanine were selected.. Subsequent isolates that were resistant to the highest concentrations of AEC in the presence of 10 g/liter of threonine and 0.5 g/liter of DAB were evaluated for lysine production and the selection process was repeated. A strain 13A52-8A66 was isolated. This strain was grown in the presence of 60 g/liter of lysine, 5 g/liter of alanine, 10g/liter of threonine and 0.5 g/liter of diaminobutyrate in MVtml broth (MVtm broth containing 0.15 mM leucine) . Growth was greatly stimulated by 0.15mM leucine. This isolate grew in the presence of 54 g/liter of lysine and 0.45 g/liter of AEC and produced as much as 37.5 g/liter of lysine in an 11 liter fed-batch fermentation (Figure 2). It is noteworthy that very little diaminopimelate was produced while a parent strain (13A52) produced as much as 12 g/liter of diaminopimelate (Figure 3) and 25 g/liter of lysine (Figure 2) in a similar fermentation run. Mutant 13A52-8A66 also produced 6-7 grams per liter of alanine, 15-17 grams of glutamate and 1 g/liter of valine (Figures 2 and 3). The lysine yields (lysine produced per gram dry wt.) varied from 0.25-3.0 during peak lysine production in strain 13A52-8A66 and was less than 1.0 for stain 13A52- (Figure 4). Lysine productivity of 13A52-8A66 was shown to be approximately 0.13 g lysine per liter per hour about twice that of strain 13A52-8 when rates of lysine production were most rapid in 1-liter batch fermenter runs although cell mass was less when strain 13A52-8A66 was used. In 11 liter fed batch fermentations, it has not been possible to obtain cell yields over 20 g/liter with strain 13A52-8A66. We believe a growth inhibitor is produced that prevents growth late in the fermentation. It has been shown in shaken flask cultures that valine, at concentrations exceeding 5mM, severely inhibited growth.

Aspartokinase acitivity was not significantly repressed by growth of strain 13A52-8A66 in rich media (trypticase soy broth) while these isoenzymes were severely repressed by growth of the wild type strain MGA3 in rich media (Table 1) Aspartokinase activity in extracts of strain 13A52-8A66 grown in a mineral salts medium was not inhibited as severely by lysine or by lysine+threonine + DAP as was activity in extracts of wild type cells (Table 1). The three amino acids in concert decreased aspartokinase activity by 90% in extracts of wild type cells but only about 39% in extracts of strain 13A52-8A66. Therefore, we conclude that the selection procedure resulted in enrichments for mutants that have altered expression of aspartokinase as well as one or more aspartokinase isoenzymes that are less sensitive to feedback inhibition than those present in wild type cells.

DISCUSSION

Development of methods to rapidly evaluate large numbers of mutants in shaken flasks that give reliable indications of lysine production in fed-batch fermentions has been a difficult problem to solve. Selection for resistance to lysine, alanine and diaminobutyrate followed by selection for resistance to continuously higher concentrations of AEC in the presence of 10 g/liter of threonine and 0.15 mM isoleucine, methionine and leucine appears to have been an effective means of selecting improved strains.

The best lysine producing strains of *B. methanolicus* selected were partially or completely dependent on leucine for growth. Patek *et al* (1994) have shown that some strains of *Corynebacterium glutamicum* bred for lysine production were leucine auxotrophs, and it has been suggested that mutations in leucine biosynthetic genes are linked to high lysine productivity. These authors have also shown that mutations in the *leu*A gene that is separated from *lys* C by an open reading frame cause increases in lysine production when leucine was growth limiting. The best lysine producing strains of *B. methanoilcus* have had the highest carbon conversion efficiencies and lysine production is inhibited less by threonine than in other strains. Lysine production of strain 13A52-8A66 is inhibited 55% by 6 g/liter of threonine in shaken flask trials. The aspartokinase activity, although not significantly repressed in rich media, remains partially sensitive to inhibtion by lysine and threonine.

Because DAP decarboxylase is competetively inhibited by lysine (Mills, Flickinger, 1993), it is expected that this enzyme will be severely inhibited by lysine when high concentrations accumulate within the cytoplasm of cells. Preliminary experiments have indicated that the intracellular lysine concentrations exceed the extracellular concentrations by a factor of at least 2 in 11-liter fermentation runs in which extracellular levels were 37 g/liter. This data indicates that for lysine production to continue, intracellular DAP levels should be high. Therefore, it is not surprizing to observe that most lysine producing strains also produced substantial amounts of DAP. We do not understand the mutation in strain 13A52-8A66 which permits the production of high levels of lysine without accumulation of DAP. Lysine excreting mutants of this bacterium that possess a feedback resistant aspartokinase also possess excretion systems driven by membrane potential or ATP hydrolysis and excrete lysine against a concentration gradient (Broer et al, 1992: Gutman et al 1992)). Selection for AEC resistance strains of *B. methanolicus* may may have selected for mutants that excreted or excluded AEC and therefore excreted lysine against a concentration gradient. This hypothesis remains to be tested.

The inability to achieve high cell mass (greater than 20 g/liter) in 11-liter fermentation runs with strain 13A52-8A66 has been an obstacle to higher lysine production. In previous runs with mutant 13A52 cell dry weights of 25 g/liter were achieved. It is possible that the accumulation of growth inhibitors cocurrently with lysine cause growth inhibition when cell densities exceed 20 g/liter. Valine is known to severely inhibit growth of strain 13A52-8A66 at concentrations of 5 mM Growth inhibition by valine can be overcome by isoleucine. It is possible that valine inhibits α-acetolactate synthase that is required for the synthesis of both isoleucine and valine as it does in

Eschericia coli K12 (E. Umbarger, 1961).

    Bacillus methanolicus offers several advantages over other methylotophs for production of amino acids. These include the presence of an NAD-linked methanol dehydrogenase that permits more adenosine triphosphate synthesis per mole of methanol oxidized than the cytochrome linked MDHs of gram negative methylotrophs. Auxotrophs and other mutants of B. methanolicus are easily isolated. This is in contrast to the gram-negative methylotrophs that have yielded few auxotrophic mutants in many attempts.

REFERENCES

Arfman, N. and L. Dijkhuizen (1993) In J.C. Murrell and D.P. Kelly (eds.) Microbial Growth on C1 Compounds. pp 267-264. Intercept, Andover U.K.

Broer, S., L. Eggeling and R. Kramer. (1992) Appl. Environ. Microbiol. **59**:316-321

Broer, S, B. Kramer (1990) J. Bacteriol. **172**:7241-7248.

Dijkhuizen, L. N. Arfman, M.M. Attwood, A.G. Brook, W. Harder and E.M. Watling. 1988. FEMS Microbiol. Lett. **52**:209-214

Gutmann, M.,C. Hoischen and R. Kramer (1992). Biochim. Biophys. Acta. 1112:115-123

Lentin, J.D. and H.G.D. Niekus (1987) In H.W. Van Verseveld and J.A. Duine (eds.) Microbial Growth on C1 Compounds. pp263-271. Martinus Nijhoff, Dordrecht, The Netherlands.

Mills, D.A. and M.C. Flickinger (1993). Appl. Environ. Microbiol. **59**:2927-2937.

Motoyama, H. H. Anazawa, R. Katsumata, K Araki, and S Teshiba (1993). Biosci. Biotech. Biochem. 57: 82-87.

Patek, M., K Krumbach, L. Eggeling, and H. Sahm (1994) Appl. Environ. Microbiol. 60: 133-140.

Sano, K, I Shiio (1970) Microbial production of L-lysine. J. Gen. Appl. Microbiol. **16**:373-391.

Schendel, F.J., C. E. Bremmon, M. C. Flickinger, M. Guettler, and R.S. Hanson. (1990) Appl. Environ. Microbiol. **56**:963-970.

Schendel F.J, Flickinger MC (1992). Appl. Environ. Microbiol. 58, 2806-2814.

Shiio, I., A. Yokota and S-i. Sugimoto. 1987. Agric. Biol. Chem. **51**:2485-2493.

Tosaka, O., H. Enei and Y. Hirose (1983). Trends Biotechnol 1:70-74.

Umbarger HE. (1961) In Cold Spring Harbor Symposium on Quantitative Biology, pp301-311, Long Island Biological Association, Cold Spring Harbor, New York

Table 1. SPECIFIC ACTIVITIES OF AND INHIBITION OF ASPARTOKINASE IN DIFFERENT STRAINS OF *Bacillus methanolicus*.

Strain	Growth medium[1]	Aspartokinase[2] (Specific Act.)	Inhibition of AK (Relative Activity)[3] in presence of		
			lys	lys+thr	lys+thr+DAP
MGA3	MVtml	36. +/- 11	0.5	0.15	0.11
MGA3	TSB	2 +/- 1	ND	ND	0.59
13A52-8A66	MVtml	33 +/- 11	0.85	0.72	0.69
13A52-8A66	TSB	23 +/- 3	ND	ND	0.92

1. The growth media used were the minimal vitamin (MV) medium (Schendell et al, 1988)supplemented with threonine, methionine and leucine (0.15 mM each) for the growth of strain 13A52-8A66 and Trypticase soy broth (TSB).

2. A unit of aspartokinase activity is reported as nmoles aspartate hydroxymate formed per minute. The specific activity is reported as units per mg. of protein in crude extracts.

3. The relative activity of aspartokinase in the presence of inhibitors (lysine, threonine and diaminopimelate) is calculated as the ratio of the specific activity in the presence of inhibitors / the specific activity in the absence of inhibitors. All inhibitors were added at a concentration of 5mM.

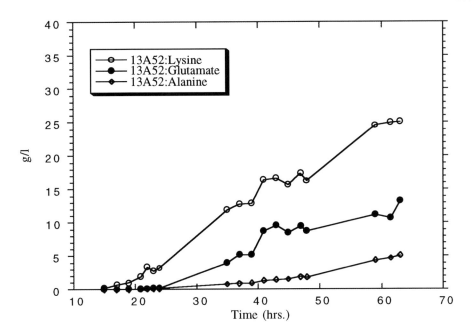

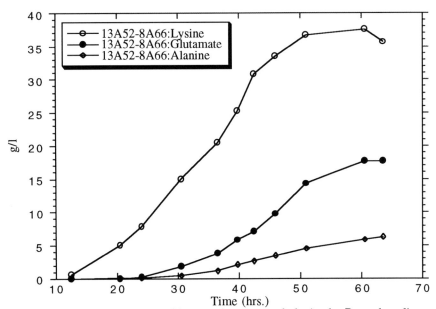

Figure 2. Production of lysine, glutamate and alanine by B. methanolicus strains 13A52 and 13A52-8A66

234

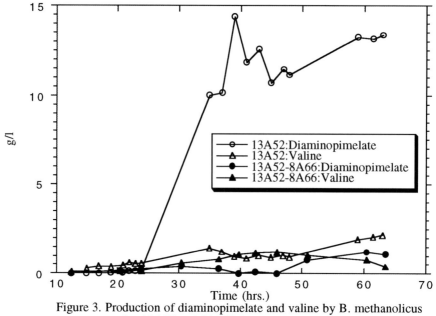

Figure 3. Production of diaminopimelate and valine by B. methanolicus
strains 13A52 and 13A52-8A66

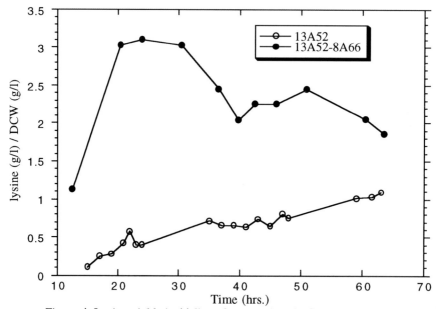

Figure 4. Lysine yields in 11 liter fermentations by B. methanolicus strains

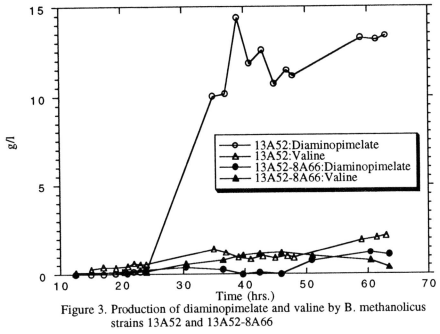

Figure 3. Production of diaminopimelate and valine by B. methanolicus strains 13A52 and 13A52-8A66

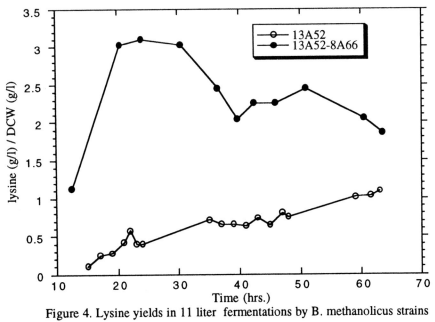

Figure 4. Lysine yields in 11 liter fermentations by B. methanolicus strains

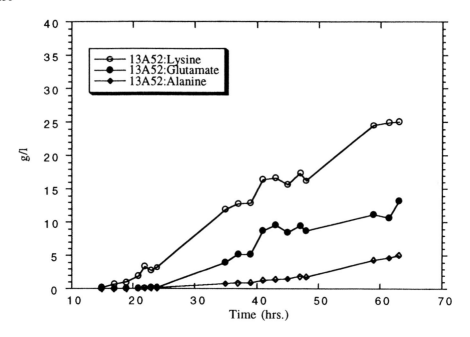

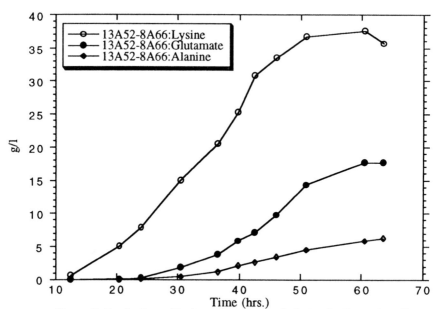

Figure 2. Production of lysine, glutamate and alanine by B. methanolicus strains 13A52 and 13A52-8A66

PHA biosynthesis, its regulation and application of C1-utilizing microorganisms for polyester production

Alexander Steinbüchel, Roman Wieczorek and Niels Krüger

Institut für Mikrobiologie der Westfälischen Wilhelms-Universität Münster, Corrensstraße 3, D-48149 Münster, Germany

INTRODUCTION

Polyhydroxyalkanoic acids (PHA) represent a storage compound for carbon and/or energy and occur as insoluble inclusions in the cells of a wide range of different bacteria (Steinbüchel 1991) mostly during cultivation of the cells in the presence of an excess of carbon source and if one other essential nutrient limits growth (Anderson, Dawes 1990). In addition to poly(3-hydroxybutyric acid), poly(3HB), which is the most well-known representative of PHA, 90 different hydroxyalkanoic acids have been detected as constituents of PHA from bacterial origin (Steinbüchel, Valentin 1995). These polyesters have been attracted much attention during the last years since they are biodegradable thermoplastics and/or elastomers which exhibit many other interesting properties and which therefore lend themselves for a wide range of different technical applications (Hocking, Marchessault 1994).

PHA can be obtained by bacterial fermentation utilizing many different carbon sources from (i) renewable resources such as, e. g., carbohydrates, organic acids and alkohols, from (ii) fossil resources such as, e. g., alkanes and from (iii) compounds provided by the chemical industry.

Bacteria are able to utilize a wide range of different C_1-compounds as carbon sources for growth and for the synthesis of various metabolic products. PHA can be also obtained from various C_1 compounds. Among those C_1-compounds listed in Fig. 1 carbon dioxide, methanol and methane are most interesting for biosynthesis and accumulation of PHA and most relevant for the biotechnological production of these polyesters, since these carbon source are not only cheap and available in large quantities but provide also good growth to several bacteria. Not all C_1-assimilation pathways seem to provide sufficient precursor substrates for the PHA biosynthesis pathways allowing the conversion of carbon dioxide, methane or methanol into PHA (Table 1). Many bacteria relying on the Calvin cycle for carbon dioxide fixation such as the aerobic chemolithotrophic bacteria and photosynthetic bacteria belonging to the purple sulfur bacteria or to the nonsulfur purple bacteria as well as facultative methylotrophs accumulate large amounts of PHA; only cyanobacteria accumulate very little PHA if they are able to synthesize these polyesters at all. Many methylotrophic bacteria relying on the serine pathway accumulate PHA, whereas those relying on the ribulosemonophosphate pathway do not. However, the ability to synthesize poly(3HB) was conferred to the latter

M. E. Lidstrom and F. R. Tabita (eds.), Microbial Growth on C_1 Compounds, 237–244.

Table 1. Occurrence of PHA in different groups of C$_1$-utilizing organisms.

Group of organisms	Representative organism	Carbon assimilation	Accumulation of PHA
Aerobic chemolithoautotrophic bacteria			
Knallgasbacteria	Alcaligenes eutrophus	Calvin cycle	yes
Photoautotrophic bacteria			
Cyanobacteria	Spirulina maxima	Calvin cycle	little, if any
Nonsulfur purple bacteria	Rhodospirillum rubrum	Calvin cycle	yes
Purple sulfur bacteria	Chromatium vinosum	Calvin cycle	yes
Green sulfur bacteria	Chlorobium limicola	reverse TCA cycle	no
Green nonsulfur bacteria	Chloroflexus auantiacus	HP pathway	little, if any
Methylotrophic bacteria			
Type I-methylotrophic bacteria	Methylophilus methylotrophus	RuMP pathway	no
Type II-methylotrophic bacteria	Methylobacterium extorquens	Serin pathway	yes
Facultative methylotrophic bacteria	Paracoccus denitrificans	Calvin cycle	yes
Anaerobic chemoautotrophic bacteria			
Methanogenic bacteria	M. thermoautotrophicum	Acetyl-CoA pathway	no
Acetogenic bacteria	Acetobacterium woodii	Acetyl-CoA pathway	no
C$_1$-utilizing eukaryotes			
Methylotrophic yeasts	Hansenula polymorpha	XuMP pathway	no
Plants	Arabidopsis thaliana	Calvin cycle	no

Abbreviations: PHA, polyhydroxyalkanoic acids; M., Methanobacterium; TCA, tricarboxylic acid; HP, hydroxypropionic acid; RuMP, ribulosemonophosphat; XuMP, xylulosemonophosphat.

by transfer and expression of the PHA biosynthesis genes from *Alcaligenes eutrophus* or *Chromatium vinosum* (Föllner et al. 1994). Anaerobic chemoautotrophic bacteria relying on the acetyl-CoA pathway such as the methanogenic and the homoacetogenic bacteria seem generally not able the synthesize PHA. The same is true for the green sulfur photosynthetic bacteria, which rely on the reverse TCA cycle. Green nonsulfur bacteria, which rely on the hydroxypropionate pathway, accumulate only very little PHA if they can synthesize it at all.

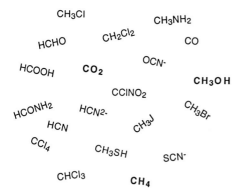

Fig. 1. C_1-compounds utilized by bacteria

It is interesting that representatives of chemolithoautotrophic and methylotrophic bacteria or recombinant organisms, which habour the PHA biosynthesis genes cloned from these bacteria, are at present the most promising candidates for the biotechnological production of PHA at an industrial scale. It is also interesting that some representatives of these groups were considered as suitable candidates for the production of single cell protein (SCP) from carbon dioxide or methanol, respectively, some decades ago (Scrimshaw, Murray 1995). When the industry had decided in the seventhies not to pursue the efforts on SCP for economical reasons, and after they had stopped the further development of these processes, mainly bacteria from these groups and plants (see below) were choosen as candidates for the biotechnological production of PHA. This was probably mainly due to the advanced knowledge of the cultivation engineering of these bacteria and also due to the fact that many of them accumulate large amounts of PHA.

RESULTS

PHA biosynthesis in Alcaligenes eutrophus. The metabolism of PHA has been most probably studied in most detail in *A. eutrophus* (Steinbüchel and Schlegel 1991). Three different laboratories cloned and sequenced the genes, which are required for the conversion of acetyl-CoA into poly(3HB). These studies revealed that the structural genes for β-ketothiolase (*phaA*), NADPH-dependent acetoacetyl-CoA reductase (*phaB*) and PHA synthase (*phaC*) constitute one single operon (*phaCAB*) which is transcribed from a promoter located approximately 300 nucleotides upstream of *phaC* and exhibiting striking similarity to the *Escherichia coli* σ^{70} consensus promotor. Other gene loci, which are required for a maximum rate of PHA accumulation and which upon inactivation by insertion of Tn5 accumulated PHA at a lower rate than the wild type, were also identified in *A. eutrophus*. The phenotype of these mutants was referred to as PHA-leaky. Among the genes affected in these mutants were the structural genes for the lipoamide dehydrogenase (*pdhL*) of the pyruvate multienzyme complex (Hein und Steinbüchel 1994) and for proteins exhibiting a striking similarity to HPr and the EI component of bacterial sugar:phosphoenol pyruvate phosphotransferase system (*phaH*, *phaI*); the latter are most probably involved in the regulation of the mobilization of accumulated PHA (Pries et al. 1991).

Properties and function of phasins. In addition to these genes we also identified the structural gene (*phaP*) for a 24-kDa protein, which is bound to the PHA granules and which represents the main component of the "membrane" surrounding the granules in *A. eutrophus* (Wieczorek et al. 1995). Like *phaL*, *phaH* and *phaI* mutants Tn5-induced *phaP* mutants accumulate less PHA than the wild type, and the phenotype was therefore also referred to as PHA-leaky. However, in contrast to *pdhL*, *phaH* or *phaI* mutants and to the wild type, cells of *phaP* mutants contained only one single large PHA granule. This GA24 protein constitutes up to approximately 5 % of the total cellular protein in cells which had accumulated PHA; after isolation of PHA granules by centrifugation in density gradients this protein is by far the most predominant protein species bound to the granules (Wieczorek et al. 1995 and Fig. 2). These observations, the location of the GA24 protein and studies on related proteins from other bacteria led to the concept of phasins. Phasins are amphiphilic proteins with a low molecular mass and with an extraordinary high affinity to PHA granules due to two hydrophobic stretches of amino acids close to the C-terminus of the protein (Fig. 3). These hydrophobic domains were not only identified in the GA24 protein, but also in the corresponding GA14 protein of the Gram-positive *Rhodococcus ruber* (Fig. 3). The phasin protein from *R. ruber* has been studied in most detail, and the involvement of the two hydrophobic domains for the binding of the protein has been clearly demonstrated employing genetically engineered phasins proteins (Pieper-Fürst et al. 1995). Phasins provide an amphiphilic layer at the surface of PHA granules which separates the hydrophobic core consisting of non crystalline PHA from the mostly hydrophobic constituents of the cytoplasm. Other putative functions of phasins were discussed recently (Steinbüchel et al. 1995).

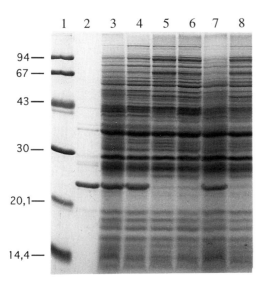

Fig. 2. Expression of the GA24-protein in the wild type and in PHA-negative mutants of different strains of *A. eutrophus*. Lanes 1, molecular mass standard proteins (from top to bottom: phosphorylase b, bovine serum albumin, ovalbumin, carbonic anhydrase, soybean trypsin inhibitor and α-lactalbumin); 2, proteins bound to poly(3HB) granules isolated from H16; 3, crude extract from H16 (a wild type); 4, crude extract from HF39 (a streptomycin-resistant mutant of H16); 5, crude extract from HF39-PSI (a spontaneous PHA-negative mutant of HF39); 6, crude extract from H16-SPI (a spontaneous PHA-negative mutant of H16); 7, crude extract from TF93 (a wild type); 8, crude extracts from TF93-PHB⁻ (a PHA-negative mutant of TF93).

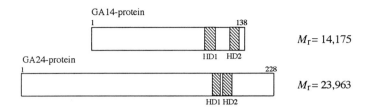

Fig. 3. Comparison of the *R. ruber* GA14 protein and the *A. eutrophus* GA24 protein. The numbers at the top indicate the position of amino acids. HD, hydrophobic domains.

Regulation of the expression of the GA24 protein in A. eutrophus. A reasonable Shine-Dalgarno sequence was identified upstream of *phaP* (Wieczorek et al. 1995). S1 nuclease protection assay identified the *phaP* promoter 143 nucleotides upstream of the translational start site (Fig. 4). Upstream of it a 2-10" and a "-35" region were identified which exhibited a high degree of homology to the corresponding regions of the *E. coli* σ⁷⁰ consensus promoter sequence (Fig. 5). From their sequences, both - the *phaP* promoter and the ribosome binding site (Fig. 5) - will probably allow transcription and translation, respectively, to occur at a high rate. However, the GA24 protein is not expressed constitutively, and the expression of the GA24 protein is subject

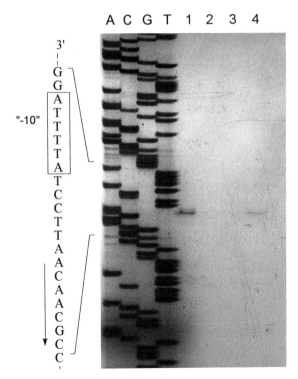

Fig. 4. S1 nuclease protection assay of the *phaP* transcript. Lanes ACGT, Standard sequencing reactions to size the mapping lanes; lane 1, strain H16 grown under nitrogen limitation in mineral salts medium with gluconate as sole carbon source (MMGlu); lane 2, strain H16 grown in nutrient broth medium; lane 3, strain H16-PHB-4 grown in MMGlu; lane 4, strain H16-SPI grown in MMGlu. The -10 region is shown. The arrow indicates the origin and direction of transcription.

242

to a very strong and effective regulation: (i) Although large amounts of the GA24 protein are synthesized in PHA accumulating cells, this protein is absent from cells which had not accumulated PHA (Fig. 2) independently whether PHA accumulation is impaired by inactivation of *phaC* due to mutagenesis or by nutritional constraints. (ii) In the wild type the transcript can be only detected in those cells which are accumulating PHA; in cells cultivated in complex nutrient broth medium and which do not accumulate PHA, only very faint amounts of the messenger RNA were detected (Fig. 4). In the PHA-negative *phaC* mutants PHB-4 and SPI the *phaP* transcript was not or only at a much lower concentration detected, respectively (Fig. 4). In addition (iii), even in PHA accumulating cells of *A. eutrophus* only that much GA24 protein is synthesized that can be bound to the granules; no soluble GA24 protein could be detected in the cytoplasm even by applying the very sensitive Western blot analysis (Wieczorek et al. 1995).

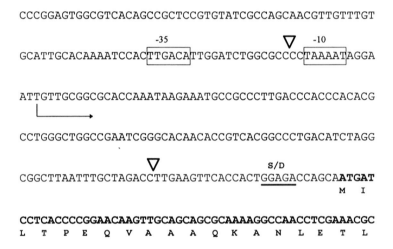

Fig. 5. *A. eutrophus phaP* 5′-upstream region. Arrow, transcriptional start site; S/D, Shine Dalgarno sequence; boxes, "-10" and "-35" region; triangles, identified insertions of Tn5-mob in two independant transposon induced PHA-leaky mutants (Wieczorek et al. 1995).

This regulation phenomenon is currently being investigated in our laboratory. Our working hypothesis proposes autoregulation of the expression of the GA24 protein (Fig. 6). Expression of the GA24 protein occurs only at a high level if PHA is synthesized and accumulated and as long as the surface of the granules provides sufficient space for the binding of this protein (Situation A). If for any reason PHA synthesis and accumulation are impaired, or if the surface of PHA granules in PHA accumulating cells is already covered with this protein (Situation B), the concentration of soluble GA24 protein in the cytoplasm is expected to increase drastically if *phaP* would be expressed constitutively. However, such a wasteful overproduction of the GA24 protein does obviously not occur in *A. eutrophus*. Instead, by a mechanism, which is yet unknown, the formation of a stable *phaP* transcript is diminished if situation B occurs. A repression might be achieved by binding of the GA24 protein to, e. g., the promoter region of *phaP*. Whether the expression of the GA24 protein is also regulated at the translational level remains to be elucidated.

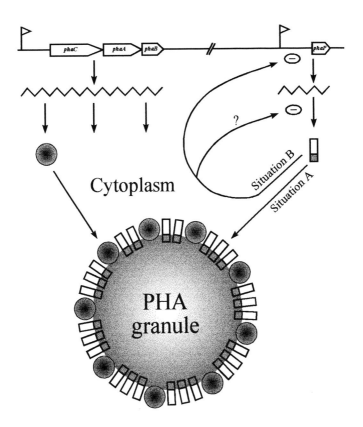

Fig. 6. Model for the regulation of the expression of the GA24 protein in *A. eutrophus*. The genes of the PHA operon (*phaCAB*) and of the GA24 protein (*phaP*) and their promoters are shown. The PHA synthase protein molecules are shown as circles, whereas the GA24 protein molecules are shown as rectangle with the C-terminus that contains the hydrophobic domains in black (for further details see text).

Phasins in other PHA accumulating bacteria. We have analyzed more than 50 different PHA accumulating bacteria from several different physiological and taxonomic groups with respect to the occurrence of granule-associated proteins. It could be demonstrated that all investigated bacteria possess at least one protein which is bound to the granules (R. Wieczorek and A. Steinbüchel unpublished results, Föllner et al. 1995, Steinbüchel et al. 1995). Most of these proteins exhibited molecular weights btween 11 and 25 kDa, and the synthesis of these proteins occurred mostly concomittantly with the synthesis of PHA. In various methylotrophuc bacteria two granule-associated proteins mostly exhibiting molecular weight of 11 and 20 kDa were identified (Föllner et al. 1995). In *C. vinosum*, the PHA biosynthesis genes of which has been analyzed in detail (Liebergesell, Steinbüchel 1992, Liebergesell et al. 1992, 1993) granule associated proteins of 14 and a 17 kDa were identified in addition to the PHA synthase (Liebergesell et al. 1992). This indicates that phasins are wide spread among PHA accumulating bacteria and that they have an important function for PHA metabolism.

CONCLUSIONS

The biosynthesis of PHA of various C_1-utilizing bacteria has been studied in much detail at a physiological, biochemical and molecular level. The availibility of PHA biosynthesis genes from various bacterial sources and knowlegde of their molecular genetics (Steinbüchel et al. 1992, 1995) prompted plant genetigists to transfer and express these genes in plants. First results obtained with transgenic cell lines of *Arabidopsis thaliana* harboring the *A. eutrophus* PHA operon composed of the genes for PHA synthase, β-ketothiolase and an NADPH-dependend acetoacetyl-coenzyme A reductase demonstrated that production of poly(3HB) will be feasible with plants (Poirier et al. 1995). This will most probably allow the economic production of considerable amounts of poly(3HB) directly from carbon dioxide. It will certainly only a matter of time when transgenic plants will be available which are also able to produce other PHA from carbon dioxide.

REFERENCES

Anderson AJ, Dawes EA (1990) Microbiol. Rev. 54,450-472.
Föllner C et al. (1994) Appl. Microbiol. Biotechnol. 40, 284-291.
Föllner C et al. (1995) Can. J. Microbiol. 41, 124-130.
Hein S, Steinbüchel A (1994) J. Bacteriol. 176, 4394-4408.
Hocking, Marchessault (1994) In Griffin GJL, ed, Chemistry and technology of biodegradable polymers, pp. 48-96, Chapman & Hall, London.
Liebergesell M, Steinbüchel A (1992) Eur. J. Biochem. 209, 135-150.
Liebergesell M et al. (1992) FEMS Microbiol. Lett. 99, 227-232.
Liebergesell M et al. (1993) Appl. Microbiol. Biotechnol. 40, 292-300.
Liebergesell M et al. (1994) Eur. J. Biochem. 226, 71-80.
Pieper-Fürst et al. (1995) J. Bacteriol. 177, 2513-2523.
Poirier Y et al. (1994) Bio/Technology 13, 142-150.
Pries, A. et al. (1991) J. Bacteriol. 173, 5843-5833.
Scrimshaw NS,Murray EB (1995) In Rehm HJ, Reed G, eds, Biotechnology, 2nd ed., VCH Publishers, Weinheim, Germany, Vol. 9, pp.221-237.
Steinbüchel A (1991) In Byrom D, ed, Biomaterials, pp 123-213, MacMillan Publishers, Basingstoke, UK
Steinbüchel A, Schlegel HG (1991) Mol. Microbiol. 5, 535-542.
Steinbüchel A et al. (1992) FEMS Microbiol. Rev. 103, 217-230.
Steinbüchel A, Valentin HE (1995) FEMS Microbiol. Lett. 128, 219-228.
Steinbüchel A et al. (1995) Can. J. Microbiol. 41, 94-105.
Wieczorek R et al. (1994) J. Bacteriol. 177, 2425-2435.

CYANOBACTERIAL SECONDARY METABOLITES

Richard E. Moore
Department of Chemistry, University of Hawaii, Honolulu, Hawaii 96822

Blue-green algae (cyanobacteria) are among the most primitive of organisms, having an evolutionary history spanning some four billion years. These procaryotic, photosynthetic microorganisms, which have the morphology and physiology of both bacteria and plants, are ubiquitous and adaptive to a wide variety of environments, including hot springs and polar ice caps. Although the most conspicuous species are found in freshwater and terrestrial habitats, it is estimated that the total number of autonomous marine species is more than double that of strictly terrestrial ones.

The first blue-green algae to be chemically investigated were the toxic species that frequently bloom in eutrophic lakes and reservoirs and account for animal and human intoxications. Neurotoxicity and hepatotoxicity are the most common types of poisoning. Strains of *Anabaena flos-aquae*, a filamentous species that elaborates predominantly neurotoxic alkaloids (anatoxins) (Devlin et al., 1977; Carmichael, Gorham, 1978), and strains of *Microcystis aeruginosa*, a colonial species that produces hepatotoxic cyclic heptapeptides (microcystins) (Bishop et al., 1959; Rinehart et al., 1988), are responsible for most of the toxic outbreaks. As more and more water supplies become eutrophic, there is growing concern about the effects of toxic algal blooms on the quality and safety of water that is used for human, as well as wildlife and domestic animal, consumption. In a recent case of human poisoning in Australia (Falconer et al., 1983), for example, hepatotoxicity coincided with the appearance of a toxic bloom of *M. aeruginosa* in the drinking-water supply. In developing countries such as China, cyanobacterial poisoning appears to be playing a significant role in the markedly higher incidence of human liver cancer in areas that are heavily dependent on surface drinking-water (Yu, 1989; Nishiwaki-Matsushima et al, 1992).

Prompted by a review (Baslow, 1977) that blue-green algae contain antibiotics, growth stimulants, and carcinogens in addition to potent toxins, we began an intensive study of this phylum as a source of new pharmaceuticals and agrochemicals about two decade ago (Moore et al., 1988). From the results to date, including those from other laboratories, there is no doubt that cyanophytes produce novel secondary metabolites and that some important leads have been generated. This phylum, however, has yet to be fully exploited as a source of potentially useful commercial products. With the exception of modest screening programs at Lilly (in collaboration with the author's laboratory) and Merck (Schwartz et al., 1990), little has been done by industry. Big industry has been reluctant, for economic reasons, to invest the capital needed for a meaningful evaluation.

Some significant leads that have been obtained from blue-green algae are cited in the next four paragraphs:

M. E. Lidstrom and F. R. Tabita (eds.), Microbial Growth on C₁ Compounds, 245–252.

In an intensive program initiated at the National Cancer Institute about nine years ago to discover new antiAIDS agents from natural sources, extracts of blue-green algae, obtained through a contract with this laboratory, were screened for anti-human immunodeficiency virus (HIV-1) properties. The first lead (actually the first lead of the entire NCI program) came as a consequence of screening extracts of the terrestrial blue-green algae *Lyngbya lagerheimii* DN-7-1 and *Phormidium tenue* CN-2-1 (Gustafson et al., 1989). These extracts protected human lymphoblastoid T cells from the cytopathic effect of HIV-1 infection. The active agents were found to be sulfoquinovosyl diacylglycerols, lipids which are structural components of chloroplasts and which occur widely in higher plants and photosynthetic microorganisms.

Several toxins from blue-green algae have been found to be important tools for pharmacological research. The aplysiatoxins (Kato, Scheuer, 1975; Mynderse et al., 1977) and lyngbyatoxins (Cardellina et al., 1979), highly inflammatory acetogenins and alkaloids, respectively, associated with a marine blue-green alga *Lyngbya majuscula* that is responsible for a severe contact dermatitis in Hawaii (Sedula et al., 1982), are potent activators of protein kinase C and powerful tumor promoters, having mechanisms of action that are essentially identical with those of the phorbol esters (Fujiki et al., 1990; Fujiki, Suganuma, 1995). The microcystins (Botes et al., 1984, 1985; Carmichael et al., 1988) and potent inhibitors of protein-serine/threonine phosphatases 1 and 2A (MacKintosh et al., 1990; Honkanen et al., 1990), exhibiting exactly the same effects at the molecular level as okadaic acid. At the cellular and whole animal levels, however, the microcystins and okadaic acid respond very differently. Okadaic acid is able to penetrate cellular membranes to inhibit protein phosphatases located inside the cells; however, microcystins are not able to get inside cells. Okadaic acid is a potent inflammatory agent that is responsible for diarrhetic shellfish poisoning; however, it is not hepatotoxic. Microcystins are among the most potent hepatotoxins known, but under normal circumstances, they do not appear to possess inflammatory or diarrhea-producing activities. Nodularin, a microcystin-related cyclic pentapeptide associated with *Nodularia spumigena*, the first blue-green alga to be implicated in animal poisoning (Francis, 1878), exhibits similar biological effects (Honkanen et al., 1991). Microcystins are tumor-promoting whereas nodularin is carcinogenic (Ohta et al., 1994). Clearly cyanobacterial toxins have become very important new probes for the study of cellular regulation and other biological processes.

A surprisingly large number of extracts of blue-green algae show broad-spectrum fungicidal activity, but not broad-spectrum antibacterial activity. Extracts of more than 1500 strains representing some 400 species of blue-green algae have been screened and nine percent of the extracts have been found to be antifungal at 1 mg/disc against one or more test organisms, viz. *Aspergillus oryzae, Candida albicans, Penicillium notatum, Saccharomyces cerevisiae,* and *Trichophyton mentagrophytes* (R. E. Moore, G. M. L. Patterson, unpublished). For most of the extracts of cyanophytes belonging to the Stigonemataceae, indole alkaloids that possess isonitrile groups such as the hapalindoles (Moore et al., 1987), fischerindoles (Park et al., 1992), ambiguines (Smitka et al., 1992), and welwitindolinones (Stratmann et al., 1994) are responsible for the antifungal activity. A fungicide with potential utility in the agricultural area is majusculamide C, a cyclic nonadepsipeptide from a deep-water variety of the marine cyanophyte *Lyngbya majuscula* (Carter et al., 1984). Potent activity was observed against a broad-spectrum of fungal plant pathogens, including resistant strains, such as *Phytophthora infestans*, the causative organism of tomato late blight, and *Plasmopora viticola*, the causative organism of grape downy mildew (Moore, Mynderse, 1982). Unfortunately adequate amounts of this substance have not become available yet for advanced testing and development.

Blue-green algae have been found to be excellent sources of new anticancer agents (Patterson et al., 1991a; Gerwick et al., 1994). Research in the author's laboratory has shown that a relatively high percentage of extracts of blue-greens collected in the field are active *in vivo*, but often the cyanophytes, specifically those of marine origin, can not be recollected in sufficient quantities or with the same activities for follow-up isolation, identification, and biological evaluation of the agents responsible for the antineoplastic activity (Moore, 1982). Six percent of the extracts of over 2000 laboratory-cultured strains, however, were found to be cytotoxic against human tumor cell lines at MIC's <20 μg/mL (Patterson et al., 1991b). Less than 1% of the extracts, however, were solid tumor selective (Valeriote et al., 1994) and/or tumor selective, but several of these showed equal cytotoxicity against drug sensitive and drug resistant cell lines. Furthermore, some of the non-cytotoxic extracts (<1%) showed significant multiple-drug-resistance (MDR) reversing activity. In contrast to field-collected blue-greens, almost all of the cultured blue-greens having anticancer activity proved to be terrestrial and freshwater species. In our hands, antitumor activity was almost always lost in the marine cultivates. Our search for a new antitumor agent with potential clinical activity has recently culminated in the discovery that cryptophycin-1, the major cytotoxin in *Nostoc* sp. GSV 224 and a potent fungicide that was first isolated from *Nostoc* sp. ATCC 53789 by researchers at Merck (Hirsch et al., 1990), shows excellent activity against drug-sensitive and drug-resistant solid tumors implanted in mice (Trimurtulu et al., 1994). A cryptophycin analog is currently heading for Phase I human clinical trials.

The literature on secondary metabolites from blue-green algae published prior to 1980 has been comprehensively reviewed (Moore, 1981). This paper briefly reviews the literature published since 1980.

Peptides and Depsipeptides

Most of the secondary metabolites that have been isolated from blue-green algae are peptides and depsipeptides and the majority of these are cyclic (Moore, 1995).

The most familiar peptides are the microcystins which belong to a class of cyclic heptapeptides comprised of approximately 50 members (Rinehart et al., 1994). Several of the microcystins have the general structure cyclo[D-Ala-L-X-D-*erythro*-β-methyl-isoAsp-L-Y-Adda-D-isoGlu-N-methyldehydroAla where X and Y are variable L-amino acids and Adda is a unique β-amino acid ($2S,3S,8S,9S$)-3-amino-9-methoxy-2,6,8-trimethyl-10-phenyl-($4E,6E$)-decadienoic acid (Namikoshi et al., 1989). For example, X and Y in microcystin-LR, the major hepatotoxin in most strains of *M. aeruginosa* from the Northern Hemisphere, are leucine and arginine. The structurally-related cyclic pentapeptide, nodularin, lacks the D-Ala and L-Leu units found in microcystin-LR. Acyclic peptides related to microcystins and nodularin have also been isolated (Choi et al., 1993). The biosynthesis of microcystin-LR has been studied (Moore et al., 1991).

Calophycin, a potent broad-spectrum fungicide from *Calothrix fusca* EU-10-1 (Moon et al., 1992), is a cyclic decapeptide containing a novel ($2R,3R,4S$)-3-amino-2-hydroxy-4-methylpalmitic acid unit (Hamp). The unusual Hamp unit has also been identified in puwainaphycin E, one of a family of cyclic decapeptides from *Anabaena* sp. BQ-16-1 (Gregson et al., 1992). The laxaphycins are a large family of cyclic undeca- and dodecapeptides, the major representative of each class being laxaphycins A and B, respectively, which together act synergistically to produce the antifungal activity of the crude extract of *Anabaena laxa* FK-1-2 (Frankmölle et al., 1992). The laxaphycins closely resemble a group of cyclic peptides known as the hormothamnins that have been isolated from the marine cyanophyte *Hormothamnion enteromorphoides* (Gerwick et al., 1992). Scytonemin A is a cyclic undecapeptide from *Scytonema* sp. U-3-3 that possesses an unusual β-amino acid unit and three substituted prolines (Helms et al., 1988).

The tantazoles are modified hexapeptides from the terrestrial cyanophyte

Scytonema mirabile BY-8-1. All of the compounds possess a sequence of four contiguous cysteine-derived Δ²-thiazoline rings attached 4,2' to one another with an isopropyl group connected to C-2 of the first thiazoline ring (ring A) and a threonine-derived oxazole ring (ring E) attached to C-4 of the fourth thiazoline (ring D) via C-2. A glycine-derived appendage is linked to C-4 of the oxazole ring. Tantazole B is a solid tumor selective cytotoxin (Carmeli et al., 1993). Structurally-related mirabazoles lack the oxazole ring (Carmeli et al., 1991).

Westiellamide is a modified cyclic hexapeptide from *Westiellopsis prolifica* (Prinsep et al., 1992). Its isolation from a blue-green alga provides circumstantial evidence for algal symbionts (*Prochloron* spp.) playing a role in the biosynthesis of closely-related cyclic peptides found in marine tunicates, e.g. the bistratamides in *Lissoclinum bistratum* (Degnan et al., 1989). Similar peptides (e.g. dolastatin 3) found in marine molluscs undoubtedly have a cyanobacterial origin (Pettit et al., 1987).

The cryptophycins form the largest depsipeptide class to date (25 members) (Golakoti et al., 1995). Structurally most cryptophycins consist of a δ-hydroxy acid unit (A), an α-amino acid unit (B), a β-amino acid unit (C), and an α-hydroxy acid unit (D), connected together in an ABCD sequence. Total syntheses of cryptophycin-1, -3 and -4 have been achieved (Barrow et al., 1995). Cryptophycin-24 from *Nostoc* sp. GSV 224 is identical with a marine sponge constituent, arenastatin A (Kobayashi et al., 1994).

Several biologically-active cyclic depsipeptides have been isolated from *Microcystis* spp., viz. a tyrosinase inhibitor, microviridin (Ishitsuka et al., 1990); the plasmin and trypsin inhibitors, micropeptins A and B (Okino et al., 1993a) and 90 (Ishida et al., 1995); a cell-differentiation-promoter, microcystilide A (Tsukamoto et al., 1993); the aeruginopeptins (Harada et al., 1993); and the cyanopeptolins (Martin et al., 1993). These depsipeptides and a serine proteinase inhibitor, A90720A, from *Microchaete loktakensis* IC-39-2 (Lee et al., 1994) possess an unusual 3-amino-6-hydroxy-2-piperidone unit that was first described in dolastatin 13, one of the cytotoxins found in the mollusc *Dollabella auricularia* (Pettit et al., 1989). The total structure of A90720A was elucidated by X-ray crystallography of the bovine trypsin-A90720A complex. Hapalosin is a multidrug-resistance reversing agent from *Hapalosiphon welwitschii* IC-52-3 (Stratmann et al., 1994a).

The angiotensin-converting enzyme inhibitor microginin (Okino et al., 1993) and the thrombin and trypsin inhibitors aeruginosins 298-A (Murakami et al., 1994) and 98-A and B (Murakami et al., 1995) are acyclic peptides. Majusculamide D (Moore, Entzeroth, 1988) and microcolins A and B are immunosuppresive, acyclic N-acylpyrrolinones where the acyl group is a peptidal group. Mirabimides A-D (Carmeli et al., 1991a) are cytotoxic, acyclic N-acylpyrrolinones where the acyl group is a depsipeptidal group.

Alkaloids

Anatoxin-a(s) (Matsunaga et al., 1989), a unique arginine-derived (Hemscheidt et al., 1995) phosphate ester of a cyclic N-hydroxyguanidine and the most potent anticholinesterase known, is one of the neurotoxic alkaloids produced by *Anabaena flos-aquae*. Cylindrospermopsin (Ohtani et al., 1992), a cyclic guanidine, is a potent hepatotoxic alkaloid associated with a strain of *Cylindrospermopsis raciborskii* that was implicated in an outbreak of hepatoenteritis on Palm Island in northern Queensland, Australia in 1979. Homoanatoxin-a (Skulberg et al., 1992) is a bicyclic neurotoxic alkaloid associated with *Oscillatoria agardhii* NOF-81. Besides the paralytic shellfish poisons, neosaxitoxin and saxitoxin, *Aphanizomenon flos-aquae* elaborates a novel tricyclic alkaloid, aphanorphine (8-hydroxy-1,3-dimethyl-2,3,4,5-tetrahydro-1,4-methano-3-benz-azepine) (Gulavita et al., 1989).

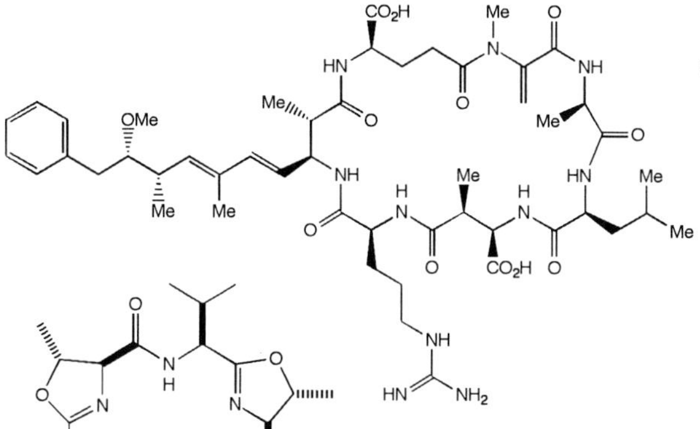

microcystin-LR

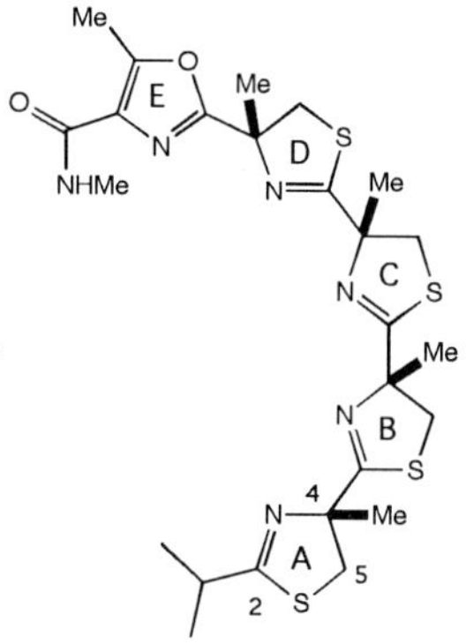

tantazole B

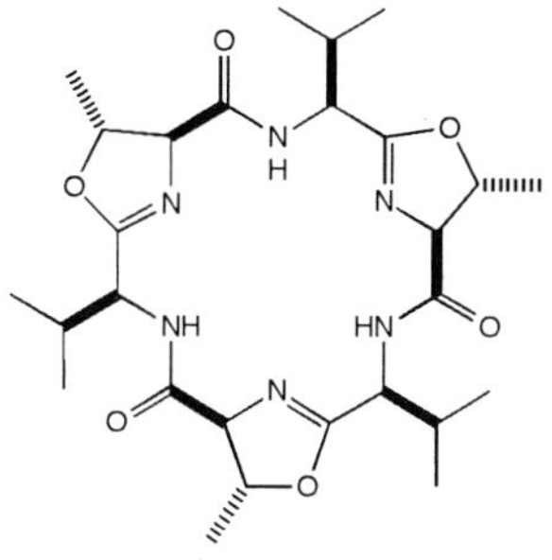

westiellamide

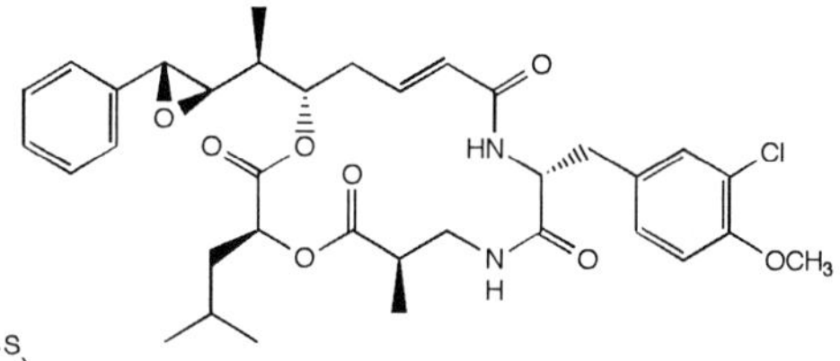

cryptophycin-1

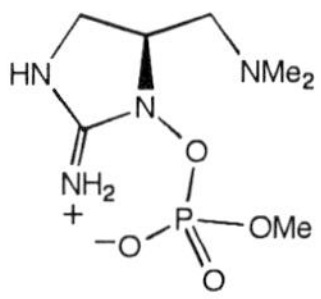

anatoxin-a(s)

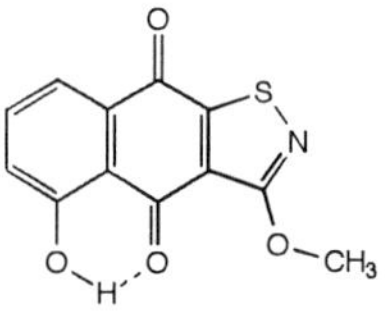

aulosirazole

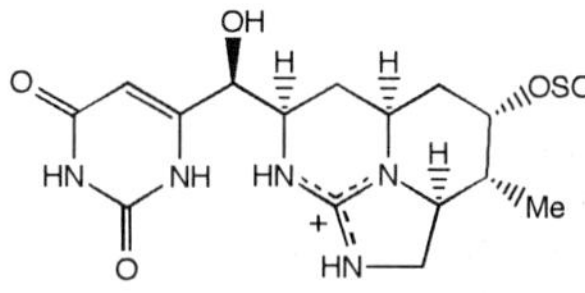

cylindrospermopsin

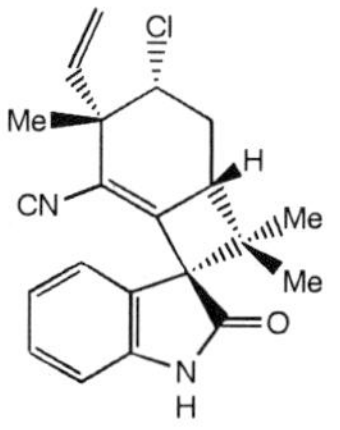

**welwitindolinone
A isonitrile**

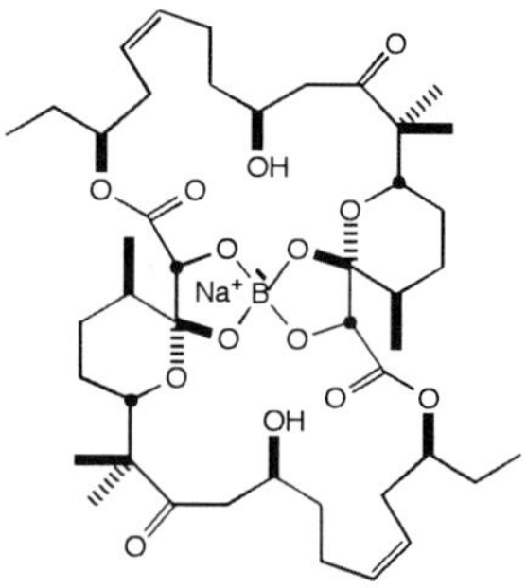

borophycin

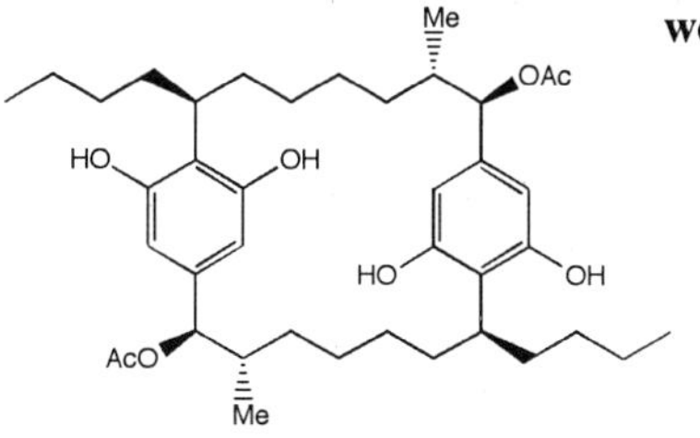

cylindrocyclophane D

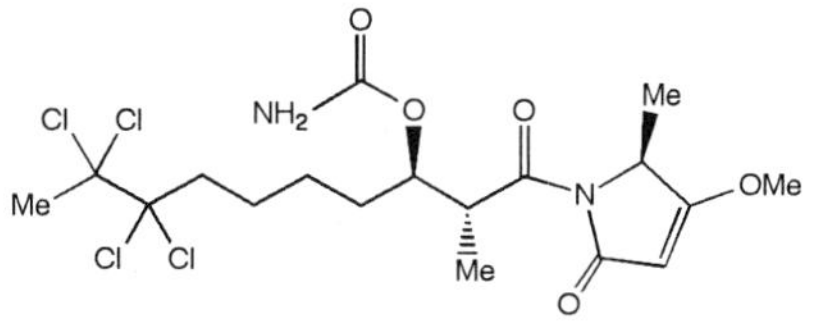

mirabimide E

Structures of some secondary metabolites from cultured blue-green algae

250

Lyngbyatoxin A is the major indole alkaloid associated with dermatitis-producing *Lyngbya majuscula* and is identical with the fermentation product teleocidin A-1 (Sakai et al., 1986). The arginine vasopressin inhibitor hapalindolinone A (Schwartz, 1987) and the MDR-reversing agent N-methylwelwitindolinone C isocyanate (Stratmann et al., 1994) are examples of the large number of indole alkaloids found in the family Stigonemataceae. Scytonemin (Proteau et al., 1993) is a dimeric indole alkaloid that functions as an effective ultraviolet sunscreen pigment in the sheaths of cyanobacteria. Cytotoxic, antiviral indolocarbazoles have been isolated from *Nostoc sphaericum* EX-5-1 (Knübel et al., 1990) and the tjipanazoles are antifungal indolo[2,3-a]carbazole alkaloids from *Tolypothrix tjipanasensis* DB-1-1 (Bonjouklian et al., 1991). The bauerines are antiviral β-carbolines from *Dichothrix baueriana* GO-25-2 (Larsen et al., 1994).

Aulosirazole (Stratmann et al., 1991b), a naphtho[2,3-d]isothiazole alkaloid from *Aulosira fertilissima* DO-8-1), is a solid-tumor selective cytotoxin. Curacin A (Nagle et al., 1995), a cyclopropane-containing thiazole alkaloid from *Lyngbya majuscula*, is an inhibitor of microtubule assembly and the binding of colchicine to tubulin. Mirabimide E (Paik et al., 1994), a N-acylpyrrolinone from *Scytonema mirabile* BY-8-1, is a solid tumor selective cytotoxin. Tolyporphin is a novel multi-drug resistance reversing porphyrin from the blue-green alga *Tolypothrix nodosa* HT-58-2 (Prinsep et al., 1992a). Unusual nucleosides such as tubercidin and toyocamycin (Stewart et al., 1988) and 9-deazaadenosine (Namikoshi et al., 1993) sometimes account for potent cytotoxicity associated with extracts of Nostocaceae.

Polyketides

Acutiphycins (Barchi et al., 1984) are cytotoxins from *Oscillatoria acutissima* B-1. Mirabilene isonitriles (Carmeli et al., 1990) are cytotoxic, acetogenic constituents of *Scytonema mirabile* BY-8-1. Oscillatoxins (Moore et al., 1984; Entzeroth et al., 1985) are inflammatory agents that are structurally and biologically related to the aplysiatoxins. Acetogenic nostocyclophanes from *Nostoc linckia* UTEX B1932 and cylindrocyclophanes from *Cylindrospermum licheniforme* ATCC 29204 are presently the only natural [m.n]paracyclophanes (Moore et al., 1990; Bobzin, Moore, 1993). Borophycin is a Boeseken complex of boric acid from *Nostoc linckia* GA-5-23 (Hemscheidt et al., 1994) that is related to the fermentation products boromycin and aplasmomycin. New malyngamides have been isolated from *Lyngbya majuscula* (Ainslie et al., 1985).

The scytophycins (Ishibashi et al., 1986) and tolytoxin (Carmeli et al., 1993a) are cytotoxic polyketides that have been isolated from several blue-greens belonging to the Scytonemataceae. Structurally-related swinholides found in the sponge *Theonella swinhoei* are believed to be produced by a cyanobacterial symbiont (Kitagawa et al., 1990).

Carbohydrates

An unusual *O*-acetyl-*O*-butyryl-*O*-carbamoyl-*O*,*O*-dimethyl-α-cyclodextrin has been found in *Tolypothrix byssoidea* (Moore et al., 1986).

Acknowledgment. Cyanobacterial research in the author's laboratory has been supported by grants from the National Cancer Institute (CA 12623) and National Science Foundation.

References

Ainslie RD et al (1985) J. Org. Chem. 50, 2859-2862.
Barchi JJ et al (1984) J. Am. Chem. Soc. 106, 8193-8197.

Barrow R et al (1995) J. Am. Chem. Soc. 117, 2479-2490.
Baslow MH (1977) Marine Pharmacology. Krieger Publishing, New York.
Bishop CT et al (1959) Can. J. Biochem. Physiol. 37, 453-471.
Bobzin SC, Moore RE (1993) Tetrahedron 49, 7615-7626.
Bonjouklian R et al (1991) Tetrahedron 47, 7739-7750.
Botes DP et al (1984) J. Chem. Soc. Perkin Trans I, 2311-2318.
Botes DP et al (1985) J. Chem. Soc. Perkin Trans I, 2747-2748.
Cardellina et al (1979) Science (Washington, D.C.) 204, 193-195.
Carmeli S et al (1990) J. Org. Chem. 55, 4431-4438.
Carmeli S et al (1991) Tetrahedron Lett. 32, 2593-2596.
Carmeli S et al (1991a) Tetrahedron 47, 2087-2096.
Carmeli S et al (1993)Tetrahedron Lett. 34, 5571-5574.
Carmeli S et al (1993a) Tetrahedron Lett. 34, 6681-6684.
Carmichael WW et al (1988) Toxicon 26, 971-973.
Carmichael WW, Gorham PR (1978) Mitt. Internat. Verein. Limnol. 21, 285-295.
Carter DC et al (1984) J. Org. Chem. 49, 236-241.
Choi BW et al (1993) Tetrahedron Lett. 34, 7881-7884.
Degnan BM et al (1989) J. Med. Chem. 32, 1354-1359.
Devlin JP et al (1977) Can. J. Chem. 55, 1367-1371.
Entzeroth M et al (1985) J. Org. Chem. 50, 1255-1259.
Falconer IR et al (1983) Med. J. Aust. 1, 511-514.
Francis G (1878) Nature (London) 18, 11-12.
Frankmölle WP et al (1992) J. Antibiotics 45, 1458-1466.
Fujiki H, Suganuma M (1995) J. Toxicol., submitted.
Fujiki H et al (1990) In Hall S and Strichartz G, eds, Marine Toxins (ACS Symposium
    Series No. 418), pp 232-240, American Chemical Society, Washington, D.C.
Gerwick WH et al (1992) Tetrahedron 48, 2313-2324.
Gerwick WH et al (1994) J. Appl. Phycol. 6, 143-149.
Golakoti, T et al (1995) J. Am. Chem. Soc. 117, submitted.
Gregson JM et al (1992) Tetrahedron 48, 3727-3734.
Gulavita N et al (1989) Tetrahedron Lett. 29, 4381-4384.
Gustafson KR et al (1989) J. Natl. Cancer Instit. 81, 1254-1258.
Harada K-i et al (1993), Tetrahedron Lett. 34, 6091-6094.
Helms GL et al (1988) J. Org. Chem. 53, 1298-1307.
Hemscheidt T et al (1994) J. Org. Chem. 59, 3467-3471.
Hemscheidt T et al (1995) J. Chem. Soc. Chem. Commun., 205-206.
Hirsch CF et al (1990) U.S. Patent 4,946,835, issued August 7.
Honkanen RE et al (1990) J. Biol. Chem. 265, 19401-19404.
Honkanen RE et al (1991) Mol. Pharm. 40, 577-583.
Ishida K et al (1995) Tetrahedron Lett. 36, 3535-3538.
Ishibashi M et al (1986) J. Org. Chem. 51, 5300-5306.
Ishitsuka MO et al (1990) J. Am. Chem. Soc. 112, 8180-8182.
Kato Y, Scheuer PJ (1975) Pure Appl. Chem. 41, 1-14.
Kitagawa I et al (1990) J. Am. Chem. Soc. 112, 3710-3712.
Kobayashi M et al (1994) Chem. Pharm. Bull. 42, 2196-2198, 2394-2396.
Koehn FE et al (1992) J. Nat. Prod. 55, 613-619.
Knübel G et al (1990) J. Antibiotics 43, 1236-1239.
Larsen LK et al (1994) J. Nat. Prod. 57, 419-421.
Lee AY et al (1994) Chem. Biol. 1, 113-117.
MacKintosh C et al (1990) FEBS Letters 264, 187-192.
Martin C et al (1993) J. Antibiotics 46, 1550-1556.

Matsunaga S et al (1989) J. Am. Chem. Soc. 111, 8021-8023.
Moon S-S et al (1992) J. Org. Chem. 57, 1097-1103.
Moore BS et al (1990) J. Am. Chem. Soc. 112, 4061-4063.
Moore RE (1981) In Scheuer PJ, ed, Marine Natural Products, Vol IV, pp 1-52, Academic Press, New York.
Moore RE (1995) J. Indust. Microbiol., submitted.
Moore RE, Entzeroth, M (1988) Phytochemistry 27, 3101-3103.
Moore RE, Mynderse JS (1982) U.S. Patent 4,342,751, issued August 3.
Moore RE et al (1984) J. Org. Chem. 49, 2484-2489.
Moore RE et al (1986) J. Org. Chem. 51, 5307-5310.
Moore RE et al (1987) J. Org. Chem. 52, 1036-1043.
Moore RE et al (1988) In Fautin DG, ed., Biomedical Importance of Marine Organisms, pp 143-150, California Academy of Sciences, San Francisco.
Moore RE et al (1991) J. Am. Chem. Soc. 113, 5083-5084.
Murakami M et al (1994) Tetrahedron Lett. 35, 3129-3132.
Murakami M et al (1995) Tetrahedron Lett. 36, 2785-2788.
Mynderse JS et al (1977) Science (Washington, D.C.) 196, 538-540.
Nagle DG et al (1995) Tetrahedron Lett. 36, 1189-1192.
Namikoshi M et al (1989) Tetrahedron Lett. 30, 4349-4352.
Namikoshi M et al (1993) J. Am. Chem. Soc. 115, 2504-2505.
Nishiwaki-Matsushima R et al (1992) J. Cancer Res. Clin. Oncol. 118, 420-424.
Ohta T et al (1994) Cancer Res. 54, 6402-6406.
Ohtani I et al (1992) J. Am. Chem. Soc. 114, 7942-7944.
Okino T et al (1993) Tetrahedron Lett. 34, 501-504.
Okino T et al (1993a) Tetrahedron Lett. 34, 8131-8134.
Paik S et al (1994) J. Am. Chem. Soc., 116, 8116-8125.
Park A et al (1992) Tetrahedron Lett. 33, 3257-3260.
Patterson GML et al (1991a) J. Appl. Phycol. 6, 151-157.
Patterson GML et al (1991b) J. Phycol. 27, 530-536.
Pettit GR et al (1987) J. Am. Chem. Soc. 109, 7581-7582
Pettit GR et al (1989) J. Am. Chem. Soc. 111, 5015-5017.
Prinsep MR et al (1992) J. Nat. Prod. 55, 140-142.
Prinsep MR et al (1992a) J. Am. Chem. Soc. 114, 385-386.
Proteau PJ et al (1993) Experientia 49, 825-829.
Rinehart KL et al (1988) J. Am. Chem. Soc. 110, 8557-8558.
Rinehart KL et al (1994) J. Appl. Phycol. 6, 159-176.
Sakai S et al (1986) Tetrahedron Lett. 27, 5219-5220.
Schwartz RE et al (1987) J. Org. Chem. 52, 3706-3708.
Schwartz RE et al (1990) J. Indust. Microbiol. 5, 113-124.
Serdula M et al (1982) Haw. Med. J. 41, 200-201.
Stewart JB et al (1988) J. Antibiotics 41, 1048-1056.
Skulberg OM et al (1992) Environ. Toxicol. Chem. 11, 321-329.
Smitka TA et al (1992) J. Org. Chem. 57, 857-861.
Stratmann K et al (1994) J. Am. Chem. Soc. 116, 9935-9942.
Stratmann K et al (1994a) J. Org. Chem. 59, 7219-7226.
Stratmann K et al (1994b) J. Org. Chem., 59, 6279-6281.
Tsukamoto S et al (1993) J. Am. Chem. Soc. 115, 11046-11047.
Valeriote F et al (1994) In Valeriote FA, Corbett TH, and Baker LH, eds., Anticancer Drug Discovery and Development, pp. 1-25, Kluwer, Norwell, Massachusetts.
Yu SZ. In Tang ZY, Wu MC and Xia SS, eds, Primary Liver Cancer; pp 30-37, Springer-Verlag, Berlin/Heidelberg.

APPLICATION OF CELLULAR FUNCTIONS OF THE METHYLOTROPHIC YEAST - metabolism, gene expression, and posttranslational process-

Yasuyoshi SAKAI, Yoshiki TANI,* and Nobuo KATO

Department of Agricultural Chemistry, Faculty of Agriculture, Kyoto University, Kitashirakawa-Oiwake, Sakyo-ku, Kyoto 606-01, Japan, and *Graduate School of Biological Sciences, Nara Institute of Science and Technology, Ikoma 630-01, Japan

INTRODUCTION

Since the first discovery of the methylotrophic eucaryote, *Candida boidinii* (initially identified as *Kloeckera*) (Ogata 1969), the methylotrophic yeast has been used extensively both in academic and applied fields. In 1970s, the overall pathway for methanol-dissimilation and assimilation was elucidated by purifying each enzyme responsible for C_1-metabolism, and some of them were found to be compartmented into single-membrane bound organelles, peroxisomes. As methanol is a cheap carbon source, production of SCP attracted much attention in the early 1970s, and these interests resulted in the establishment of a high-cell density cultivation method. In 1980s, recombinant DNA technology expanded and invaded into this field of studies. A heterologous gene expression with *Pichia pastoris* and *Hansenula polymorpha* has now become a basic technology in molecular biology and for the production of pharmaceutical proteins (Cregg 1993; Gellissen et al. 1992). Several processes have also been developed with *Candida boidinii* for the production of useful chemicals through unique C_1 metabolic functions (Tani 1991). And in the last decade, the mechanism of peroxisome assembly and protein sorting has begun to emerge at molecular level using these methylotrophic yeasts as model organisms (Subramani 1993). These studies have medical importance in connection to the human peroxisomal genetic disorder (e.g. Zellweger syndrome). Here, we summarize our recent studies with the methylotrophic yeast, *C. boidinii,* based on a concept "application of cellular functions to produce useful chemicals", and also try to prospect future development in the field. These cellular functions include 1) metabolism, 2) gene expression, and 3) posttranslational processes, e.g. organelle assembly and protein translocation.

RESULTS

EXPLORING NEW METABOLIC FUNCTIONS

1) MULTI-STEP BIOCONVERSION WITH PURIFIED METHANOL-ASSIMILATING ENZYMES

The first stage in the methanol-assimilating pathway of the methylotrophic yeast is catalyzed by alcohol oxidase (AOD), catalase (CAT), dihydroxyacetone synthase (DHAS), and dihydroxyacetone kinase (DHAK). In the present study, this multi-enzyme system was applied to produce $[1, 3-^{13}C]$dihydroxyacetone phosphate (Yanase et al. 1995). Besides these four enzymes, the production reaction utilizes purified preparation of another exogenous enzyme, adenylate kinase (ADK) to regenerate ATP (Fig. 1A). The product, ^{13}C-labeled dihydroxyacetone phosphate,

M. E. Lidstrom and F. R. Tabita (eds.), Microbial Growth on C₁ Compounds, 253–260.
© *1996 Kluwer Academic Publishers. Printed in the Netherlands.*

A

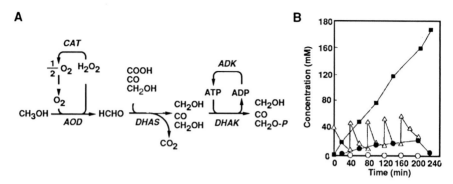

B

Fig. 1. Enzymatic preparation of [1, 3-^{13}C]dihydroxyacetone phosphate (DHAP) from [^{13}C]methanol and hydroxypyruvate using the methanol-assimilating system of the methylotrophic yeast. (A) Reaction scheme. (B) Fed-batch reaction. The starting reaction mixture contained potassium phosphate buffer (pH 7.0, 100 mM), MgCl$_2$ (5 mM), thiamin pyrophosphate (0.5 mM), ATP (40 mM), [^{13}C]methanol (40 mM), hydroxypyruvate (80 mM), H$_2$O$_2$ (50 mM), DHAS (2 U/ml), DHAK (4 U/ml), ADK (10 U/ml), and CAT (200 U/ml). Methanol (40 mM), hydroxypyruvate (80 mM), H$_2$O$_2$ (50 mM), and ATP (40 mM) were fed at 40-min intervals. Closed square, DHAP. Open triangle, methanol. Closed circle, dihydroxyacetone. Open circle, formaldehyde.

is a convenient raw material for preparation of various kinds of ^{13}C-labeled sugars, which are useful for ^{13}C-NMR diagnosis (Syrota, Jehensen 1991). In the production, [^{13}C]methanol and hydroxypyruvate were used as the starting materials. DHAS catalyzes a transketolase reaction between formaldehyde and D-xylulose 5-phosphate *in vivo*, and it can also utilize hydroxypyruvate as an acceptor of formaldehyde *in vitro*. The latter reaction leads to the formation of dihydroxyacetone and CO$_2$, the benefit of which is that the generated CO$_2$ can be spontaneously eliminated from the reaction system avoiding the product inhibition of the DHAS-catalyzed reaction. Catalase was added to support oxidation of [^{13}C]methanol to formaldehyde through its peroxidative function and supply oxygen to AOD. After extensive optimization of each reaction step, the production system was established as described in the legend to Fig. 1. Under the optimum reaction conditions, a fed-batch reaction afforded 185 mM [1, 3-^{13}C] dihydroxyacetone phosphate from [^{13}C]methanol; and the molar yield of the ester relative to methanol added was 92.5% (Fig. 1B).

2) METHYLFORMATE SYNTHASE: A KEY ENZYME IN A NOVEL FORMALDEHYDE OXIDATION PATHWAY

Formaldehyde is the key intermediate which occupies the central position in the C$_1$-metabolism. In the dissimilatory pathway of the methylotrophic yeast, formaldehyde is oxidized to formate in a glutathione (GSH)-dependent reaction catalyzed by GSH-dependent formaldehyde dehydrogenase (EC 1.2.1.1.) and S-formylglutathione hydrolase (EC 3.1.2.12). Subsequently, formate is oxidized to CO$_2$ by formate dehydrogenase (EC 1.2.1.2.).

Recently, we found that a considerable amount of methylformate accumulated in the culture media of methylotrophic yeasts. When *P. methanolica* was

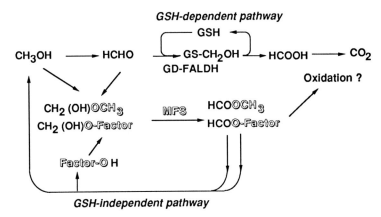

Fig. 2. GSH-independent and GSH-dependent formaldehyde oxidation pathways in the methylotrophic yeast. GD-FALDH, GSH-dependent formaldehyde dehydrogenase.

grown on a medium containing 2% (v/v) methanol, methylformate accumulated until the cells reached mid-exponential phase, at which point maximum methylformate (ca. 0.5 mM) was observed. Further studies revealed that methylformate formation was found to be catalyzed by a new type of alcohol dehydrogenase, which was named methylformate synthase (MFS). MFS was induced on methanol-medium, and methylformate accumulation and MFS activity were enhanced by the addition of formaldehyde in the growing medium (Sakai et al. 1995a). MFS was purified to an apparent homogeneity from cell-free extract of *P. methanolica*. MFS was composed of four identical subunits of 42 kDa, and included two zinc atoms per subunit. The reaction of MFS required formaldehyde, NAD^+, and a high concentration (0.5 M) of methanol in the assay mixture. Although a high concentration of methanol was required to express enzyme activity *in vitro*, reduction of NAD^+ was observed when a unknown heat stable factor was added in the place of methanol (Y. Sakai, unpublished). Thus, MFS showed a factor-dependent formaldehyde dehydrogenase activity similar to the enzyme activity described with a gram positive bacterium, *Amycolatopsis methanolica* (van Ophem et al. 1992). We propose a novel GSH-independent formaldehyde oxidation pathway in the methylotrophic yeasts (Fig. 2). In the pathway, formaldehyde is oxidized in a factor-dependent manner or in a methanol-dependent manner to methylformate.

In a biotechnological aspect, MFS could provide a new approach to produce water soluble esters in an aqueous reaction mixture. Methylformate accumulated up to the concentration of 138 mM under aqueous conditions when cells having high MFS activity were used in the reaction (Y. Sakai, unpublished).

GENE EXPRESSION AND METABOLIC ENGINEERING

The yeast gene expression system is considered advantageous in the following respects: 1) yeast cells are easy to handle and to bring up the production system to an industrial scale; 2) eucaryotic intracellular structures, which are important for secretion, modification and folding of the proteins exist in yeast cells as well as in the higher organisms. And also, *C. boidinii* shows potential as a producer of useful metabolites such as citric acid, ATP, and aldehydes (Tani 1991). Gene expression

	ADK activity (U / mg protein)
MeOH	170
MeOH + Glycerol	320
Glycerol	25
Glucose	0.40
MeOH + Glucose	0.50
EtOH	0.40
MeOH + EtOH	0.29

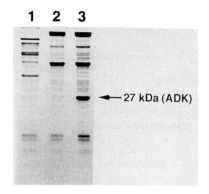

Fig. 3. Expression of *S. cerevisiae ADK1* under *AOD1* promoter in *C. boidinii*. Regulation of ADK by carbon sources and SDS-PAGE analysis of intracellular soluble proteins (15 μg). Lane 1, glucose-grown transformant; lane 2, methanol-grown wild type; lane 3, methanol-grown transformant.

coupled with molecular breeding system could be considered as a basic technology for metabolic engineering, and as a new type bioconversion system.

A transformation system for *C. boidinii* was established by deriving a uracil auxotroph strain (*ura3*), and constructing a chromosomal integrative vector and an autonomous replicating vector (Sakai et al. 1991; Sakai et al. 1993). And any specific gene in *C. boidinii* could be efficiently deleted by one-step gene disruption, if it is not lethal (Sakai, Tani 1992a). The alcohol oxidase gene (*AOD1*), which is efficiently expressed during the growth on methanol but repressed on glucose or ethanol, was cloned from the *C. boidinii* genome (Sakai, Tani 1992b). Heterologous gene expression in *C. boidinii* under the *AOD1* promoter was investigated using the *S. cerevisiae* adenylate kinase (*S. cerevisiae ADK1*)-expression vector pTRex. The single integrant strain was grown on various carbon sources, and the ADK activity was followed to characterize the transcriptional control of the *C. boidinii AOD1* promoter (Sakai et al. 1995b). The pattern of ADK regulation in the transformant was essentially the same as the *C. boidinii AOD1* regulation., i.e., (a) cells grown on methanol or methanol plus glycerol medium showed highly induced levels of ADK activity, (b) their activities were repressed to low levels when they were grown in the presence of glucose or ethanol in the medium, (c) glycerol-grown cells represent a middle-level expression. The produced ADK protein

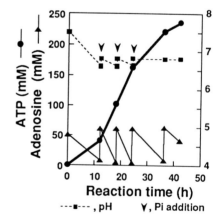

Fig. 4. ATP production from adenosine with cells of *C. boidinii* transformant. The reaction was performed at a 20-ml scale (Sakai et al. 1995). Arrows show the addition of potassium phosphate for pH adjustment.

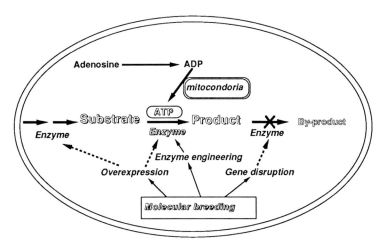

Fig. 5. A new metabolic engineering and bioconversion system with *C. boidinii*.

accumulated up to 28% of the total soluble proteins on glycerol plus methanol medium, and all of the produced ADK was in a soluble, active, and modified form as was expressed in *S. cerevisiae* (Fig. 3).

Previous studies of ATP production with *C. boidinii* cells suggested that the phosphorylation of AMP to ATP catalyzed by ADK was the rate-limiting step (Tani et al. 1994). As methanol-induced transformants had ca. 10,000-fold enhanced levels of ADK activity, ATP production with these transformants was investigated to seek whether the enhancement in ADK activity would increase ATP productivity in *C. boidinii* cells. The wild type strain was low in ADP and ATP productivities. In contrast, cells of transformant strain phosphorylated AMP to ADP efficiently, reflecting the high activity of ADK in the cells. Consequently, transformant had very high ATP productivity (Sakai et al. 1994). ATP production was performed with successive feeding of adenosine under controlled pH. As shown in Fig. 4, after 42 h, the ATP concentration in the reaction mixture was 230 mM (117 g/l), at an 88% conversion efficiency.

As many bioconversion reactions require ATP as a substrate or energy source, simultaneous expression of ADK and other heterologous enzymes should enable us to establish a bioconversion system with *C. boidinii* that does not require continuous addition of ATP. Such a system would provide a novel strategy for further metabolic engineering in *C. boidinii* (Fig. 5).

POSTTRANSLATIONAL PROCESSES INVOLVED IN THE EXPRESSION OF PEROXISOMAL ENZYME ACTIVITY

Intracellular production of enzyme protein is the basic technology to prepare biocatalytic cells to produce useful chemicals. Under highly methanol-induced conditions, much portion of intracellular volume (up to 80%) is occupied by single membrane organelles, peroxisomes. From a biotechnological point of view, peroxisome is an attractive organelle "to pack" the produced heterologous proteins where they are protected from the degradation by cytosolic proteases. Most peroxisome matrix proteins are destined for peroxisomes by a 3-amino acid sequence, -SKL and its derivatives (Peroxisome Targeting Signal 1: PTS1), located at an extreme carboxyl end (Subramani 1993). And so, an enzyme protein can be easily targeted to peroxisomes just by adding 3 amino acids, when the addition does not affect its activity.

Recently, molecular bases for peroxisomal protein transport began to appear with the use of yeast genetics. So far, at least 15 *PAS* (Peroxisome assembly) genes are identified with *S. cerevisiae*. Methylotrophic yeasts, *P. pastoris, H. polymorpha,* and *C. boidinii,* are currently used as model systems in these peroxisomal studies because of their ability to proliferate massive peroxisomes (Subramani 1993).

During the investigation on the function of a peroxisome membrane protein of *C. boidinii* (McCammon et al. 1990), we found that a putative peroxisomal transporter, Pmp47, is responsible for transport and folding of a specific matrix enzyme (Y. Sakai, unpublished). The knock out strain (*pmp47Δ*) of *C. boidinii* could not grow on methanol. Biochemical and immunoelectron microscopic observations showed that DHAS protein aggregated in the cytosol as an inclusion body while AOD was active and was imported into peroxisomes. Pmp47 was also depleted from the two peroxisome assembly mutants of *C. boidinii* (Sakai et al. 1995c). In these strains, DHAS was in an active form in the cytosol (or nucleus).

Fig. 6 summarizes our working hypothesis for peroxisomal transport in methanol-induced cells of *C. boidinii*. At least four components are necessary for the translocation and folding of DHAS: DHAS, Pmp47, an unknown substrate transported by Pmp47, and a peroxisomal translocation-folding machinery (PTFM). In *pmp47Δ*-cells, DHAS protein could not be folded properly and accumulated as aggregates in the cytoplasm. According to our model, this is explained as follows: although both DHAS and the solute transported by Pmp47 are present in the cytoplasm of *pmp47Δ*, PTFM is absent in the cytoplasm, so DHAS could not be folded properly. This problem might be amplified by the high concentration of DHAS in the cells, hindering folding in the cytoplasm. However, peroxisome-deficient strains allowed

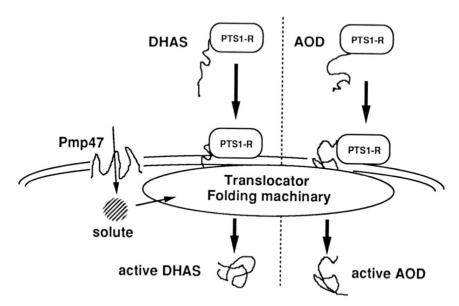

Fig. 6. Suggested pathway of two PTS1-peroxisomal enzymes, AOD and DHAS. PTS1-R, PTS1-receptor.

folding of DHAS in the cytoplasm since PTFM was also mislocalized in the cytosol.

Understanding of the mechanism for translocation and folding of peroxisomal proteins will give us a novel strategy for the overproduction of foreign proteins within peroxisomes, and then, the system may be applied to bioconversion reactions in a more sophisticated and economical manner in the future.

CONCLUSION

The methylotrophic yeast, *Candida boidinii*, was initially isolated during screening for methyl acetate esterase. The project had started from purely applied backgrounds, and at that time, nobody could expect such a great expansion of the field of the methylotrophic yeast.

After extensive biochemical, genetical, and cellular studies of methylotrophic yeasts, various biological functions have been utilized so far and many more may be utilized in the future. At first, its unique C_1-metabolism will provide a novel reaction not found in other organisms. Here, as an example, a multi-step C_1-assimilating system was developed to produce ^{13}C-labeled DHAP. Although C_1-metabolism in the methylotrophic yeast had been studied extensively, current studies on methyl formate formation led to a surprising finding of a new pathway for formaldehyde oxidation. These new findings will lead to the production of new metabolites. Second, the strong and inducible gene expression system of the methylotrophic yeast can be directly applied for the production of useful proteins. Third, when the molecular breeding system is coupled with gene expression, the system can be considered a novel bioconversion system. In other words, intracellular overexpression of enzymes means amplification of the reaction involved in bioconversion, and by-product formations can be eliminated by the disruption of the specific gene. Fourth, an ATP-production system shown here will provide a possibility for an energy-coupled bioconversion with *C. boidinii* cells. Fifth, using methylotrophic yeast cells, we will be able to "optimize" the localization of foreign proteins or to pack foreign proteins into peroxisomes. Thus, the future of the methylotrophic yeast in the field of biotechnology is prospected to be quite bright.

The methylotrophic yeast could be "a bridge" between applied and basic sciences. As basic sciences could contribute to applied sciences, applied sciences could also contribute to a feed back on basic sciences. In both fields, a novel concept leads to a new approach, and will open a new field of studies. We hope biotechnological studies using the methylotrophic yeasts could also help understanding complex cellular functions and their mechanisms at molecular level.

REFERENCES

Cregg, JM (1993) Bio/Technology 11, 905-910.
Gellissen G et al. (1992) Trends in Biotechnology 10, 413-417.
McCammon MT et al. (1990) J. Biol. Chem. 265, 20098-20105.
Ogata K et al. (1969) Agric. Biol. Chem. 33, 1519-1520.
Sakai Y et al. (1991) J. Bacteriol. 173, 7458-7463.
Sakai Y et al. (1993) J. Bacteriol. 175, 3556-3562.
Sakai Y et al. (1994) Bio/Technology 12, 291-293.
Sakai Y et al. (1995a) FEMS Microbiol. Lett. 127, 229-234.
Sakai Y et al. (1995b) Appl. Microbiol. Biotechnol. 42, 860-864.
Sakai Y et al. (1995c) Biosci. Biotech. Biochem. 59, 869-875.
Sakai Y, Tani Y (1992a) J. Bacteriol. 174, 5988-5993.
Sakai Y, Tani Y (1992b) Gene 114, 67-73.
Subramani S (1993) Annu. Rev. Cell. Biol. 9, 445-478.

Tani Y (1991) In Goldberg I and Rokem JS, eds, Biology of Methylotrophs, pp 253-270, Butterworth-Heinemann, Boston, USA.
Syrota A, Jehensen P (1991) Eur. J. Nucl. Med. 18, 897-923.
Tani Y et al. (1994) Proc. Jpn. Acad. 70 Ser. B, 53-57.
van Ophem PW et al. (1992) Eur. J. Biochem. 206, 511-518.
Yanase H et al. (1995) Appl. Microbiol. Biotechnol. 43, 228-234.

Evolution of Dichloromethane Utilization

Thomas Leisinger, Andreas Mägli, Monika Schmid-Appert, Kurt Zoller and Stéphane Vuilleumier

Mikrobiologisches Institut, Swiss Federal Institute of Technology, ETH-Zentrum, CH-8092 Zürich, Switzerland

INTRODUCTION

To date, some 2000 naturally occurring organohalogen compounds have been identified. Many of these are of biogenic origin, i.e. they are produced by bacteria, fungi, algae, plants and animals. Other halogenated compounds are discharged into the environment as the result of natural physical processes, such as forest fires and eruptions of volcanoes (Gribble 1994). The industrial solvent dichloromethane falls into the latter category. The compound is not known to be produced by organisms, but reportedly occurs in gasses emitted by volcanoes (Isidorov 1990). One thus would assume that volcanism, a process whose relevance for dichloromethane emission is unknown, and industrial chemical synthesis, which occurs at an estimated rate of 3 x 10^5 tons per year, are the main sources of dichloromethane in the environment.

The dichloromethane carbon atom is at the oxidation state of zero and therefore contains a considerable amount of energy. In contrast to trichloromethane and tetrachloromethane, the compound can be utilized as the sole carbon and energy source by both aerobic and anaerobic bacteria. Aerobic utilization of dichloromethane is restricted to methylotrophs containing the enzyme dichloromethane dehalogenase (Bader, Leisinger 1994). Anaerobic utilization of the compound is independent of a dichloromethane dehalogenase type enzyme and proceeds by an as yet unknown pathway which so far has only been observed in a novel acetogenic bacterium (Mägli et al. 1995). The present communication addresses three topics relating to the evolution, in the broadest sense, of microbial dichloromethane utilization. We report on efforts to elucidate the pathway for anaerobic dichloromethane utilization, on protein engineering of dichloromethane dehalogenase and on the characterization of insertion sequences associated with the dichloromethane utilization genes of methylotrophic bacteria.

RESULTS

Isolation of a dichloromethane-utilizing acetogenic bacterium in pure culture.
Two strictly anaerobic mixed cultures capable of growth with dichloromethane as the sole carbon and energy source have previously been described (Freedman et al. 1991; Stromeyer et al. 1991). Both of them originally produced acetate and methane from dichloromethane. As a prerequisite for biochemical studies on the dehalogenation mechanism we have isolated a pure culture from one of them in a three-stage procedure (A.Mägli, unpublished). The original dichloromethane utilizing mixed culture was first subjected to serial transfer through minimal

M. E. Lidstrom and F. R. Tabita (eds.), Microbial Growth on C₁ Compounds, 261–268.
© *1996 Kluwer Academic Publishers. Printed in the Netherlands.*

medium with 5 mM dichloromethane, and we obtained the homoacetogenic mixed culture DM. The two major components of this culture were isolated and characterized, but neither of them was capable of anaerobic growth with dichloromethane in pure culture (Braus-Stromeyer et al. 1993). At the second stage, culture DM was subcultured with dichloromethane, leading to a spontaneous increase in the growth rate of the system to 0.035 h^{-1} and to the appearance of a major new organism which we termed DMC. First attempts to isolate organism DMC in pure culture were unsuccessful. Strain DMC, however, was able to acetogenically grow with dichloromethane when associated with *Acetobacterium woodii*, *Methanobacterium hungatei* or with a *Desulfovibrio* sp.. This observation was taken to indicate that organism DMC is responsible for both the dehalogenation of dichloromethane and acetogenesis and that its dependence on a partner during growth with dichloromethane stemmed from the need for a growth factor produced by the associated organism (Mägli et al. 1995). At a third stage we were able to grow strain DMC in pure culture on minimal medium containing 5 mM dichloromethane plus 50% (v/v) of filter-sterilized spent medium from a culture of *Desulfovibrio* sp. strain DMB. Addition of spent medium was discontinued after a few transfers, and the organism was then able to grow on minimal medium with 5 mM dichloromethane at a growth rate of 0.025 h^{-1}.

Table 1. Selected properties of the strictly anaerobic, dichloromethane utilizing bacterium strain DMC

Non-motile, spherical rod with Gram-positive cell wall
42.7 % G+C
89% 16S rDNA identity with *Desulfitobacterium dehalogenans*
and *Desulfotomaculum orientis*
Optimum growth at 34°C and pH 7.0
Growth with dichloromethane
No growth with H$_2$/CO$_2$, formate, methanol, sugars, methoxylated amines
Presence of CO dehydrogenase, hydrogenase and formate dehydrogenase activities

---

Table 1 lists some of the properties of strain DMC. Its distinct phylogenetic position and the lack of high 16S rDNA sequence identity to any taxa previously described indicate that strain DMC represents a new genus. The characteristics of anaerobic growth, Gram-positive cell wall and relatively low G+C content are in agreement with its position within the radiation of the *Clostridium/Bacillus* subphylum. The most striking feature of strain DMC is its apparent inability to grow with carbon sources other than dichloromethane. Forty potential substrates were tested in addition to the compounds listed in Table 1, but none of them supported growth. Since the organism uses a C$_1$-substrate for growth, forms acetate as the major product from dichloromethane (A.Mägli, unpublished) and possesses CO dehydrogenase activity, the acetyl-CoA pathway (Drake 1994) is presumably operative. The inability of DMC to grow with H$_2$/CO$_2$, however, contrasts with this assumption and stresses the need for enzymological studies to reach an understanding of the metabolism of strain DMC.

Our main interest is with the dehalogenation of dichloromethane by strain DMC. To explore the conditions for establishing a cell-free dehalogenation system, preliminary experiments were performed with cell suspensions of a coculture of strain DMC with *Desulfovibrio* sp. strain DMB (Mägli et al. 1995). The degradative capacity of cells was fully

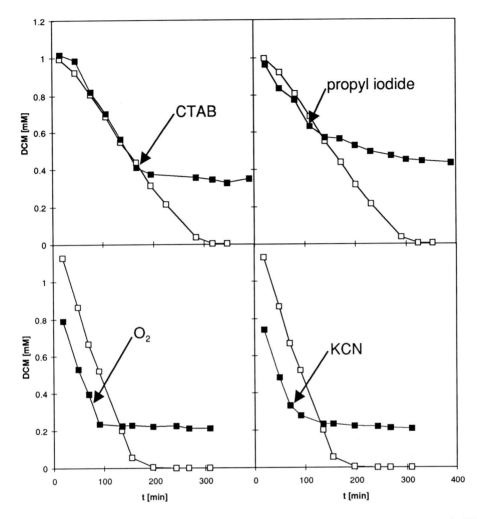

Figure 1: Cell suspension experiments with the coculture of strain DMC with *Desulfovibrio* sp. strain DMB. Experiments were done anaerobically at a protein concentration of 0.9 mg/ml in a total volume of 3 ml. At the time points indicated by arrows, 0.008% CTAB (hexadecyltrimethylammonium bromide), 50 μM propyl iodide or 1 mM potassium cyanide were added. Oxygen treatment was done by exchanging the gas phase (9 ml) with air. Open squares represent dichloromethane degradation in control experiments.

maintained upon storage in liquid nitrogen, whereas cells that had been frozen at -20°C or -80° C were completely inactive. The effect of some inhibitors on dichloromethane degrading cell suspensions is shown in Figure 1. Degradative activity was destroyed by the detergents CTAB (Fig.1) and Triton X-100. Since sodium and proton ionophores such as CCCP (carbonylcyanide-3-chlorophenyl- hydrazone), sodium ionophore II and monensin did not affect the degradation process (not shown), we conclude that dehalogenation is not dependent on an intact membrane potential. Rather, cell permeabilization by detergents appears to deplete a cofactor involved in dichloromethane dehalogenation. However, addition of potential cofactors such as various thiols (including glutathione), coenzyme B_{12} and tetrahydrofolate to either permeabilized cell suspensions or cell extracts did not restore dehalogenation activity. These preliminary results demonstrate that dichloromethane dehalogenation in strain DMC is mechanistically different from the glutathione-dependent dehalogenation reaction observed in aerobic methylotrophs.

According to our present view, dichloromethane dehalogenation in strain DMC may be catalyzed by an enzyme transferring dichloromethane onto tetrahydrofolate, thus yielding methylenetetrahydrofolate and inorganic chloride. As an intermediate of the acetyl-CoA pathway, methylenetetrahydrofolate yields electrons by oxidation of the methylene group to carbon dioxide on the one hand, and is reduced to methyltetrahydrofolate on the other hand, thereby providing the methyl group of acetate. A mechanism based on a dehalogenase/methyltransferase which transfers the methyl group of CH_3Cl to tetrahydrofolate has recently been demonstrated in a homoacetogenic bacterium growing with chloromethane (Messmer et al. 1993). By analogy, a dehalogenase/methylenetransferase reaction may be involved in dichloromethane utilization by strain DMC. Such a reaction might rely on a Co[I] corrinoid as the nucleophile for the dehalogenation of dichloromethane. The inhibition of dichloromethane dehalogenation in cell suspensions by cyanide, propyl iodide and oxygen (Fig. 1) is compatible with such a mechanism. Alternatively, a sulfhydryl group, which would be propylated and inactivated by propyl iodide, could serve as a nucleophile. A third possible mechanism would feature the use by the dehalogenase of the N_5-nitrogen of tetrahydrofolate as the nucleophilic moiety, thereby leading to the formation of chloromethyltetrahydrofolate, which spontaneously rearranges to methylenetetrahydrofolate.

Protein engineering of dichloromethane dehalogenase/glutathione S-transferase
The key enzyme in the degradation of dichloromethane by aerobic methylotrophic bacteria is dichloromethane dehalogenase (Leisinger et al. 1994). This enzyme catalyzes the glutathione (GSH) -dependent dehalogenation of dichloromethane to formaldehyde and inorganic chloride:

$$CH_2Cl_2 + GSH \rightarrow [GS\text{-}CH_2Cl] + HCl$$
$$[GS\text{-}CH_2Cl] + H_2O \rightarrow [GS\text{-}CH_2OH] + HCl$$
$$[GS\text{-}CH_2OH] \rightarrow HCOH + GSH$$

Dichloromethane dehalogenases, functionally and by their sequence, belong to the Theta-class of the large and ubiquitous superfamily of glutathione S-transferases (GST), which are enzymes involved in the binding and degradation of xenobiotics. However, dichloromethane dehalogenases display a number of properties which distinguish them from the majority of GST enzymes described so far. Most intriguing is their apparently restricted substrate range, dihalomethanes being their only known substrates. Considering the observed differences in protein sequences (56% identity between *Methylobacterium* sp. DM4 and *Methylophilus* sp. DM11 dichloromethane dehalogenases, and sequence identities below 25% between the dehalogenases and Theta-class GST enzymes), it seems quite improbable that these enzymes would have evolved recently in response to anthropogenic contamination of the environment with dihalomethanes. The absence of closely similar GST sequences, however, makes it difficult to relate dichloromethane dehalogenases to some specific GST precursor of known function. The high affinity of dichloromethane dehalogenases for glutathione, compared to GST enzymes of the eukaryotic Alpha, Mu and Pi classes, is another interesting feature. Somewhat paradoxically, dichloromethane dehalogenases and other members of the Theta-class cannot be purified on commercially available glutathione affinity supports. The cofactor binding site of Theta-class enzymes may therefore turn out to be somewhat different from that of the structurally well-characterized enzymes of the Alpha-, Mu- and Pi-classes (reviewed in Dirr et al. 1994).

In the absence of detailed structural information on dichloromethane dehalogenases or related enzymes, we have initiated protein engineering studies in order to investigate the role and importance of different residues for substrate binding, GSH affinity, and catalytic activity of the enzyme. The cloned dichloromethane dehalogenase gene of *Methylophilus* sp. DM11 can be expressed at high level in *E.coli* (Bader, Leisinger 1994). PCR-mediated site-directed mutagenesis has yielded a first series of mutants carrying single-residue changes within the N-terminal region of the enzyme. Numerous protein engineering studies on GST enzymes of the Alpha, Mu and Pi classes have previously defined a conserved N-terminal tyrosine residue to be essential for GST activity, the side chain hydroxyl group of tyrosine apparently acting as an enhancer of GSH nucleophilicity (Dirr et al. 1994). Inspection of sequence alignments indicates that the position of this tyrosine is not well-conserved in Theta-class enzymes (Fig. 2). Some of these enzymes have two tyrosine residues at the N-terminus,

```
Methylophilus sp. DM11      .........STKLRYLHHPASQPCRAVHQFMLENNIE
Methylobacterium sp. DM4    SPNPTNIHTGKTLRLLYHPASQPCRSAHQFMYEIDVP
Escherichia coli            ...........MKLFYKPGAC.SLASHITLRESGKD
Drosophila Dmgst27          ...........MDFYYHPCSAPCRSVIMTAKALGVD
Human erythrocyte Theta 1   .........MGLELYLDLLSQPCRAVYIFAKKNDIP
Human liver Theta 2         .........MGLELFLDLVSQPSRAVYIFAKKNGIP
Rat liver Theta 1           ..........VLELYLDLLSQPCRAIYIFAKKNNIP

Blowfly Theta X-ray         ...........MDFYYLPGSAPCRSVLMTAKALGIE
Rat liver Mu X-ray          ..........PMILGYWNVRGLTHPIRLLLEYTDSS
Human liver Alpha X-ray     .........AEKPKLHYFNARGRMESTRWLLAAAGVE
Pig lung Pi X-ray           .........PPYTITYFPVRGRCEAMRMLLADQDQS
```

Figure 2. Alignment of N-terminal sequences from selected GSTs of the Theta-class with sequences from GSTs of known three-dimensional structure. Residues corresponding to the N-terminal tyrosine and serine (Y6 and S12, respectively, in the dichloromethane dehalogenase from *Methylophilus* sp. DM11) are highlighted in boldface (see text).

while others lack a tyrosine residue altogether. In the case of strain DM11 dichloromethane dehalogenase the N-terminal tyrosine (Tyr-6) is present, but shifted towards the N-terminal end, suggesting that it may not be playing the same role as in GST enzymes from other classes. We therefore have produced mutations of the enzyme at Tyr-6 as well as at other positions of the N-terminal end of the protein.

Our results (Table 2) indicate that the N-terminal tyrosine is not required and indeed entirely dispensable with respect to enzyme activity, as is His-8 which, in sequence alignments, is the residue corresponding to the N-terminal residue of the Alpha-, Mu- and Pi-class enzymes. In contrast, a mutation at serine 12, a residue conserved across the entire Theta-class of GST enzymes but not in other classes (Fig. 2), essentially abolished enzyme activity (Table 2), despite high level expression of the mutant enzyme and properties very similar to the wild type enzyme during purification. Thus, in the case of Theta-class enzymes, the serine side chain corresponding to Ser-12 in the dichloromethane dehalogenase of *Methylophilus* sp. DM11 may be the functional equivalent of the N-terminal tyrosine residue of other GST enzymes.

Our data have been independently confirmed by a series of recent studies. Replacement of the N-terminal tyrosine by phenylalanine in GST from *E.coli* did not affect enzyme activity (Nishida et al. 1994). In the case of the Theta-class GST D27 from *Drosophila*, both adjacent N-terminal tyrosine residues have been shown to be involved in, but not essential for catalytic activity (Lee, Tu 1995). Finally and most importantly, the first X-ray structure of a Theta-class GST provides clear evidence that the serine corresponding to Ser-12 in the DM11 dichloromethane dehalogenase, and not an N-terminal tyrosine residue, is in close contact with the thiol group ot the glutathione cofactor. (Wilce et al. 1995).

Table 2. Kinetic properties of mutants of the strain DM11 dichloromethane dehalogenase

Activity of the mutant proteins was assayed spectrophotometrically at 340 nm by measuring the formation of NADH in a coupled enzyme assay with formaldehyde dehydrogenase at 30°C in 100 mM phosphate buffer pH 8.5, at [GSH] = 1 mM (in the determination of K_m (CH$_2$Cl$_2$) and k_{cat} (CH$_2$Cl$_2$) and at [CH$_2$Cl$_2$] = 1 mM (in the case of K_m (GSH)), respectively.

Enzyme	K_m (CH$_2$Cl$_2$) (μM)	k_{cat} (CH$_2$Cl$_2$) (s^{-1})	K_m (GSH) (μM)
wild-type	59 (± 6)	3.3 (± 0.1)	66 (± 3)
Y6F	32 (± 3)	2.6 (± 0.1)	78 (± 3)
H8Y	55 (± 4)	3.0 (± 0.1)	88 (± 5)
S12A	ND	< 0.1	ND

Identification of insertion elements associated with the dichloromethane utilization genes of methylotrophic bacteria. In most dichloromethane utilizing methylotrophs, *dcmA*, the structural gene of dichloromethane dehalogenase and *dcmR*, the gene encoding a putative regulatory protein governing expression of *dcmA*, are encoded on a 4.2 kb *BamH1* fragment. We have examined the regions flanking this conserved *BamH1* fragment by restriction analysis and have found that sequence homology in the *dcm* region extends over at least 10 kb in a number of dichloromethane utilizing strains. Nucleotide sequence analysis of the *dcm* region in *Methylobacterium* sp. DM4 has now led to the identification of three types of insertion sequences (M.Schmid-Appert et al., unpublished).

As shown in Figure 3, two identical direct repeats of insertion sequence IS*1354* flank the conserved *BamH1* fragment. IS*1354* extends over 1431 bp. Both of its copies have 48 bp imperfect terminal inverted repeats which are flanked by 8 and 9 bp direct repeats. The major open reading frame (*orf*) of IS*1354* encodes a putative transposase of 421 amino acids with 21% to 27% amino acid identity to the IS*256* class of insertion elements (Byrne et al, 1989). IS*1355*, another insertion element, is located directly adjacent to the left hand copy of IS*1354* (Fig. 3). This 975 bp element carries 9 bp perfect terminal inverted repeats, but no sequence duplication was observed in the flanking DNA. Two overlapping *orfs* of IS*1355* encode proteins of 153 and 179 amino acids which display sequence identities between 43% and 48% to IS*Rm4* (Soto et al. 1992) from *Rhizobium meliloti* and to IS*1031* (Coucheron 1993) and IS*1032* (Iversen et al. 1994) from *Acetobacter xylinum*. The third insertion element detected in the *dcm* region of *Methylobacterium* sp. DM4 is IS*1357*. It is positioned at the right hand end of IS*1355* (Fig.3). Since IS*1355* has apparently inserted into IS*1357*, the ends of the latter element could not be determined. IS*1357* contains an incomplete *orf* lacking a start codon but encoding a 369 amino acid polypeptide with 19% identity to the putative transposase of IS*1070* from the cyanobacterium *Calothrix* sp. PCC7601 (Mazel et al. 1991).

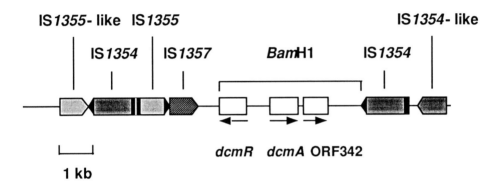

Figure 3. Organization of the of the *dcm* region in *Methylobacterium* sp. DM4. A core region encompassing the dichloromethane utilization genes *dcmA* and *dcmR* as well as *orf342* is flanked by insertion sequences whose characteristics are described in the text. *orf 342* encodes a putative protein with 27% amino acid identity to a subunit of 7α-cephem-methoxylase (Coque et al. 1995). The insertion sequences designated IS*1354*-like and IS*1355*-like have not been fully sequenced.

     The occurrence in methylotrophic bacteria of the three novel insertion elements described here was tested by Southern blot analysis. Ten dichloromethane utilizing facultative methylotrophs possessing a type A dichloromethane dehalogenase (Leisinger et al, 1994) contained at least one copy of each IS element. In *Hyphomicrobium* sp. GJ21, another strain with a type A dichloromethane dehalogenase, only IS*1355* and IS*1357* were present. None of the three insertion elements was detected in the genomes of *Methylobacterium extorquens* AM1 and *M.organophilum* XX, two organisms unable to grow with dichloromethane, and the elements were also undetectable in *Methylophilus* sp. DM11, a facultative methylotroph possessing a type B dichloromethane dehalogenase. It thus appears that their occurrence is restricted to methylotrophic bacteria relying on a type A dehalogenase for dichloromethane utilization. However, a possible function of these elements in transferring the dichloromethane utilization genes among methylotrophic bacteria has yet to be demonstrated.

CONCLUSIONS

The adaptation of bacteria to dichloromethane has resulted in organisms capable of utilizing this compound as the sole carbon and energy source, and three aspects related to the evolution of this metabolic trait have been investigated.

- A novel, strictly anaerobic acetogenic bacterium growing with dichloromethane as the apparently exclusive carbon and energy source has been isolated in pure culture and characterized. Further studies now aim at elucidating the mechanism of dichloromethane dehalogenation in this organism.

- A system for mutagenesis and expression of dichloromethane dehalogenase/glutahione S-transferase, the key enzyme of dichloromethane utilization in aerobic methylotrophs has been established. A first series of mutagenesis studies at the N-terminal end of the enyzme, demonstrates that a serine residue conserved across the Theta-class of GST enzymes is essential for the function of DCM dehalogenase, and that the N-terminal tyrosine shown to be essential for the function of GST enzymes of the Alpha, Mu and Pi classes of GSTs is dispensable.

- The dichloromethane utilization genes of methylotrophs possessing a type A dichloromethane dehalogenase were found to be associated with insertion sequences. Three novel insertion sequences, IS1354, IS1355 and IS1357, were identified by nucleotide sequence analysis. It remains to be explored whether these elements, by themselves or in conjunction with the dichloromethane utilization genes, move by transposition.

ACKNOWLEDGEMENTS

The work reported here was supported by grants from the Swiss Federal Institute of Technology, Zürich and by the Swiss Priority Programme Biotechnology.

REFERENCES

Bader R and Leisinger T (1994) J.Bacteriol. 176, 3466-3473.
Braus-Stromeyer SA et al (1993) Appl.Environ.Microbiol. 59, 3790-3797.
Byrne ME et al (1989) Gene 81, 361-367.
Coque JJR et al (1995) J.Bacteriol. 177, 2230-2235.
Coucheron DH (1993) Mol.Microbiol. 9, 211-218.
Dirr H et al. (1994) Eur.J.Biochem. 220, 645-661.
Drake HL, ed (1994) Acetogenesis, Chapman and Hall, New York, USA.
Freedman DL and Gossett JM (1991) Appl.Environ.Microbiol. 57, 2847-2857.
Gribble GW (1994) Env.Sci.Technol. 28, 310A-319A.
Isidorov VA (1990) Organic chemistry of the earth's atmosphere, Springer-Verlag, Berlin, Germany.
Lee HC and Tu CP-D (1995) Biochem.Biophys Res Commun 209, 327-334.
Leisinger T et al (1994) Biodegradation 5, 237-248.
Mazel D et al (1991) Mol.Microbiol. 5, 2165-2170.
Mägli A et al (1995) Appl.Environ.Microbiol. 61,2943-2949.
Messmer M et al (1993) Arch.Microbiol. 160, 383-387.
Nishida M et al (1994) J Biol.Chem. 269, 32536-32541.
Soto MJ et al (1992) Gene 120, 125-126.
Stromeyer SA et al (1991) Biodegradation 2, 129-130
Wilce CJ et al (1995) EMBO J. 14, 2133-2143.

Microbial Ecology of PCB Transformation in the Environment: A Niche for Methanotrophs?

Angela S. Lindner and Peter Adriaens[*]
Department of Civil and Environmental Engineering
The University of Michigan, Ann Arbor, MI 48109-2125

ABSTRACT
       Whereas anaerobic and aerobic heterotrophic activity against polychlorinated biphenyls (PCBs) has been well established in the laboratory and in the field, the potential niche for methanotrophic bacteria in these environments has received only scant attention. We recently demonstrated that methanotrophic groundwater isolates were able to hydroxylate *ortho* - and *ortho -/para* - substituted chloro- and chlorohydroxybiphenyls and that heterotrophs present in mixed cultures may have been responsible for subsequent ring-fission. The oxidation of 2-chlorobiphenyl by the Type II methanotroph CSC1, present in the groundwater mixture used, was shown to proceed via a 1,2-intramolecular rearrangement, resulting in migration of the chlorine to the *meta* -position. Preliminary evidence from incubations of strain CSC1 with other substituted aromatics indicates that this mechanism may be ubiquitous and is affected by the substituent, based on the appearance of both hydroxylated transformation products. Analysis for products effecting chlorine migration in PCB (or haloaromatic) contaminated environments may thus provide a fingerprint for *in situ* cometabolic methanotrophic activity and should be included in investigations on natural or enhanced bioremediation processes.

INTRODUCTION
       Natural intrinsic biotransformation of polychlorinated biphenyls (PCBs) has been observed in river and lake sediments, as well as in soils. Based on the distribution of products observed, both anaerobic and aerobic biodegradative processes have been recognized. For example, natural reductive dechlorination of industrial (Aroclor) mixtures in anaerobic methanogenic sediments results in the accumulation of predominantly the *ortho*- chlorinated mono- and dichlorobiphenyls, 2- and 4-chlorobiphenyls, and 2,2'-, 2,6-, 2,4-, 2,3-, and 2,4'-dichlorobiphenyls (Brown et al., 1987; Quensen et al., 1988; Mohn, Tiedje, 1992). Recently, naturally occurring aerobic PCB degradation was demonstrated by analysis of Hudson River sediment cores within and outside PCB-contaminated regions (Flanagan and May, 1993). Mono- and dichlorobenzoate intermediates (2-, 4-, 2,4-, and 2,5-substitution) corresponding with PCB depth profiles and to the PCB isomers which accumulate from reductive dechlorination of Aroclors, were detected in the contaminated sediment cores. In bench-scale and field experiments, aerobic PCB-degrading populations in sediments

269

M. E. Lidstrom and F. R. Tabita (eds.), Microbial Growth on C₁ Compounds, 269–276.
© 1996 *Kluwer Academic Publishers. Printed in the Netherlands.*

were found to be stimulated by oxygen, hydrogen peroxide and biphenyl amendments, resulting in the overall decrease of mono- to trichlorobiphenyls and the accumulation of corresponding chlorobenzoate intermediates (Anid et al., 1993; Harkness et al., 1993). Several isolates were found to exhibit cometabolic activity against tetra- and hexachlorinated biphenyls. Although the level of activity of aerobic heterotrophs can be expected to be low under the prevailing conditions of low oxygen tensions, the evidence for stimulation of cometabolism testifies to the potential for sequential anaerobic/aerobic treatment strategies.

However, these same environments harbor methanotrophic bacteria, which frequently are microaerophiles and live on the interface between aerobic and anaerobic zones where stable sources of methane are present (such as those generated by methanogens). Since these microorganisms play an important role in the carbon and electron flux in the environment, their potential role in PCB cometabolism and, in particular, in oxidation of substituted aromatic compounds was investigated (Higgins et al., 1980).

RESULTS
a.) Transformation of Chloro- and Chlorohydroxybiphenyls by Methanotrophic Groundwater Isolates. The strains used in this study were (i) a mixed culture (MM1) containing a type II methanotroph and several heterotrophs, (ii) the pure Type II methanotroph CSC1 isolated from (i), and (iii) an axenic type I methanotrophic culture, *Methylomonas* sp. MM2. These cultures were previously isolated from a relatively pristine aquifer (Moffett Naval Air Station, California) and were shown to oxidize trichloroethylene (TCE) (Henry, Grbic'-Galic', 1990).

Both mixed culture MM1 and the type II strain CSC1 were found to transform 2- and 4-chlorobiphenyl (CB), 4-hydroxy-2-chlorobiphenyl (4OH2CB), 2-hydroxy-3-chlorobiphenyl (2OH3CB), and 2,4- and 2,4'-dichlorobiphenyl (DCB). Except for 4CB, 30% and 17% of which disappeared upon incubation with CSC1 and MM1, respectively, substrate disappearance was generally found to be higher in the mixed culture incubations (30-51%) than in incubations with strain CSC1 (24-42%) alone. The primary products found in both the pure and mixed culture incubations were identified as hydroxylated chlorobiphenyls (UV-visible spectrum; λ_{max} 246 nm; pH 7), based upon mass spectral analysis. The formation of hydroxylated intermediates was the result of methanotrophic oxidation, since (i) these metabolites were found in both the mixed and pure culture incubations, (ii) the range of chlorobiphenyl substrates oxidized was nearly identical for both inocula, and (iii) hydroxybiphenyl monooxygenases have been shown to be incapable of hydroxylating chlorinated biphenyl analogues. Incubations of these congeners with the type I methanotrophic *Methylomonas* sp. MM2, however, did not result in the formation of hydroxylated intermediates, which is consistent with the findings that the particulate MMO (pMMO) has a more restricted enzyme specificity.

Upon addition of 0.1 N NaOH to the neutral extracts from the 3CB, 4CB, and 4OH2CB incubations, highly water-soluble, yellow-pigmented compounds (λ_{max} 418-442, pH 10) formed, and a hypochromic effect (i.e., absence of wavelength shift and decrease in absorbance) was observed. Since *ortho* - quinones have been described to exhibit such hypochromic shifts, these intermediates are hypothesized to have a quinone-type structure (March, 1985). Acidic extracts of mixed culture incubations with 2- and 4-CB and with chlorinated hydroxybiphenyls contained both chlorinated and non-chlorinated products, resulting from ring cleavage presumably due to heterotrophic activity. This observation may indicate a role for methanotrophs to activate aryl halides via hydroxylation to facilitate further metabolism by heterotrophs.

b.) Transformation of Hydroxybiphenyls and Biphenyl by Methanotrophs. Since an hydroxylated CB was detected in 2CB incubations, incubations of the mixed culture MM1 and the pure methanotroph CSC1 with biphenyl, 2-hydroxybiphenyl (2OHB), 3-hydroxybiphenyl (3OHB), and 2,2'-dihydroxybiphenyl (2,2'-OHBP) were conducted to determine the metabolic activities of the strains and the effects of substituent type and position on the oxidation. Results from the biphenyl and monohydroxybiphenyls are provided in Table 1. Oddly, substrate disappearances followed a trend in direct opposition to those obtained with the chloro- and chlorohydroxybiphenyl incubations discussed above. The CSC1 incubations yielded, on average, a 138% greater disappearance of substrate than the mixed culture incubations. One logical explanation for this behavior is that the presence of chlorine may, perhaps, inhibit the degradation potential of the CSC1 but not of the heterotrophs present in the mixture. For both strains used, an hydroxylated metabolite as well as the as yet unidentified yellow product were detected. Gas chromatographic/ mass spectral analysis of the biphenyl incubation extract showed the appearance of two distinct peaks, identified as mono-hydroxylated biphenyls. The loss of a proton resulting from a benzylic C-H cleavage (169, [M-1]+) and of hydrogen due to molecular thermolysis (169,

Table 1: Results from incubations of mixed culture MM1 and CSC1 with biphenyl and monohydroxylated biphenyls

Substrate	Treatment	Substrate Recovered (mg/L)		Metabolites
		Mix MM1	CSC1	
BP	Live	2.6 ± 1.9	1.9 ± 1.4	Yellow, OHB
	Killed	9.5	10.5	
	Chemical	10.5	ND*	
2OHB	Live	5.2 ± 3.8	0.8 ± 0.2	Yellow, diOHB
	Killed	9.0	10.2	
	Chemical	9.2	11.1	
3OHB	Live	5.4 ± 0.2	3.1 ± 1.0	
	Killed	5.9	10.4	
	Chemical	6.7	11.9	

*Not determined

[M-2]+) as well as strong ions resulting from the loss of CO and CHO are characteristic for alcohols. These potential products could possibly have been formed either by an arene oxide intermediate or by a direct loss mechanism as to be discussed in detail below. Incubation of CSC1 with 2-hydroxybiphenyl (λ_{max} 286 nm) yielded a UV spectrum showing complete disappearance of this compound and formation of a metabolite with a maximum wavelength value at 315 nm (Figure 1). Incubations of 3-hydroxybiphenyl and 2,2'-dihydroxybiphenyl resulted in similar product formation with shifted maximum wavelengths in the 315-320 nm range; however, without complete disappearance of the substrates. After each OHBP incubation, acidification

272

yielded no shift in wavelengh, only an decrease in intensity (hypochromic shift). Such a shift in maximum wavelength would be expected if the product were a dihydrodiol, since acidification would result in the release of H_2O and the formation of a monohydroxylated compound. This provides strong evidence for the formation of a quinone-type compound, which, upon acidification, would not result in a shift in maximum wavelength.

    c.) Mechanism of Oxidation of Substituted Biphenyls by Methanotrophs. Formation of the hydroxylated intermediates could result either from direct hydroxylation of the aromatic ring in a previously unsubstituted position or by an NIH shift rearrangement--an hydroxylation-induced intramolecular migration of a substituent on an aromatic ring--of an arene oxide intermediate. Because the transformations catalyzed by soluble methane monooxygenases (sMMO) resemble those of cytochrome P-450, which has been demonstrated to form hydroxylated intermediates via an NIH shift, a similar mechanism has been implicated in sMMO-mediated oxidations of compounds.

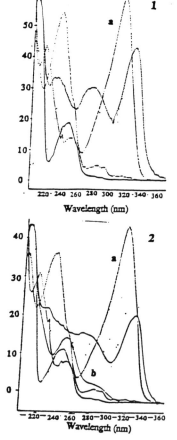

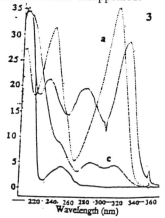

Figure 1: UV absorption spectra of 2OHBP (1), 3OHBP (2), and 2,2'OHBP (3) metabolites after incubation with CSC1. Peaks labeled (a) are the hypothesized quinone-type metabolites. Peak b is 3OHBP; peak c is 2,2'-OHBP. All of 2OHBP disappeared.

The NIH shift was fully characterized in the late 1960's and early 1970's by researchers at the National Institute of Health using the enzymes phenylalanine hydroxylase, tryptophan-5-hydroxylase, aryl hydroxylase, and tyrosine hydroxylase. It was determined that the migration of the substituent followed a "stepwise" mechanism (Figure 1), where the enzyme plays no role beyond the addition of the hydroxyl to a specific position on the molecule (Jerina et al., 1967).The common intermediate formed in the direct loss pathway (a), the migration pathway (b), and the spontaneous isomerization pathway (c) is an arene oxide, whose formation by both higher organisms and lower organisms and plants has been well-documented (Jerina et al., 1970; Daly et al., 1972). It has been found in the pathway shown above that the formation of the arene oxide is the rate-limiting step, whereas the isomerization of the arene oxide to the phenolic compound is the non-rate-limiting step (Boyd et al., 1972; Jerina, Daly, 1973 ; Kurata et al., 1988; and Nasir et al.,1992). The rapid formation of the ketone intermediates was recently reported by Nashed et al. (1993)except for when methyl substituents are present. These researchers were also able to conclude that the energy of aromatization is the major driving force for the enolization of the ketone intermediates. Formation of the phenolic intermediates was found to be slow in relation to the formation of the ketone from the arene oxide, and its rate is affected by steric factors, such as the nonplanarity of the aromatic system or steric interference between the phenolic hydroxyl group and a peri substituent.

As shown in Figure 2, a charged intermediate results from the stepwise mechanism; hence, the charge distribution in this intermediate determines the extent to which migration occurs. As a result, the degree of substituent migration and retention can be predicted upon the basis of the type of substituent present. In general, it was found that compounds with electron-donating substituents, such as phenols and aniline, exhibit very low degrees of migration and retention, favoring the direct loss pathway (Daly et al., 1967; Kurata et al., 1988). However, electron-withdrawing substituents, such as chlorine and bromine, cause no delocalization of charge in a cationoid

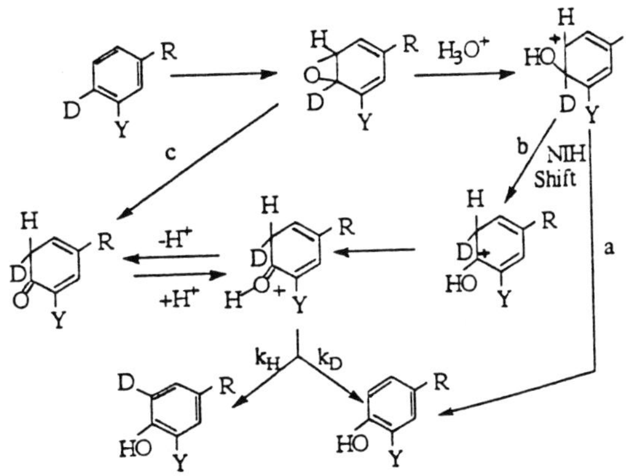

Figure 2 : Pathways consistent with the stepwise mechanism (Y=aryl substituent; R=H, Cl, NH$_2$, etc.) (Jerina and Daly, 1973)

intermediate, thus, promoting migration and retention. Location of the hydroxylation in relation to the substituent highly influences the degree of migration and retention. Migrations and retentions of deuterium during *ortho* - and *para* -hydroxylation are comparable, whereas *meta* -hydroxylation of monosubstituted benzenes has been shown to result in lower migration and retention of deuterium(Daly et al., 1967). Environmental conditions, namely pH, have also been shown to affect the extent of NIH shift occurring. As pH increases, retentions on compounds with electron-donating substituents have been shown to decrease, while pH has no effect on compounds with electron-withdrawing substituents present. In experiments on the hydroxylation of 1,4-dimethyl benzene, it was found that the extent that methyl migrates is a function of pH, with greater migrations under neutral and basic conditions.

Evidence for the migration of chlorine during oxidation of 2-chlorobiphenyl by an axenic methanotrophic groundwater isolate, CSC1, expressing the soluble form of methane monooxygenase, was recently reported by Adriaens (1994). Based upon GC, MS, and ^{1}H NMR results, 88% of the original 2-chlorobiphenyl was found to be oxidized to 2-hydroxy-3-chlorobiphenyl, consistent with the previous findings that the more electronegative the substituent, the higher the degree of migration occurring. A proposed mechanism for this oxidation follows pathway b in Figure 2, with the initial formation of an arene oxide, the spontaneous opening of the arene oxide ring, subsequent formation of the corresponding ketone with concurrent NIH shift, and, finally, enolization to the product. In the case of hydroxy biphenyls, the electron-donating phenolic substituent would favor the direct loss mechanism over the arene oxide pathway. However, the formation of quinone-type compounds may indicate that both pathways exist.

CONCLUSIONS

The niche for methanotrophic transformation of PCBs in the environment has long been overlooked in favor of studying the roles of anaerobic and aerobic heterotrophs on such transformations. However, because *ortho-* chlorinated PCB congeners have been shown to be recalcitrant to both aerobic heterotrophs and anaerobic methanogenic populations, further study using methanotrophs is warranted. This study shows that Type II methanotrophs, harboring sMMO, hydroxylated o-, p-, and o,p-chlorinated congeners preferentially, possibly by means of an arene oxide intermediate. The hydroxylated metabolites formed are more water soluble than their parent compounds and, hence, potentially pose more of an environmental threat because of their greater mobility in water environments. When in the presence of dioxygenase-harboring heterotrophs, the hydroxylated metabolites are transformed to catechols and ring-cleavage products, thus, offsetting this problem of greater mobility. An overall schematic describing the potential for anaerobic and aerobic PCB transformation is proposed in Figure 3. The type of substituent present on the ring was found to influence the oxidation mechanism followed by methanotrophs. Whereas, 2CB was shown to undergo an NIH shift mechanism upon incubation with sMMO, hydroxylated biphenyls are believed to form quinone-type compounds by means of a direct loss mechanism.

Further efforts will be directed in determining the ubiquity of the NIH shift during oxidation of substituted aromatic compounds. Use of a matrix composed of various deuterated substrates (with either electron-donating or electron withdrawing substituents) and several different methanotrophs will enable achieving this goal. The

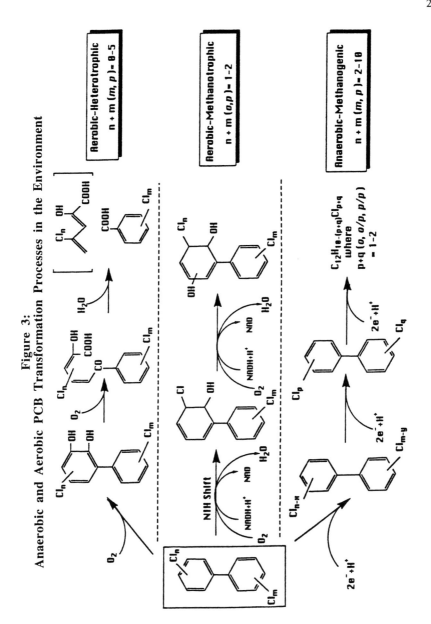

Figure 3:
Anaerobic and Aerobic PCB Transformation Processes in the Environment

identity of the intermediates as well as the effects of the type of substituent on migration may help establish the role that methanotrophs play in contaminated environments.

REFERENCES

Adriaens, P et al. (1989) Appl. Environ. Microbiol. 55, 887-892.
Adriaens, P et al. (1991) Progress report. Research and Development Program for the Destruction of PCBs, General Electric Co., Schenectady, N.Y.
Adriaens, P and Hickey, WJ (1993) In Stoner, DL, editor, Biotechnology for Hazardous Waste Treatment, pp. 97-137, Lewis Publishers, Boca Raton, FL.
Adriaens, P (1994) Appl. Environ. Microbiol. 60, 1658-1662.
Adriaens, P and Grbic'-Galic', D (1994) Environ. Sci. Technol. 28, 1325-1330.
Anid, PJ et al. (1993) Biodegradation 4, 241-248.
Boyd, DR et al. (1972) Biochemistry 10, 1961-1966.
Brown, JF et al. (1987) Science 236, 709-712.
Daly, JW et al. (1967) In Mayo, F., Gen. Chairman, Oxidation of Organic Compounds. Vol. III. Ozone Chemistry, Photo and Singlet Oxygen and Biochemical Oxidations, pp. 279-289, American Chemical Society, Washington, D.C.
Castle, L et al. (1980) J. of Molc. Catalysis 7, 235-243.
Daly, JW (1972) Experientia 28, 1129-1264.
Flanagan, W and May, R (1993) Environ. Sci. Technol. 27, 2207-2212.
Guroff, G et al. (1967) Science 157, 1524-1530.
Harkness, M et al. (1993) Science 259, 503-507.
Henry, SM and Grbic'-Galic', D. (1990) Microb. Ecol. 20, 151-169.
Higgins, IJ et al. (1980) Science 286, 561-564.
Jerina, DM et al. (1967) J. Am. Chem. Soc. 89, 5488-5489.
Jerina, DM et al. (1970) Biochemistry 9, 147-154.
Jerina, DM and Daly, JW (1973) In King, TE; Mason, HS; and Morrison, M, editors, Oxidases and Related Redox Systems, pp. 143-161, University Park Press, Baltimore, Maryland.
Kurata, T et al. (1988) J. Am. Chem. Soc. 110, 7472-7478.
March, J (1985) Advanced Organic Chemistry--Reactions, Mechanisms, and Structure, 3rd edition, John Wiley & Sons, New York, NY.
Mohn, WW and Tiedje, JM (1992) Microbiol. Rev. 56, 482-507.
Nasir, MS et al. (1992) J. Am. Chem. Soc. 114, 2482-2494.
Sondossi, M et al. (1991) J. Ind. Microbiol. 7, 77-88.
Quensen, FJ et al. (1988) Science 242, 752-754.
Uchiyama, H et al. (1992) Appl. Environ. Microbiol. 58, 3067-3071.

Metabolic aspects of plant interaction with commensal methylotrophs

S.K. Freyermuth, R.L.G. Long, , S. Mathur, M.A. Holland,[1] T.P.
Holtsford,[2] N.E. Stebbins, R.O. Morris and J.C. Polacco

Department of Biochemistry and [2]Division of Biological Sciences,
University of Missouri-Columbia, USA. [1]Department of Biology, Salisbury
State University, Salisbury, MD.

INTRODUCTION
      Pink-pigmented, facultatively methylotrophic bacteria (PPFM's) are
normally associated with seeds and leaves of soybean and all other plants we
have examined. We became aware of PPFM's as covert colonizers of callus
cultures. PPFM's produce a constitutive urease which appears to contribute
to the 'background' urease of callus cultures and unifoliate leaves of a mutant
defective in the structural gene (*Eu4*) for the plant-encoded ubiquitous urease
isozyme (Polacco et al., 1989; Holland and Polacco, 1992). A second class (class
II) of soybean mutants, which have pleiotropically lost the activities of both
the ubiquitous urease and the embryo-specific ureases (Meyer-Bothling et al.,
1987) harbors PPFM's which are deficient in urease and hydrogenase activities
in planta. The bacteria are transiently urease- and hydrogenase-negative
when cultured away from the plant, and re-acquisition of these activities is
accelerated by supplemental nickel (Holland and Polacco, 1992).
      We report here our progress in elucidating the nature of the class II
genes: *Eu2* and *Eu3*. Mutation in either gene results in the simultaneous loss
of activity of both the plant ureases and the urease of associated PPFM's.
Further, plant-PPFM metabolic associations are explored with respect to a
potential PPFM role in biosynthesis of the plant hormone, cytokinin.

RESULTS
      Surface sterilization does not remove PPFM's from soybean seeds. We
make soybean seeds 'tissue culture-clean' for germination to axenic seedlings,
the source of explants for callus induction. This involves soaking 5 min. in
70% ethanol, extensive washing in sterile distilled water (st. dH$_2$O) and
soaking 5 min. in 1.3% NaOCl (25% strength commercial bleach). Seeds are
then germinated in st. dH$_2$O in petri dishes, a condition in which they easily
manifest contamination (failure to germinate and overgrowth by bacteria or
fungi). We almost invariably find PPFM's as covert colonizers of callus
derived from all sections of "pristine" seedlings: hypocotyl, radicle, plumule,

M. E. Lidstrom and F. R. Tabita (eds.), Microbial Growth on C$_1$ Compounds, 277–284.
© 1996 *Kluwer Academic Publishers. Printed in the Netherlands.*

cotyledon. "Finding PPFM's" involves grinding callus in st. dH$_2$O and plating on methylotroph selection media containing either methylamine or methanol (ATCC media formulations 396 or 784, respectively) as sole C/energy sources. In addition to pink colonies, other methylotrophs appear as white or yellow colonies. Cells of the former have the morphology of *Hyphomicrobium* spp., reported by Horsch and King (1983) to be a "covert contaminant" of *Datura innoxia* cultures and to contribute to the "leakiness" of an adenine auxotrophic line (King et al., 1980).

In one recent experiment we sampled bacteria in 14 day seedlings derived from sterilized and unsterilized seeds (germinated in germination paper rolls in which non-sterile seeds are usually not overgrown by microbial contaminants). While sterilization eliminated all or most bacteria forming colonies on rich medium (Luria Broth) at 37° C, it reduced the PPFM titer at this stage by only about 75% (to about 1.4 x 10^6 cfu·g fw^{-1}). We have not yet distinguished between two possibilities: that PPFM's are within the seed coat and therefore escape sterilization, or that they are unusually resistant to ethanol/NaOCl. There is precedent for seedborne bacteria; Dunleavy (1989) reported that the yellow bacterium, *Curtobacterium plantarum*, is found within immature and mature soybean seeds.

PPFM's cluster with isolates of Methylobacterium spp. Most of our characterization of PPFM's employed a single isolate from soybean leaf (soy leaf #2). We called this isolate *Methylobacterium* sp. and suggested that it was most likely *M. mesophilicum* (Holland and Polacco, 1992) although subsequent nutritional tests (Green and Bousfield, 1983) indicated much closer relatedness to *M. extorquens* .

Since 16S rRNA sequence comparison is a more reliable measure of bacterial relatedness than physiological criteria (Olsen et al., 1994) we PCR-amplified 16S rDNA (Weisberg et al., 1991) from 11 strains and compared their restriction fragment patterns. We compared four identified species (ATCC type cultures for *M. extorquens, M. mesophilicum* and *M. organophilum*, plus *Methylobacterium* sp. DM4 obtained from Gälli and Leisinger [1985]) with 7 isolates recovered from 5 different plants (2 monocots and 3 dicots). Most of the PPFM isolates cluster among the known *Methylobacterium* spp. The best Fitch-Margoliash (1967) tree of Nei and Li (1979) genetic distances places most of the PPFM's within the *Methylobacterium* group (Fig. 1A). This tree fits the data at least twice as well (by minimizing least squares deviations) as any tree which puts the PPFM's outside the *Methylobacterium* group. The PPFM isolates were also very similar (or identical) to known *Methylobacterium* spp. in nine nutritional assays (Fig. 1B). Four isolates were identical to *M. extorquens* and *M. organophilum*. No PPFM isolate had more than 2 nutritional differences separating it from a known *Methylobacterium* isolate. Obviously, more phylogenetic data and experimentation are needed to address plant host specificity among the PPFM's.

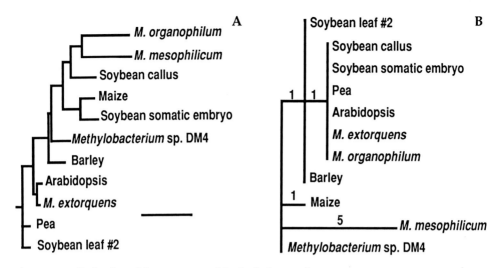

Figure 1. Relationships among Methylobacteria. (A) Fitch-Margoliash (1967) tree of Nei and Li (1979) distances based of restriction fragments of 16S rDNA. The scale bar represents a distance of 0.04 substitutions per nucleotide position. (B) Parsimony tree based on nine nutritional assays (growth on tryptic soy broth, sensitivity to erythromycin, polymixin B and nalidixic acid, utilization of arabinose, rhamnose and sucrose as C source, nitrate as N source and methylamine as C/N source). The numbers above the branches denote the number of nutritional traits that differ among the isolates separated by that branch.

The nature of the soybean Eu2 and Eu3 genes. Mutational elimination of soybean urease activity has unmasked plant controls on urease expression in associated PPFM's. Four urease genes have been identified (Table I).

Table I. Patterns of loss of urease activity as a result of lesions in urease loci of soybean (summarized in Polacco and Holland, 1993).

LOCUS	FUNCTION	DEFECTIVE ALLELE	UREASE ACTIVITY	
			EMBRYO	UBIQUITOUS
Eu1 :	STRUCTURAL	*eu1-sun*	-	+
Eu4 :	STRUCTURAL	*eu4*	+	-
Eu2 :	Ni Insertion?	*eu2*	-	-
Eu3 :	Ni Insertion?	*eu3-e1*	-	-

Eu1 and *Eu4* are the structural genes for the embryo-specific and tissue-ubiquitous ureases, respectively (Meyer-Bothling and Polacco, 1987; Torisky et al., 1994). The pleiotropic *Eu2* and *Eu3* genes are formally identical to bacterial urease accessory genes, whose products are necessary for insertion of nickel into the urease active site (Lee et al., 1992). Mutation in *Eu4*, *Eu2* and *Eu3* has revealed PPFM-plant interactions in production of 'plant' urease.

First, elimination of the ubiquitous urease in *eu4/eu4* plants did not eliminate urease expression in all tissues. As shown in Table II (highlighted entries) there was much higher urease activity in unifoliate leaves and callus than expected from activity in trifoliate leaves. (A pair of unifoliates is the first true leaves; all subsequent leaves are trifoliates. As *eu4/eu4* unifoliates expand, their urease activity drops to the trifoliate level.) Although the urease of NH_4Cl-supported *eu4/eu4* callus was fully 40% that of wild type callus (Table I), it was insufficient to confer on *eu4/eu4* callus the ability to utilize urea N or sensitivity to ≥ 50 mM urea (Polacco et al., 1989). These observations were explained by demonstrating that most *eu4/eu4* callus urease is bacterial (Holland and Polacco, 1992).

Table II. Whole tissue normalized urease levels in class II (*eu2/eu2* and *eu2/eu2*) and class III (*eu4/eu4*) mutants (summarized from data in Meyer-Bothling et al. 1987; Polacco et al. 1989). Callus activities are given for growth on either of two N sources.

	LEAF		CALLUS	
GENOTYPE	unifoliate (young)	trifoliate	NO_3^-/NH_4^+	NH_4^+
Williams 82 (progenitor)	1.0	1.0	1.0	1.0
eu2/eu2	0.002	0.0002	nd	0.02
eu3-e1/eu3-e1	0.001	0.002	0.0001	0.005
Eu3-e3/Eu3-e3	0.003	0	nd	0.003
eu4/eu4	**0.2**	0.02	**0.14**	**0.39**

If bacteria generate the background urease in *eu4/eu4* callus and unifoliates, then why are these tissues urease-negative in *eu2/eu2* and *eu3-e1/eu3-e1* mutants (Table II)? The short answer is that PPFM's on these mutants are urease-negative. They are transiently urease-negative in free-living culture and the re-acquisition of urease is accelerated by extra nickel supplementation (Table III).

Table III. Urease activity of PPFM's isolated from leaves of urease-negative soybean mutants (taken from Holland and Polacco, 1992).

SOURCE OF PPFM's	UREASE[a] (nmol urea·min^{-1}·OD$_{550}$)	
	Minus Ni	Added Ni[b]
Williams 82 (progenitor)	21.0 ± 0.8	17.9 ± 2.4
eu2/eu2	3.9 ± 0.4	26.8 ± 3.3
eu3-e1/eu3-e1	4.6 ± 1.1	19.9 ± 0.5
eu4/eu4	16.3 ± 4.2	16.3 ± 1.4

[a] Cultures were initiated from single colonies that appeared on AMS plates 5 d after plating leaf macerates and were grown overnight in tryptic soy broth ± added Ni.

[b] Added as 10 µM $NiSO_4$ in the presence of 10 mM Kcitrate, pH 6.0.

We do not know which plant signal is altered in the class II *eu2/eu2* and *eu3-e1/eu3-e1* mutants resulting in a urease-negative (and hydrogenase-negative) phenocopy in associated PPFM's. We are confident, however, that it is neither Ni nor urea. By the criteria of tissue Ni content, seedling uptake of ^{63}Ni, Ni delivery from mutant root stocks and Ni delivery from mutant seed coats to wild type embryos Ni uptake and movement appear normal in class II mutants (Holland and Polacco, 1992). Class II mutants accumulate urea (Stebbins et al., 1991) and it is possible that efflux of urea or altered efflux/production of other N compounds represses PPFM urease expression. This does not seem likely since PPFM urease is constitutive *in vitro* and since *eu4/eu4* mutants also accumulate urea but harbor urease-positive PPFM's.

To understand the plant signal altered in the class II mutants, it behooves us to understand the nature of the *Eu2* and *Eu3* genes. Our approach is to clone plant urease accessory genes and to test their identity to *Eu2* or *Eu3* by using them to correct *eu2/eu2* and *eu3-e1/eu3-e1* mutants in callus culture. Since it is possible that a cloned urease accessory gene may be a class II gene not yet identified by mutation we are employing a parallel approach: production of antisense transgenic plants and cell lines. Are these lines urease-negative? Do they harbor urease-negative PPFM's?

Urease and urease accessory genes have been studied extensively in the bacterium *Klebsiella aerogenes*. The *K. aerogenes* gene cluster contains three structural genes, *ureA, B* and *C* and four accessory genes, *ureD, E, F* and *G*. The three structural genes encode subunits of the urease apoenzyme. These three subunits appear to be "fused" in plant ureases: UreA is 59% identical to residues 1-101 in jackbean, UreB is 52% identical to residues 132-237, and UreC is 60% identical to residues 271-840 (Mulrooney and Hausinger, 1990). Besides the three structural genes which encode the apourease, all four accessory genes in *K. aerogenes* are necessary for the functional incorporation of the nickel-containing urease metallocenter (Lee et al., 1992). UreD is proposed to be a chaperone protein that stabilizes a urease apoprotein conformation making it competent for nickel incorporation (Park et al., 1994). The UreE ORF contains potential metal-binding sites, including a carboxy-terminus in which 10 of 15 residues are histidines. This structure agrees with the hypothesized role of UreE as binding intracellular nickel and functioning as a nickel donor during urease metallocenter assembly (Lee et al., 1993). While nothing is known about the UreF function, the deduced UreG sequence contains a nucleotide binding motif and is believed to be involved in coupling ATP hydrolysis to the nickel incorporation process (Lee et al., 1993).

We have recovered plant (soybean and *Arabidopsis*) homologs of bacterial urease accessory gene *ureG*. Both plant cDNA clones are partial; the largest, from *Arabidopsis*, is 780 bp and contains a 400 bp region with 56 to 66% nucleotide identity to bacterial *ureG*. Like UreG protein, it contains a deduced P-loop nucleotide binding domain (Fig. 2). This region of the

Arabidopsis clone also shows homology to partial *ureG* clones from soybean and barley, 78 and 77% identity, respectively.

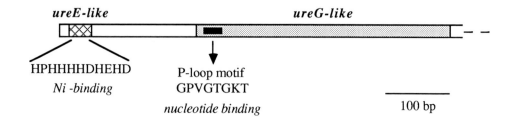

Figure 2. Accessory gene homolog from *Arabidopsis*. The incomplete cDNA (lacking the C-terminus) has domains from *K. aerogenes* urease accessory genes *ureE* and *ureG*, including their respective deduced binding sites for Ni and nucleotides.

Just as the urease structural genes in bacteria are fused into one plant structural gene, it appears that the accessory genes may be fused as well. The plant *ureG* homolog also contains a region of homology to bacterial UreE. The deduced amino-terminal end of the *Arabidopsis* urease accessory protein contains an 11 amino acid histidine-rich region of homology (73.5% identical at the nucleotide level) to the C-terminus of the *K. aerogenes* UreE protein. This is the region of *K. aerogenes* UreE that is believed to be involved in binding and donating nickel during metallocenter assembly. The *ureG/ureE* homolog maps to a single locus on *Arabidopsis* chromosome II (H. M. Goodman, personal communication). Transgenic antisense plants and anti-UreG/UreE antibody will soon be available to help us test the identity of this plant accessory gene to soybean *Eu2* and *Eu3*.

A possible explanation for PPFM persistence in plants-- cytokinin production by PPFM's. Cytokinins are plant hormones which stimulate cell division and are important during seed fill and germination. The cytokinin/auxin ratio is routinely increased to induce shoot formation in cell culture. While biosynthetic genes have been isolated for other plant hormones (e.g., auxin, gibberellins, ethylene, abscisic acid, jasmonate) none has yet been recovered for cytokinins. We are pursuing two observations suggesting that plant cytokinins are microbial in origin, with PPFM's being at least one donor.

Corpe and Basile (1982) showed that callus from *Streptocarpus prolixis*, a flowering plant, regenerated plantlets within 15 days when cultured "co-biotically" with PPFM's. PPFM-free controls exhibited no development after 30 days. However, the authors did not indicate whether cytokinins mimicked the PPFM effects or whether PPFM's produce cytokinins.

The second observation arose from attempts to improve the germination of heat-treated seeds. A 50° C/48 h treatment of soybean seeds lowered PPFM titer by >90% (Holland and Polacco, 1992) and germination frequency by about 30% (Holland and Polacco, 1994). Imbibition of heated seeds with cytokinins (benzyl adenine + zeatin [0.5 mg·L^{-1}]) or with washed PPFM's was equally effective in partially restoring germination. Imbibition with spent, but not fresh, PPFM medium, was nearly as effective as cytokinins in restoring germination (Holland and Polacco, 1994). PPFM rescue of aged or moribund seeds extends to other plants (Holland et al., unpublished).

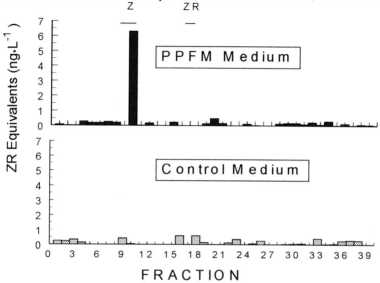

Figure 3. Cytokinin RIA assay of HPLC (C-18)-resolved components of spent PPFM medium. Eluate fractions (1 ml) and standard zeatin riboside (ZR) were used as a competitors against a [^{3}H]zeatin derivative. The monoclonal antibody recognizes the substituted dimethylallyl side chain of zeatin (Z) or ZR, whose elution positions are shown.

The ability of PPFM's and their spent medium to mimic cytokinin effects prompted us to determine cytokinins in spent AMS/methanol-C medium (from 8-day cultures of soybean leaf isolate #2 grown). Isolation and analyses were as described (Jameson and Morris, 1989): Radioimmunoassay (RIA) of HPLC-resolved compounds removed from medium by immobilized monoclonal antibody to zeatin/zeatin riboside (ZR). There was a reproducible RIA signal in an HPLC fraction eluting close to zeatin (Fig. 3). The peak shape, however, suggested that this compound was a zeatin derivative ('ZR equivalent'). Its RIA-derived value (6.3 ng·L^{-1}) is a minimum since: (1) losses during recovery were not corrected; (2) the derivative most likely had reduced affinity, based on HPLC peak height, for the monoclonal antibody; and (3) cytokinin production was not optimized for phase of batch culture or for medium components (such as plant inducers).

Cytokinin in spent PPFM medium appears low compared to levels reported in cultures of plant pathogenic bacteria (Akiyoshi et al., 1987; Stevens and Berry, 1988; Lichter et al., 1995). However, we did not attempt to isolate isopentenyl adenine/adenosine which often make up the bulk of bacterial cytokinins (Stevens and Berry, 1988). Zeatin-like compounds produced by ubiquitous plant-associated bacteria may contribute to the overall hormone balance of the plant (Taller and Wong, 1989; Scholz-Seidel and Ruppel, 1992).

CONCLUSIONS

Several observations point to a tight, perhaps mutualistic, plant-PPFM association. (1) We have found PPFM's in all plants we have examined. (2) PPFM's are common covert colonizers of cell culture and it is not yet clear whether a plant can be cured of PPFM's. (3) The inability of some host plant mutants to produce active ureases is reflected in the urease-negative phenocopy of associated PPFM's. (4) At least one PPFM isolate produces a cytokinin, a trait associated with bacteria associated with plant roots.

REFERENCES
Akiyoshi DE et al., (1987) J. Bacteriol. 169, 4242-4248.
Corpe WA, Basile DV (1982) 23, 483-493.
Dunleavy JM (1989) Int. J. System. Bacteriol. 39, 240-249.
Fitch WM, Margoliash A (1967) Science 155, 279-284.
Gälli R, Leisinger T (1985) Conservation and Recycling 8, 91-100.
Green PN, Bousfield IJ (1983) Int. J. Syst. Bacteriol. 33, 875-877.
Holland MA, Polacco JC (1992) Plant Physiol. 98, 942-948.
Holland MA, Polacco JC (1994) Ann. Rev. Pl. Phys. Pl. Mol. Biol. 45, 197-209.
Horsch RB, King J (1983) Plant Cell Tiss. Org. Cult. 2, 21-28.
Jameson PE, Morris RO (1989) J. Plant Physiol. 135, 385-390.
King J et al. (1980) Planta 149, 480-484.
Lee MH, et al. (1992) J. Bacteriol. 174, 4324-4330.
Lee MH et al., (1993) Protein Sci. 2, 1042- 1052.
Lichter A et al., (1995) Molec. Plant Micr. Interact. 8, 114-121.
Meyer-Bothling LE, et al. (1987) Mol. Gen. Genet. 209, 432-438.
Meyer-Bothling LE, Polacco JC (1987) Mol. Gen. Genet. 209, 439-444.
Mulrooney SB, Hausinger RP (1990) J. Bacteriol. 172, 5837-5843.
Nei M, Li WH (1979) Proc. Nat. Acad. Sci. U.S.A. 76, 5269-5273.
Olsen GJ et al. (1994) J. Bacteriol. 176, 1-6.
Park, I-S et al. (1994) Proc. Nat. Acad. Sci. U.S.A. 91, 3233-3237.
Polacco JC et al. (1989) Mol. Gen. Genet. 217, 257-262.
Polacco JC, Holland MA (1993) In KW Jeon, M Friedlander, eds, International Review of Cytology (Vol 145), pp 65-103, Academic Press, San Diego.
Scholz-Seidel C, Ruppel S (1992) Zentralb. Mikrobiol. 147, 319-328.
Stebbins N, et al. (1991) Plant Physiol. 97, 1004-1010.
Stevens GA, Berry AM (1988) Plant Physiol. 87, 15-16.
Taller BJ, Wong T-Y (1989) Appl. Environ. Microbiol. 55(I), 266-267.
Torisky RS, et al. (1994) Mol. Gen. Genet. 242, 404-414.
Weisberg WG et al. (1991) J. Bacteriol. 173, 697-703.

CO_2 Fixation in Chemoautotroph-Invertebrate Symbioses: Expression of Form I and Form II RubisCO

Colleen M. Cavanaugh and Jonathan J. Robinson

Department of Organismic and Evolutionary Biology, Harvard University, Cambridge, Massachusetts, USA

INTRODUCTION

The discovery of thriving communities of invertebrates at deep-sea hydrothermal vents gave us our first insights into a new community based on chemosynthesis. A major contributor to the base of the food chain was sulfur-oxidizing chemoautotrophic bacteria found within the invertebrates, including vestimentiferan tubeworms, vesicomyid clams, and mytilid mussels (see reviews Cavanaugh 1994, Fisher 1990). In these associations, the bacterial endosymbionts are hypothesized to provide the animal host with a source of nutrition via production of organic compounds from CO_2. The main evidence that the symbionts are autotrophic is the co-occurrence of Gram-negative bacteria and activities of ribulose-1,5-bisphosphate carboxylase/oxygenase (RubisCO), the CO_2-fixing enzyme of the Calvin-Benson cycle, in specific animal tissues. Although the bacterial partners have not yet been grown in pure culture from any of the invertebrate hosts, these data suggest that they are indeed chemoautotrophs since RubisCO is only known to occur in autotrophic organisms, i.e., plants, algae, photo- and chemoautotrophic bacteria (Tabita 1988, Tabita 1995). The symbionts thus afford the host the opportunity to utilize carbon, nitrogen, and energy sources which are otherwise unavailable to a metazoan. The host, in turn, provides the symbionts simultaneous access to substrates from both reducing and oxidizing environments.

It is now clear that these chemoautotroph-invertebrate symbioses have evolved between a diverse group of marine invertebrates and chemoautotrophic bacteria (see reviews Cavanaugh 1994, Fisher 1990). Initially discovered in the giant hydrothermal vent vestimentiferan *Riftia pachyptila* (Cavanaugh et al. 1981, Felbeck 1981), chemoautotrophic bacteria, occuring as either endo- or episymbionts, have now been described in association with animal species from five phyla including vestimentiferans, pogonophorans, bivalves, gastropods, nematodes, and oligochaetes. Environments where these symbioses occur include not only hydrothermal vents, but also hypersaline cold seeps, tectonic subduction zones, and shallow-water reducing muds. Each host has been shown to associate with a unique symbiont species based on phylogenetic analyses of the bacterial symbiont gene encoding 16S rRNA, with almost all of the symbionts examined to date belonging to the γ-subdivision of the *Proteobacteria* (Distel et al. 1994, Distel et al. 1988, Eisen et al. 1992). In contrast, studies of the phylogenetic relationships of epibionts occuring as a monoculture on the vent shrimp *Rimicaris exoculata* indicate they fall within the ϵ-subdivision of the *Proteobacteria* (Polz, Cavanaugh 1995).

M. E. Lidstrom and F. R. Tabita (eds.), Microbial Growth on C_1 Compounds, 285–292.
© 1996 *Kluwer Academic Publishers. Printed in the Netherlands.*

Our recent studies indicate that the symbionts are also diverse in their carbon fixation enzymes. Carbon isotopic compositions of host tissues, which have been used to demonstrate the dependence of vent symbiont species on chemosynthetically-derived organic material (see reviews Rau 1985, Van Dover, Fry 1989), provided the clue that different forms of RubisCO may be used in different symbioses (Robinson, Cavanaugh 1995). Here evidence is reviewed and presented indicating that symbionts from different hosts encode and express either a form I or a form II RubisCO. These findings are discussed with regard to the interpretation of stable carbon isotope ratios in the environment.

Stable carbon isotopes as trophic indicators: The use of stable carbon isotopes as elemental tracers has been fundamental in the study of carbon sources utilized by hydrothermal vent species (Rau 1985, Van Dover, Fry 1989). Symbiont-containing hydrothermal vent organisms have $\delta^{13}C$ values that cluster around two distinct ranges (Figure 1), referred to as the -30‰ group and the -11‰ group (Childress, Fisher 1992, Van Dover, Fry 1994). The -30‰ group include all of the mollusk symbioses discovered to date. The -11‰ group includes all the vestimentiferans and species associated with epibiotic bacteria, the alvinellid polychaetes, and the shrimp *Rimicaris exoculata*. The $\delta^{13}C$ values of some bacterial mats collected from the Gorda Ridge also fall closer to the -11‰ group (Van Dover, Fry 1994). The typical range of $\delta^{13}C$ values of marine organic carbon is between -12‰ and -22‰ (Figure 1) (Rau, Hedges 1979, Rau 1985); therefore it was concluded that photosynthetically-fixed carbon does not play a role as a carbon source in hydrothermal vent communities (Rau 1985). Rather, carbon fixed at the vents by chemoautotrophic bacteria was considered to be the primary carbon source, and it was later shown that chemoautotrophic symbioses also act as primary producers in these environments (Fisher 1990). The stable isotope composition of the bivalve symbioses, which belong to the -30‰ group, has been explained by the fact that sulfur-oxidizing bacteria in the genera *Thiomicrospira* and *Thiobacillus* also discriminate strongly against ^{13}C (Ruby et al. 1987) (Figure 1). However, the isotopic enrichment observed in the -11‰ group has never been adequately explained (Childress, Fisher 1992).

Several theories discussing the carbon isotope enrichment in the vestimentiferan tubeworm *R. pachyptila* have been proposed, but they have failed to fully account for the differences observed between these two groups. The bivalve and vestimentiferan symbioses inhabit similar environments in which dissolved inorganic carbon (DIC) differs by less than 3‰ (Childress, Fisher 1992), effectively ruling out any possibility that they utilize unusual sources of inorganic carbon. A C_4 hypothesis was proposed based on the observation that radiolabelled CO_2 is fixed into four carbon compounds in the plume of the vestimentiferans which could serve as a carbon transport system in this symbioses resulting in a stable isotope signature similar to that observed in C_4 plants (Felbeck 1985). Further experimentation has not corroborated this however, as radiolabelled C_4 compounds are not incorporated in host tissue to a significant degree, indicating that this is probably a minor pathway for carbon transport to the symbionts (Childress et al. 1993). Carbon limitation within the -11‰ group, initially proposed by Rau (1981), has gained support from an observed correlation between increasing animal wet weight and isotopically heavier signatures, showing a definite relationship (Fisher 1990). However, the smallest animal examined (~10mg) had a relatively enriched $\delta^{13}C$ value of -15‰, rather than closer to -30‰, suggesting that other processes are involved. Furthermore, blood ΣCO_2 concentrations in *R. pachyptila* are very high, between 29-31 mM (Childress et al. 1993), orders of magnitude higher than the measured Km for CO_2 uptake by the symbionts (~30μM) (Scott et al. 1994). This indicates that the symbionts are fully saturated with substrate assuming adequate diffusion of CO_2 from the blood space into the bacteriocytes and the symbionts themselves. The high levels of carbonic

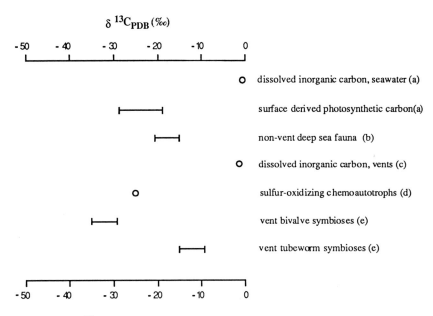

Figure 1. $\delta^{13}C$ values observed for vent and non-vent deep sea fauna, free-living chemoautotrophs, and surface plankton. The isotopic fractionation occuring during fixation and subsequent growth by sulfur-oxidizing chemoautotrophs (*Thiomicrospira* sp. str. L12 and *Thiobacillus neopolitanus*) was determined experimentally (Ruby et al. 1987), and presented here relative to reported $\delta^{13}C$ values for dissolved inorganic carbon at hydrothermal vents. Data compiled from: (a)-(Goericke et al. 1994); (b)-(Van Dover, Fry 1989) (c)-(Childress et al. 1993); (d)-(Ruby et al. 1987); (e)-(Fisher 1990).

anhydrase activities associated with the trophosome tissue (Kochevar et al. 1993) should facilitate this process allowing rapid interconversion of bicarbonate ion to CO_2, the species taken up by the symbionts (Scott et al. 1994), and fixed by RubisCO (Tabita 1988).

Stable carbon isotope ratios and form I and form II RubisCO: We hypothesized that the differences observed in the symbiosis stable carbon isotope values are due to carbon fixation by different forms of RubisCO in the two fractionation groups. RubisCO exists in two known forms, called form I and form II (reviewed by Tabita 1988). The form I type, found in the majority of autotrophs including cyanobacteria, algae, plants, and many aerobic chemoautotrophs is well characterized with a hexadecameric structure containing eight large and eight small subunits (L_8S_8) (Tabita 1988, Schneider et al. 1990). The form II type, which is more limited in its distribution, contains only large subunits arranged dimerically (the L_x form) (Tabita 1988, Schneider et al. 1990) which share ~25% amino-acid sequence identity with the form I large subunit (Somerville, Somerville 1984). The two forms of RubisCO catalyze the same reaction but differ in quaternary structure (Schneider et al. 1990) and in their kinetic isotope effect with respect to CO_2 (Estep, Tabita 1978, Roeske, O'Leary 1984, Roeske, O'Leary 1985). Carbon fixation by either a form I or form II RubisCO could thus account for the differences observed in the stable carbon isotope ratios between the two groups of symbioses.

RESULTS

To test our hypothesis, the RubisCO was characterized from representative symbioses from the -11‰ and -30‰ groups (Robinson, Cavanaugh 1995). Analysis of RubisCO expression by immunoblot and enzyme inhibition studies indicate that a form I RubisCO is expressed in the -30‰ symbioses, while a form II is expresssed in the -11‰ symbioses (Robinson, Cavanaugh 1995; summarized in Table 1). Cross reactivity with antiserum against form I RubisCO (*Nicotiana tabacum* (tobacco) large subunit) was detected only in protein extracts of the symbiont-containing tissues of the hydrothermal vent mussel *B. thermophilus* and the Atlantic coast clam *S. velum*, while the form II antiserum (against *R. rubum* RubisCO) cross-reacted only with protein from symbiont-containing tissues of the vent tubeworms *R. pachyptila* and *Tevnia jerichonana*. Recent immunoblot analyses similarly show that the bacterial symbionts which colonize the carapace of the vent shrimp *Rimicaris exoculata* (Polz, Cavanaugh 1995) also express a form II RubisCO (Table 1; J.J. Robinson, C.M.Cavanaugh, unpublished). For all of the symbionts only one cross-reacting band was found with an approximate mobility corresponding to an estimated molecular weight of ~50-55 kDa, consistent with the reported molecular weight of the form I large subunit and form II RubisCO (Gibson, Tabita 1977, Tabita 1988). No cross-reactivity was observed between groups indicating that these forms are expressed as the primary enzyme and both forms (in symbionts which may encode both) are not expressed at the same time.

Inhibition experiments using 6-phosphogluconate, which strongly inhibits the form I enzyme but has virtually no effect on the form II molecule (Gibson, Tabita 1977, Tabita 1972), corroborate our immunoblot results (Robinson, Cavanaugh 1995). RubisCO activity was reduced in *S. velum* extract at concentrations ranging from 0.4 mM to 1.0 mM 6-phosphogluconate with activity equaling ~40% of the control at the highest concentration tested. Virtually no effect was observed in extract from *R. pachyptila* at any of the inhibitor concentrations tested. This is consistent with published inhibition values for other form I and form II enzymes (Tabita 1972).

Table 1. Summary of form I and form II RubisCo expression in representative chemoautotrophic symbioses from the -11‰ and -30‰ stable carbon isotope fractionation groups[a].

Host species	$\delta^{13}C$ (‰)	RubisCO Form expressed
Solemya velum	-30 to -33.9	I
Bathymodiolus thermophilus	-30 to -37	I
Riftia pachyptila	-9.0 to -15.6	II
Tevnia jerichonana	-8.8 to -15.2	II
Rimicaris exoculata	-11.6 to -12.1	II

[a] The RubisCO form expressed was characterized by immunoblot analysis and enzyme inhibition experiments (see text). Compiled from Robinson (1995), and unpublished immunoblot data for *R. exoculata*. References for tissue δC^{13} values cited in Robinson (1995).

Southern blot and sequence analysis indicate that the members of the -30‰ and -11‰ groups encode a form I and form II RubisCO gene, respectively. RubisCO probes for the form I RubisCO large subunit gene cloned from *Synechococcus* PCC6301 (*'Anacystis nidulans'*) hybridized with DNA extracted from the gills of both *S. velum* and *B. thermophilus,* while probes for the form II gene cloned from *Rhodospirllum rubrum* hybridized with DNA extracted from the trophosome of both *R. pachptila* and *T. jerichonana* (B. Laue, D. Nelson, unpublished; J. Robinson and C. Cavanaugh, unpublished). Previous Southern blot analyses, indicating that *R. pachyptila* also hybridized with a form I RubisCO probe (Stein et al. 1989, Williams et al. 1988), have not been able to be repeated by the investigators above. This may be explained by a difference in stringency conditions during hybridization.

Sequence analysis confirms the presence of a form I gene in the symbionts of the bivalves *S. velum* and *B. thermophilus* (C. Cavanaugh, M. Fontana, and D. Distel, unpublished). PCR primers, designed based on the RubisCO genes of *Synechococcus* PCC6301 and the gastropod *Alvinoconcha hessleri* symbiont were used to amplify the symbiont RubisCO gene. Direct sequence analysis indicates that the PCR products from both bivalves each contained a single detectable sequence. The inferred amino acid sequences, unique and invariant for each host, were aligned with other form I RubisCO large subunit (cbbL) genes. The bivalve symbiont RubisCO large subunits share the highest sequence identity with each other at both the nucleotide and amino acid levels. The next most similar taxon based on amino acid sequence, is the free-living sulfur/iron-oxidizing chemoautotroph *Thiobacillus ferrooxidans* 19859, a member of the β-*Proteobacteria,* followed by the gastropod *Alvinoconcha hessleri* symbiont, and other free-living bacteria including *Proteobacteria* of the α- (*Nitrobacterr vulgaris)* and γ- (*Chromatium vinosum*) subdivisions.

To confirm that the antisera against *R. rubrum* RubisCO was cross-reacting with a form II RubisCO in *R. pachyptila*, we purified the symbiont enzyme by anion exchange chromatography essentially according to Chene (Chene et al. 1992). Briefly, trophosome tissue was homogenized in extraction buffer with 1 mM phenylmethylsulfonyl fluoride added, sonicated 3 times for 15 seconds on ice, and centrifuged for 20 minutes at 16,000 rpm in a Sorvall SS34 rotor. The supernatant was then desalted on a Pharmacia Biotech PD-10 column and loaded onto a Pharmacia Biotech Mono Q HRS/S anion exchange resin. The column was developed with a linear gradient of 0.1M NaCl to 0.6 M NaCl. The fraction containing the form II RubisCO eluted between 0.2 M to 0.25 M NaCl. Partially pure protein was concentrated in a microcon 50 microconcentrator and subjected to SDS-PAGE. Following SDS-PAGE, the band corresponding to the form II RubisCO (determined by immunoblotting) was transferred to a PVDF (Biorad Trans Blot Transfer Medium) membrane, stained with Ponceau S, excised and the first 25 amino acids determined by N-terminal sequence analysis performed by the Harvard Microchemistry Facility. The *R. pachyptila* RubisCO sequence is easily aligned with the amino acid sequences of other form II RubisCOs at the N-terminus sharing ~50% sequence identity with these enzymes (Figure 2). The lack of a methionine at the N-terminus suggests, as in many plant RubisCO enzymes (Houtz et al. 1989), that post-translational modification occurs in the *R. pachyptila* symbiont RubisCO. Efforts are currently underway to clone and express the *Riftia* symbiont RubisCO, for the determination of the kinetic isotope effect of this enzyme.

These data strongly indicate that the representative symbioses tested from the -30‰ group both encode and express a form I RubisCO while those from the -11‰ fractionation group encode and express a form II RubisCO. Although the potential for the genetic basis for both forms may exist in certain symbioses, as shown for some free-living bacteria, e.g., *Rhodobacter sphaeroides* (Tabita 1988), the gene form actually expressed,

```
                              1          10         20
                              |          |          |
Riftia pachyptila        ALDQTNRYSD LXLKEDELIA SGDYV
Rhodospirillum rubrum    ~MDQSSRYVN LALKEEDLIA GGEHV
Thiobacillus denitrificans ~MDQSARYAD LSLKEEDLIK GGRHI
Rhodobacter sphaeroides  ~MDQSNRYAR LDLQEADLIA GGRHV
Hydrogenovibrio marinus  ~MDQSNRYAD LTLTEEKLVA DGNHL
```

Figure 2. N-terminal sequence of the purified form II RubisCO from the *R. pachyptila* symbiosis. Amino acid sequence is shown aligned with form II RubisCO sequences from free-living autotrophic bacteria. X=S or L. Genbank acccession numbers: *R. rubrum* K01999, *T. denitrificans* L37437, *R. sphaeroides* M68914, *H. marinus* D28135.

i.e., transcribed and translated into protein, is the important parameter in determining which RubisCO form is actually being utilized for primary carbon fixation.

The observed $\delta^{13}C$ values of the two symbiosis fractionation groups thus may be due to the expression of the two different forms of RubisCO. These data are consistent with differences in the kinetic isotope effect with respect to CO_2 determined for purified form I and form II RubisCO by high precision methods (Guy et al. 1993, Roeske , O'Leary 1984, Roeske , O'Leary 1985). The fractionation factor for form I RubisCO purified from *Spinacia oleracae* (spinach) is 29‰ (Roeske, O'Leary 1984) and from the cyanobacterium *Synechococcus PCC6301* is 22‰ (Guy et al. 1993). The fractionation factor for form II RubisCO, determined for the *R. rubrum* enzyme in two separate investigations (Guy et al. 1993, Roeske, O'Leary 1985), is as low as -17.8‰ (Roeske, O'Leary 1985) and can range up to 23‰ depending on Mg^{+2} concentration (Guy et al. 1993). Given the extremes measured *in vitro* in the kinetic isotope effect of the two RubisCO forms, carbon fixed by symbionts using form II RubisCO would be isotopically enriched relative to that fixed using form I RubisCO. This discrimination coupled with other processes, such as diffusion, reaction rate, and heterotrophic carbon fixation (Goericke et al. 1994), could result in the heavier isotopic composition observed in the -11‰ symbioses.

Unusually heavy $\delta^{13}C$ values for autotrophic organisms are not restricted to the intracellular symbioses at hydrothermal vent ecosystems. Heavy isotope signatures are characteristic of some vent bacterial mats (Van Dover, Fry 1994) and invertebrates associated with epibiotic bacteria such as the polychaete worms within the genus *Alvinella* (Childress, Fisher 1992), suggesting that the form II enzyme may be more prevalant among bacterial autotrophs than once suspected. Furthermore, the recent discovery of a form II RubisCO in eukaryotic photoautotrophs, the dinoflagellates *Gonyaulax polyedra* (Morse et al. 1995) and *Symbiodinium* sp. (Whitney et al. 1995) indicates that this enzyme type is not restricted to prokaryotes. Corals, which harbor dinoflagellates in the genus *Symbiodinium* as symbionts, typically have $\delta^{13}C$ values ranging from -17.1‰ to -19.8‰ (Land et al. 1975), but can be as heavy as -10‰ in shallow water (Muscatine et al. 1989). Typically ascribed to high rates of photosynthesis and subsequent carbon limitation, the expression of a form II enzyme may provide an alternate explanation. Enriched carbon isotope values observed for other algae in nature, e.g., diatom blooms on Georges Bank ($\delta^{13}C$=-15‰ to -19‰) (Fry, Wainright 1991) or diatoms on the surfaces of chironomid tubes ($\delta^{13}C$=-19‰) (Peterson et al. 1993), may be due to the expression of a form II RubisCO under these conditions. The increased observations of the form II enzyme, both within the prokaryotes and the new discovery within the eukaryotes, suggests widening the survey for form II RubisCOs. Accurately resolving the distibution of RubisCO forms among primary producers and the kinetic isotope effect of different forms of RubisCO will be necessary for interpretation of $\delta^{13}C$ values encountered in the environment.

CONCLUSION

Our studies show that in representative invertebrate-chemoautotroph symbioses, bivalve symbionts encode and express a form I RubisCO, while vestimentiferan tubeworm symbionts express a form II RubisCO. Since purified form II RubisCO from free-living bactera has previously been reported to have a reduced kinetic isotope effect with respect to CO_2 relative to form I, the different RubisCO forms appear to be at least partially responsible for reported differences in stable carbon isotope ratios of the intact bivalve ($\delta^{13}C$ = -27 to -35‰) and tubeworm ($\delta^{13}C$ = - 9 to -16‰) symbioses. The increased occurrence of the form II enzyme among autotrophs has implications for the interpretation of stable carbon isotope data in ecological studies in general, and may provide an alternate explanation for other ^{13}C-enriched isotope values observed in the environment.

ACKNOWLEDGMENTS

We thank the captain and crew of the R/V Atlantis II and DSV ALVIN for their skillful help in obtaining deep-sea specimens. We thank R. John Collier for the use of his FPLC and Philip Hanna for assistance in protein isolation. RubisCO antisera and gene probes were generously donated by Lawrence Bogorad, F. Robert Tabita, George Lorimer, and Jessup Shively. This work was supported by grants from the National Science Foundation and the Office of Naval Research.

REFERENCES

Cavanaugh CM (1994) Amer. Zool. 34, 79-89.
Cavanaugh CM et al. (1981) Science. 213, 340-342.
Chene C et al. (1992) J. Mol. Biol. 225, 891-896.
Childress JJ, Fisher CR (1992) Oceanogr. Mar. Biol. Annu. Rev. 30, 337-441.
Childress JJ et al. (1993) Nature. 362, 147-149.
Distel DL et al. (1994) J. Mol. Evol. 38, 533-542.
Distel DL et al. (1988) J. Bacteriol. 170, 2506-2510.
Eisen JA et al. (1992) J. Bacteriol. 174, 3416-3421.
Estep MF, Tabita FR, Parker PL, Van Baalen, C. (1978) Plant Physiol. 61,680-687.
Felbeck H (1981) Science. 213, 336-338.
Felbeck H (1985) Physiol. Zool. 58, 272-281.
Fisher CR (1990) Rev. Aquat. Sci. 2, 399-436.
Fry B, Wainright SC (1991) Mar. Ecol. Prog. Ser. 76, 149-157.
Gibson JL, Tabita RF (1977) J. Biol. Chem. 252, 943-949.
Goericke R et al. (1994) In Lajtha K and Michener RH, eds, Physiology of Isotopic Fractionation in Algae and Cyanobacteria, pp 187-221, Blackwell Scientific Publications, Boston.
Guy RD et al. (1993) Plant Physiol. 101, 37-47.
Houtz RL et al. (1989) Proc. Natl. Acad. Sci. 86, 1855-1859.
Kochevar RE et al. (1993) Molec. Mar. Biol. and Biotech. 2, 10-19.

Land LS et al. (1975) Limnol. Oceanogr. 20, 283-287.
Morse D et al. (1995) Science. 268, 1622-1624.
Muscatine L et al. (1989) Marine Biology. 100, 185-193.
Peterson BJ et al. (1993) Ecology. 74, 653-672.
Polz MF, Cavanaugh CM (1995) Proc. Natl. Acad. Sci. USA. 92, 7232-7236.
Rau GA, Hedges JI (1979) Science. 203, 648-649.
Rau GH (1981) Science. 213, 338-340.
Rau GH (1985) In Jones ML, ed, $^{13}C/^{12}C$ and $^{15}N/^{14}N$ in Hydrothermal Vent
    Organisms: Ecological and Biogeochemical Implications, pp 243-248, Bulletin
    of the Biological Society of Washington, Washington, D.C.
Robinson J, Cavanaugh CM (1995) Limnol. Oceanogr. in press.
Roeske CA, O'Leary MH (1984) Biochem. 23, 6275-6284.
Roeske CA, O'Leary MH (1985) Biochem. 24, 1603-1607.
Ruby EG et al. (1987) Appl. Environ. Microbiol. 53, 1940-1943.
Schneider G et al. (1990) EMBO. 9, 2045-2050.
Scott KM et al. (1994) Physiol. Zool. 67, 617-638.
Somerville CR, Somerville SC (1984) Mol. Gen. Genet. 193, 214-219.
Stein J et al. (1989) In Nardon P et. al., eds, Diversity of Ribulose 1,5 Bisphosphate
    Carboxylase Genes in Thioautotrophic Symbioses, pp 343-347, Institute National
    de la Recherche Agronomique, Paris.
Tabita FR, McFadden, BA (1972) Biochem. Biophys. Res. Comm. 48, 1153-1159.
Tabita FR (1988) Microbiol. Rev. 52, 155-189.
Tabita FR (1995) Blankenship RE et al., eds, The Biochemistry and Metabolic
    Regulation of Carbon Metabolism and CO_2 Fixation in Purple Bacteria,
    Kluwer Academic Publishers, The Netherlands.
Van Dover CL, Fry B (1989) Mar. Biol. 102, 257-263.
Van Dover CL, Fry B (1994) Limnol. Oceanogr. 39, 51-57.
Whitney SM et al. (1995) Proc. R. Soc. Lond. B. 259, 271-275.
Williams CA et al. (1988) FEMS Microbiol. Lett. 50, 107-112.

Sulfur Metabolism of Autotroph-Invertebrate Symbioses

Douglas C. Nelson and Sarah C. McHatton

Section of Microbiology, University of California, Davis, CA, USA

INTRODUCTION

Two years after geologists initially examined the Galapagos Rift using the deep submergence vehicle, ALVIN (Corliss et al. 1979), these deep-sea hydrothermal vents were first visited by biologists in 1979. The discovery of symbiotic, presumably mutualistic, associations between chemolithoautotrophic sulfur bacteria (thioautotrophs) and invertebrates followed shortly thereafter (Cavanaugh et al. 1981; Felbeck et al. 1981). The taxonomic range of hosts known to associate with putatively thioautotrophic bacteria continues to expand and currently includes well over 100 species in five families of bivalves (clams and mussels) and several phyla of worms (Fisher 1990; Nelson, Fisher 1995 and references therein). The largest associations are endemic to hydrogen sulfide-rich hydrothermal vents or marine seeps, but many other smaller bivalves and worms occur in marine sediments containing high concentrations of biogenic sulfide. Globally, approximately half the oxidation of organic matter reaching marine sediments is performed by sulfate-reducing bacteria (Jorgensen 1983), and hydrogen sulfide, the waste product of this latter group, still contains most of the potential energy stored in typical organic matter (Howarth 1984). Hydrogen sulfide equilibrates between H_2S, HS^- and $S^=$ depending on pH, but the sum of these soluble forms is designated "sulfide" hereafter.

Numerous lines of evidence support the conclusion that the symbionts are thioautotrophic. These include analyses of symbiont-containing tissues for: (1) stable carbon isotope ratios, (2) activity and immunochemical localization of enzymes diagnostic for the Calvin cycle of autotrophic CO_2 fixation, (3) activities of enzymes which are diagnostic, to varying degrees, for sulfur-based energy metabolism, (4) rates of CO_2 incorporation by symbionts, (5) genes diagnostic of thioautotrophic metabolism, and (6) phylogenetic clustering of thioautotrophic symbionts (Cavanaugh et al. 1988; Cavanaugh, Robinson 1996; Distel et al. 1995; Fisher 1990; Laue, Nelson 1994; Nelson, Fisher 1995). Nonetheless, in these symbioses the energy-yielding sulfur metabolism of the symbionts is only partly understood. This paper will summarize current knowledge concerning catabolic sulfur enzymes in the thioautotrophic symbionts of worms and bivalves and their use of diverse sulfur compounds. Three lines of evidence will be discussed: (1) the stimulation of CO_2 incorporation in intact symbionts by various reduced sulfur compounds, (2) the activities of potentially catabolic enzymes detected in lysates of symbiont-containing tissues, and (3) the emergence of probes for genes encoding these catabolic, sulfur-oxidizing enzymes.

M. E. Lidstrom and F. R. Tabita (eds.), Microbial Growth on C_1 Compounds, 293–300.

RESULTS

Effects of Various Sulfur Compounds on Carbon Dioxide Incorporation by Symbiont Preparations: There is currently no evidence that any symbiotic, thioautotrophic bacterium has been successfully cultured apart from its host. Therefore, shipboard and laboratory studies with fresh preparations must provide insights into symbiont physiology. To date, these studies have focused on CO_2 incorporation by symbionts of the hydrothermal vent tube worm (*Riftia pachyptila*), the vent mussel (*Bathymodiolus thermophilus*), the vent clam (*Calyptogena magnifica*), and four sediment-dwelling, non-vent bivalves (*Solemya reidi, S. velum, Lucinoma aequizonata,* and *Lucina floridana*). The studies reported here were performed with crude homogenates of symbiont-containing tissues, i.e. gills (bivalve) or trophosome (tube worm), or preparations enriched in symbionts by differential centrifugation. The very limited number of studies based on higher levels of host-symbiont integration are discussed elsewhere (Nelson, Fisher 1995).

Symbiont-enriched preparations are likely to also contain host mitochondria due to their similar size. The mitochondria of *S. reidi* have been shown to couple oxidation of sulfide to oxygen consumption. This results in: ATP production (Powell, Somero 1986b), stoichiometric oxidation of sulfide to thiosulfate (O'Brien, Vetter 1990) and accumulation of thiosulfate in the host's blood at levels up to 1-3 mM (Anderson et al. 1987). Thus, putative stimulation of CO_2 incorporation by soluble sulfide in *S. reidi* symbiont preparations must be interpreted with caution. The symbionts might actually be metabolizing thiosulfate ($S_2O_3^=$) produced by contaminating mitochondria. Thiosulfate also accumulates in the blood of *C. magnifica* and *B. thermophilus*, but it is not known whether this results from mitochondrial oxidation of sulfide.

The symbiont of the tube worm *R. pachyptila* is a sulfide specialist, and the host's two distinct hemoglobins both contain sulfide-binding sites (Somero et al. 1989). Soluble sulfide, but not thiosulfate, stimulates CO_2 fixation (Belkin et al. 1986; Fisher et al. 1989), and the host does not accumulate significant concentrations of thiosulfate in its blood (Childress et al. 1991a). It has been calculated that homogenates of *R. pachyptila* trophosome fix CO_2 at a rate that is consistent, assuming complete nourishment of the host by thioautotrophs, with whole animal rates of O_2 consumption (Fisher et al. 1989). Elemental sulfur (S^o) globules are abundant in this and other symbionts (Wilmot, Vetter 1990; Vetter 1985) and probably represent an internal oxidizable substrate, but this has not been confirmed experimentally. Sulfite ($SO_3^=$) is a presumed universal intracellular intermediate in the oxidation of sulfide by thioautotrophs (Kelly 1989; Nelson, Hagen 1995), and the *R. pachyptila* symbiont contains enzymes implicating the APS pathway of sulfite oxidation (Felbeck et al. 1981). Therefore, it can be assumed that endogenous sulfite stimulates CO_2 incorporation. However, lack of stimulation of O_2 consumption by exogenous sulfite (Powell, Somero 1986a) suggests that sulfite is not transported.

Studies of symbionts of the vent mussel demonstrated a linear relationship of approximately 0.36 mol CO_2 incorporated per mol $S_2O_3^=$ consumed (Nelson et al. 1995), consistent with molar growth yields for free-living thioautotrophs (Kelly 1982; Nelson et al. 1986). In experiments with symbionts purified four- to ten-fold from 10 separate mussel gills, the stimulation was typically strong (mean = 48-fold, n = 7) and associated with a low control rate (e.g. Fig. 1A). However, for symbionts purified from gills of some mussels, stimulation by thiosulfate was modest (mean = 2.3-fold, n = 3) but associated with a high control rate (e.g. Fig. 1B). If the high control rates observed for these washed pellets represent autotrophic incorporation, calculations detailed elsewhere (Nelson et al. 1995) suggest insoluble S^o as the oxidizable substrate probably driving the incorporation. Exogenous sulfite was not stimulatory.

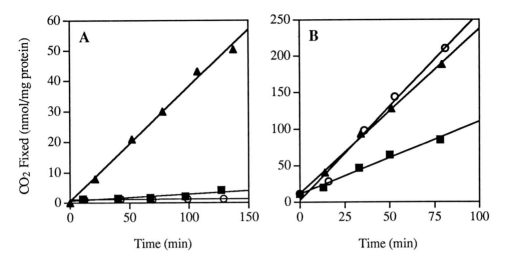

Figure 1. CO_2 incorporation as a function of time for resuspended symbiont high-speed pellets from *B. thermophilus*. Symbols: (squares) unsupplemented; (triangles) thiosulfate, 1mM; (open circles) sulfide, 200µM (A) or 100µM (B). **A.** Mussel 2011-M1 at 12°C. **B.** Mussel 2017-M1 at 5°C.

For *B. thermophilus* symbiont preparations, sulfide usually stimulated CO_2 incorporation to roughly the same extent as thiosulfate (Fig. 1B; Nelson et al. 1995), but occasionally it was inhibitory (Fig 1A; Belkin et al. 1986). When sulfide was stimulatory the universal pattern was for modest concentrations, e.g. 100µM, to be more stimulatory than 20µM (Fig. 1B). For preparations inhibited by sulfide, the lowest sulfide concentration tested (50µM, data not shown) and a higher concentration (200µM, Fig. 1B) were roughly equally inhibitory. This variable influence of sulfide but predictable stimulation by thiosulfate might be explained in two different ways. First, it is possible that sulfide can be oxidized by the host's mitochondria, as in *S. reidi*, but not by the symbionts. Thus, if purified symbiont preparations proved to be differentially contaminated with mitochondria, inhibition by sulfide would be expected only for uncontaminated preparations. However, based on electron micrographs, mitochondrial contamination was minimal and no higher in sulfide-stimulated preparations (Nelson et al. 1995). A second explanation may lie in differing exposure of individual mussels to sulfide-containing vent fluids. If a symbiont's ability to generate energy from thiosulfate oxidation is constitutive while its sulfide-oxidizing capacity is inducible, differences in habitat exposure to sulfide could explain the observed symbiont differences.

The hydrothermal vent clam accumulates thiosulfate in its blood at approximately 0.1 to 1.0mM when presented with sulfide at 0.01 to 1.0mM (Childress et al. 1991b), and in this respect it resembles the vent mussel. Additionally, it has a nonheme serum protein which allows it to concentrate sulfide (Somero et al. 1989) and contains considerable elemental sulfur in the gills (Fisher et al. 1988b). Because some or all of these possible energy sources are undoubtedly present in crude homogenate of a fresh clam gill, it is not surprising to see a high rate of incorporation in such unsupplemented controls (mean = 0.25 nmol $CO_2 \, min^{-1} \, mg^{-1}$ protein; n=3) and at most a modest stimulation by reduced sulfur compounds (Fig. 2A). For crude homogenates or symbiont-enriched fractions from gills of freshly collected clams the enhancement of incorporation by exogenous compounds averaged 1.45-fold (n = 4) for 50 to 200μM sulfide and 1.25-fold (n=3) for 1mM thiosulfate. A higher concentration of sulfide was inhibitory (Fig. 2A).

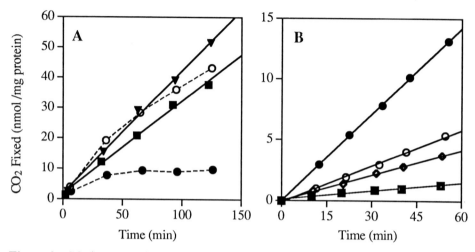

Figure 2. CO_2 incorporation as a function of time for gill symbionts of *C. magnifica*. A. Clam 2007-C1, crude homogenate at 10°C. Symbols: (squares) unsupplemented; (triangles) thiosulfate, 1mM; (open circles) sulfide, 200μM; (filled circles) sulfide, 800μM. **B.** Clam 2029-C3, resuspended high-speed pellet at 10°C. Symbols: (squares) unsupplemented; (diamonds) 20% serum; (open circles) 20% serum + 20μM sulfide; (filled circles) 20% serum + 100μM sulfide.

Some clams were maintained for one to five days in high pressure aquaria (100 atm, flowing seawater) without an exogenous reduced-sulfur source, a procedure which dramatically reduces blood concentrations of sulfide and thiosulfate (Childress et al. 1991b). Crude and enriched symbiont preparations from these bivalves showed a much lower unsupplemented rate and a greater capacity for sulfide stimulation (mean = 10.5-fold, n = 5) by the most effective concentration in the 20 to 100 μM range. The blood serum from *C. magnifica* still contains some sulfide which will partition between bound and unbound phase, imparting a low free-sulfide concentration to symbiont preparations (Childress et al. 1991b). Thus, the three "serum" or "serum plus sulfide" treatments (Fig. 2B) show the influence of increasing concentrations of free sulfide on symbiont metabolism. It is not known what the actual free sulfide concentrations were for the various treatments. The symbionts of similarly maintained clams were assayed for the influence of temperature on their sulfide-stimulated CO_2 incorporation. The results (Fig. 3) strongly suggest that they, along with the *B. thermophilus* symbiont (Nelson et al. 1995), are the only known examples of psychrophilic chemolithoautotrophic sulfur bacteria.

Due to the relative difficulty of obtaining samples from the hydrothermal vents, several researchers have attempted studies of invertebrate-thioautotroph symbioses more readily accessible from nearshore environments (Fisher 1990; Nelson, Fisher 1995). Our lab focused on *S. reidi* and *S. velum* which were collected routinely from their nearshore, sulfide-rich environments. Both bivalves are small so that pooling of gills from multiple organisms was often necessary to achieve quantities appropriate for $^{14}CO_2$ incorporation experiments. In addition, the homogenate produced by grinding the gill was thick and viscous. This made purification of the large, possibly fragile, symbionts difficult and may account for erratic time course data obtained in roughly half of the experiments with both species. In successful experiments with partially purified *S. reidi* symbionts, specific activity of CO_2 incorporation was stimulated 1.4-6.2 fold (n=7) in the presence of 500μM thiosulfate, while sulfide additions (20μM) increased CO_2 incorporation 1.8-3.1 fold (n=2). Specific activity of CO_2 incorporation for *S. velum* symbionts was at least an order of magnitude greater than for *S. reidi* and showed higher stimulation with sulfur compounds (11-fold for 20μM sulfide and 50μM thiosulfate; Fig. 4)

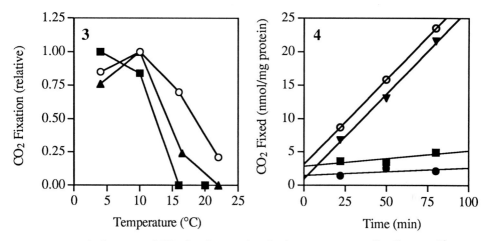

Figure 3. Relative rate of CO_2 fixation vs. incubation temperature for *C. magnifica* symbionts in 20% serum + 20μM sulfide. Symbols: (squares) crude homogenate of 2013-C3 (5 days maintenance) corrected for unsupplemented rate; (circles) crude homogenate of 2024-C1 (1 day maintenance) uncorrected; (triangles) high speed pellet of 2029-C1 (1 day maintenance) uncorrected.

Figure 4. CO_2 incorporation as a function of time for resuspended symbiont high-speed pellet from *S. velum* at 22°C. Symbols: (squares) unsupplemented; (triangles) thiosulfate, 50μM; (open circles) sulfide, 20μM; (closed circles) sulfide, 200μM.

Activities of Catabolic Sulfur Enzymes in Symbiont-Containing Tissues: Free-living, thioautotrophic bacteria, taken as a group, posses a variety of enzymatic pathways for the energy yielding oxidation of sulfide, thiosulfate, sulfite and elemental sulfur. A full description of the multiple enzymatic mechanisms known or postulated to start with each of these substrates (Kelly 1989; Nelson, Fisher 1995; Nelson, Hagen 1995) is beyond the scope of this paper. One energy-yielding mechanism of sulfite oxidation employed by some, but not all, free-living thioautotrophs is the "APS pathway". The enzymes comprising this pathway catalyze the following reactions (where APS denotes adenosine 5' phosphosulfate):

(1) $SO_3^- + AMP \leftrightarrow APS$ \hspace{2cm} (APS reductase)

(2) $APS + PP_i \leftrightarrow SO_4^= + ATP$ \hspace{2cm} (ATP sulfurylase)

These are the only enzymes reasonably diagnostic of sulfur-based energy generation that have been sought or detected in symbiont-containing tissues. Although one or both has been detected in at least 30 separate symbiont-containing tissues (Fisher 1990), the specific activities are often extremely low, and it was not always stated whether rates were corrected for appropriate controls. Additionally, incomplete release of symbiont enzymes due to poor cell breakage can underestimate activities by a factor of ten (Nelson et al. 1995). A final shortcoming in this approach is that host lytic enzymes in a homogenous preparation of B. thermophilus gill appear to have degraded exogenous RubisCO with a half life of 11-17 minutes (Nelson et al. 1995). This may also be reflected in low activities of sulfur catabolic enzymes assayed in crude homogenates. No comparable lytic activity was detected in symbiont tissues of R. pachyptila, C. magnifica (Nelson et al. 1995), S. reidi or S. velum (S. McHatton, unpublished).

In addition to its role in energy generation, ATP sulfurylase is involved in ATP-driven sulfate assimilation in a variety of heterotrophic prokaryotes, fungi and photoautotrophic eukaryotes. Even in animal tissue this sulfate-activating enzyme is a required first step in certain biosynthetic reactions (Segel et al. 1987). Problematically, ATP sulfurylase activities measured for symbiont tissues (i.e. first four entries of Table 1) do not always exceed those of non-autotrophic sulfate assimilators. Although interpretation problems are fewer than with ATP sulfurylase, a modest specific activity of APS reductase is also equivocal support for the presence of thioautotrophic symbionts. Anaerobic sulfate-reducing bacteria utilize both enzymes of the APS pathway in the "reverse" direction for reducing $SO_4^=$ to $SO_3^=$. Because of the high sulfate content of seawater, any anaerobic organic-rich niche (such as a closed bivalve) may provide an enrichment for these bacteria. Their contamination of gill surfaces alone might be responsible for the low activities detected in some symbiont containing tissues (Table 1).

Table 1. Specific Activities of ATP Sulfurylase and APS Reductase for Selected Symbiont-Containing Tissues and Controls. Values in μmol product·min^{-1}·gm^{-1} wet wt.

Organism/tissue	ATP sulfurylase	APS reductase	References
R. pachyptila trophosome	67-180	23-30	Felbeck et al. 1981 Renosto et al. 1991
B. thermophilus gill	0.21[a] ± 0.06	not detected	Nelson et al. 1995 Fisher et al. 1988a
Thyasira flexuosa gill	0.22 - 0.39	0.004 - 0.017	Dando, Southward 1986
Myrtea spinifera gill	0.20 - 0.29	0.017 - 0.043	Dando et al. 1985
Sulfate-reducing bacteria	160 - 370[b]	49[b]	Kramer, Cypionka 1989 Akagi, Campbell 1962
Rat liver	1.5	–	Renosto et al. 1991

[a] Converted from Nelson et al. (1995) assuming that protein comprises 7.1% of gill wet weight (Fisher et al. 1988a).
[b] Converted assuming that protein comprises 10% of wet weight of a typical bacterium.

The following enzymes: (1) sulfite:acceptor oxidoreductase, (2) siroheme sulfite reductase, and (3) flavocytochrome c sulfide dehydrogenase might be expected to be present in some thioautotrophic symbionts (Nelson, Fisher 1995) but neither these nor the enzymes potentially diagnostic of energy-yielding thiosulfate utilization (Kelly 1989) have been sought in symbiont-containing tissues.

Probes for Genes Encoding Catabolic Sulfur Enzymes: Following the purification of the ATP sulfurylase protein from the symbiont of *R. pachyptila* (Renosto et al. 1991), the corresponding gene (*sopT*) was cloned and sequenced (Laue, Nelson 1994). In Southern hybridizations *sopT* probes showed positive results only with DNA from thioautotrophs--including free-living bacteria of the genera *Thiobacillus* and *Beggiatoa* (Laue and Nelson, 1994) and the symbionts of other tube worms (B. Laue, unpublished). Under low stringency hybridization was negative with DNA from assimilatory and dissimilatory sulfate- reducing bacteria (Laue, Nelson 1994). The *sopT* probes are not universally diagnostic for a catabolic ATP sulfurylase gene because they failed to hybridize with DNA from *S. reidi, S. velum, C. magnifica*, and *B. thermophilus* (B. Laue, unpublished) most of which possess reasonable ATP sulfurylase activity. The failure of the *sopT* probe to hybridize might be explained by the fact that its base composition is 58 mol% G+C while those known for bivalve symbionts are less than 38 mol% (Belkin et al. 1986; Nelson, unpublished).

In addition to the *sopT Riftia* symbiont probe, a number of possibly useful heterologous probes have recently become available. The enzymes these encode are: the APS reductase of *Archaeoglobis fulgidus* (Spiech et al. 1994), flavocytochrome c sulfide dehydrogenase of *Chromatium vinosum* (Dolata et al. 1993), and siroheme sulfite reductase of *A. fulgidus* and *Desulfovibrio vulgaris* (Dahl et al. 1993; Karkhoff-Schweizer et al. 1995). Because these bacteria grow either as anaerobic sulfate reducers or sulfide-dependent photoautotrophs, there is no guarantee that one can proceed from their genes to probes that are diagnostic for thioautotrophic energy generation. The need for such probes is most clearly underscored by instances in which distinct methylotrophic and thioautotrophic endosymbionts appear to inhabit the same mussel gill (Cavanaugh et al. 1992; Fisher et al. 1993; Distel et al. 1995). Evidence suggests that in at least one of these associations only the methylotrophic symbiont may be active (Cavanaugh et al. 1992). This eliminates symbiont physiological studies and enzyme assays as possible diagnostic tools, leaving gene probes as perhaps the only way to define the energy metabolism of the putatively thioautotrophic symbiont.

CONCLUSIONS

Physiological studies with partially purified thioautotrophic symbionts offer continued prospects for understanding the spectrum of reduced sulfur compounds that can serve as energy sources to drive CO_2 incorporation. However, a method for inhibiting conversion of sulfide to thiosulfate by contaminating mitochondria (or factoring the contribution out) would strengthen this approach. Most progress in understanding symbiont physiology has been made with associations from hydrothermal, deep-sea vents; however, more available "model" systems should be sought among the numerous near-shore symbiotic associations. Of the enzymes known from free-living thioautotrophs to be diagnostic for oxidation of diverse inorganic sulfur compounds, only a small number have been sought via assays of symbiont-containing tissues. This approach should be expanded beyond those of the APS pathway. Additionally, gene probes for particular thiotrophic pathways should be sought. Although in the early stages of development, this approach may ultimately prove more definitive than enzyme assays.

ACKNOWLEDGMENTS

This work was supported in part by grants from the Biological Oceanography Program of the National Science Foundation.

REFERENCES
Akagi JM, Campbell LL (1962) J. Bacteriol. 84, 1194-1201.
Anderson AE et al. (1987) J. Exp. Biol. 133,1-31.
Belkin S et al. (1986) Biol. Bull. 170, 110-121.
Cavanaugh CM et al. (1981) Science. 213, 340-342.
Cavanaugh CM et al. (1988) Proc. Natl. Acad. Sci. USA. 85, 7786-7789.
Cavanaugh CM et al. (1992) Appl. Environ. Microbiol. 58, 3799-3803.
Cavanaugh CM, Robinson JJ (1996) In Lidstrom ME and Tabita FR, eds, Microbial
    Growth on C1 Compounds, Kluwer Academic Publishers, Dordrecht, The
    Netherlands. In press.
Childress JJ et al. (1991a) Biol. Bull. 180, 135-153.
Childress JJ et al. (1991b) Physiol. Zool. 64, 1444-1470.
Corliss JB et al. (1979) Science. 203, 1073-1083.
Dahl C et al. (1993) J. Gen. Microbiol. 139, 1817-1828.
Dando et al. (1985) Mar. Ecol. Prog. Ser. 23, 85-98
Dando PR, Southward AJ (1986) J. Mar. Biol. Assoc. U.K. 66, 915-920.
Distel DL et al. (1995) Proc. Natl. Acad. Sci. USA. 92, in press.
Dolata MM et al. (1993) J. Biol. Chem. 268, 14426-14431.
Felbeck H et al. (1981) Nature 293, 291-293.
Fisher CR (1990) Rev. Aquat. Sci. 2, 399-436.
Fisher CR et al. (1988a) Deep-Sea Res. 35, 1769-1791.
Fisher CR et al. (1988b) Deep Sea Res. 35, 1811-1831.
Fisher CR et al. (1989) Biol. Bull. 177, 372-385.
Fisher CR et al. (1993) PSZNI Mar. Ecol. 14, 277-289.
Howarth RW (1984) Biogeochem. 1, 5-27.
Jorgensen BB (1983) In Bolin B and Cook RC, eds, The Major Biogeochemical Cycles
    and their Interactions, pp 477-509, SCOPE, Stockholm.
Karkhoff-Schweizer RR et al. (1995) Appl. Environ. Microbiol. 61, 290-296.
Kelly D (1982) Phil. Trans. R. Soc. Lond. B. 298, 499-528.
Kelly D (1989) In Schlegel HG and Bowien B, eds, Autotrophic Bacteria, pp 193-217,
    Science Tech Publishers, Madison, WI.
Kramer M, Cypionka H (1989) Arch. Microbiol. 151, 232-237.
Laue BE, Nelson DC (1994) J. Bacteriol. 176, 3723-3729.
Nelson DC et al. (1986) Appl. Environ. Microbiol. 52, 225-233.
Nelson DC, Fisher CR (1995) In Karl DM, ed, The Microbiology of Deep-Sea
    Hydrothermal Vents, pp 125-167, CRC Press, Boca Raton, FL.
Nelson DC, Hagen KD (1995) Amer. Zool. 35, 91-101.
Nelson DC et al. (1995) Mar. Biol. 121, 487-495.
O'Brien J, Vetter RD (1990) J. Exp. Biol. 149, 133-148.
Powell MA, Somero GN (1986a) Biol. Bull. 171, 274-290.
Powell MA, Somero GN (1986b) Science. 233, 563-566.
Renosto F et al. (1991) Arch. Biochem. Biophys. 290, 66-78.
Segel IH et al. (1987) In Jakoby WB and Griffith O, eds, Methods in Enzymology
    Volume 143: Sulfur and Sulfur Amino Acids, pp 334-349, Academic Press,
    Orlando, FL.
Somero GN et al. (1989) Rev. Aquat. Sci. 1, 591-614.
Speich N et al. (1994) Microbiol. 140, 1273-1284.
Vetter RD (1985) Mar. Biol. 88, 33-42.
Wilmot DB, Vetter RD (1990) Mar. Biol. 106, 273-283..

Molecular Approaches to Studying Natural Communities of Autotrophs

J.H. Paul and S.L. Pichard

Department of Marine Science, University of South Florida, St. Petersburg, FL 33701

Introduction

Ribulose 1,5-diphosphate carboxylase (Rubisco) forms the major link between the inorganic and organic carbon pools in the biosphere (Hartman and Harpel, 1994). Most ecosystems employ this enzyme in the process of photosynthetic carbon fixation (PCF), although certain ecosystems devoid of light (ie. hydrothermal vents on the ocean floor) use Rubisco in chemosynthetic carbon fixation (Jannasch et al., 1985). Rubisco enjoined with the enzymes of the Calvin cycle provide the net synthesis of 3 phosphoglyceric acid from 3 CO_2's (Tabita, 1994).

Rubisco as an enzyme has been found in nature to exist in two molecular forms: Type I and Type II. The Type I, the first to be discovered and by far the most common in nature, is an enzyme consisting of small and large subunits in the form of L_8S_8. The sequence of the large subunit is relatively conserved, and is believed to be the main catalytic site, while small subunits show more diversity in sequence. In higher plants and chlorophytic algae, the large subunit is encoded by the chloroplast genome, while the small subunit is nuclear encoded. In cyanobacteria the rbcL-rbcS genes are adjacent on the genome and are co-transcribed. In chromophytic algae (diatoms, cryptomonads) the large and small subunit genes are both encoded by the chloroplast. The type II enzyme, found in certain photosynthetic bacteria and recently in a dinoflagellate (Morse et al., 1995) is formed only of large subunits, in the form L_2 (Tabita, 1988)

Rubisco is believed to be the major carbon assimilatory enzyme in PCF in the phytoplankton of the world's oceans. A critical concern of ocean scientists has been quantifying the rate of carbon fixation in the ocean, termed primary production. Estimates of the rate of primary production have been made since the 1950's with the advent of the ^{14}C-technique (Steemen-Nielsen, 1952). However, the development of methodology to accurately measure this process has provided an unending challenge to ocean researchers (Carpenter and Lively, 1980;

301

M. E. Lidstrom and F. R. Tabita (eds.), Microbial Growth on C₁ Compounds, 301–309.
© *1996 Kluwer Academic Publishers. Printed in the Netherlands.*

Falkowski et al., 1992).

Another challenge to ocean scientists has been to understand the composition of the primary producers and their distribution in the water column. The types of organisms responsible for autotrophic CO_2-fixation in the world's oceans include cyanobacterial picoplankton (Waterbury et al., 1979), eucaryotic ultraphytoplankton (Shapiro et al., 1989), prochlorophyte-like organisms (Chisholm et al., 1988), and larger eucaryotic phytoplankton (chain-forming diatoms, dinoflagellates, etc.). The distribution of these types of organisms have been shown to vary over space (vertically in the water column and geographically) and time (daily, seasonally, and interannually). Understanding these variations and the factors controlling the distribution of the various types of organisms are major goals of the marine science community.

We have become interested in the regulation of photosynthetic carbon fixation in the water column of the oceans. Little is known concerning regulatory mechanisms in algae in culture, and nothing is known about this process with natural phytoplankton populations *in situ*.

The activity of Rubisco is regulated through a myriad of mechanisms in plants. Arguably central to regulation in higher plants, but also functioning in all rubisco-containing forms, is the carbamylation of lysine 191 by activator CO_2 (Hartman and Harpel, 1994). Uncarbamylated Rubisco is incapable of carbon fixation and is inhibited by the binding of one of the substrates, RuDP. The carbamylated from uses the substrate RuDP and CO_2 for fixation. Carboxy arabinatol phosphate (CA1P) also binds to the carbamylated form of Rubisco, and is believed to be a dark regulatory mechanism of carbon fixation in higher plants. Other phosphorylated intermediates that form as a result of epimerization of the substrate such as xylose 1,5-bisphosphate (XuDP) and 3-ketoarabinatol 1,5-bisphosphate (KABP) bind to the activated and unactivated forms of the enzyme, respectively (Tabita, 1994). The enzyme rubisco activase catalyzes the removal of phosphorylated intermediates that bind and inhibit activity, and is another form of regulation (Hartman and Harpel, 1994).

Another regulatory mechanism and important factor for the correct catalytic activity of transcribed Rubiscos is the folding proteins or molecular chaperonins (Hartmen and Harpel, 1994). In higher plants, the chaperonin Chcpn 60 has been identified as the key molecule that enables nuclear encoded small subunits to be folded with plastid encoded large subunits.

Transcriptional control of Rubisco expression is also a key regulatory mechanism, previously described in higher plants and some chlorophytic algae (Tabita, 1988). The overriding regulatory feature influencing enzyme synthesis is light, at least for higher plants and most algae. Light induction of synthesis of Rubisco has been shown to be caused by increases in both large and small subunit mRNA (Tabita, 1988). How stoichiometrically equivalent amounts of large and small subunits are synthesized is a mystery.

In cyanobacteria, it was previously thought that levels of rubisco, and thus *rbcL-rbcS* mRNA were fairly invariant over time (Tabita, 1988). However, recent evidence suggests that transcriptional regulation of Rubisco occurs in cyanobacteria. In the unicellular nitrogen fixing *Synechococcus* RF-1, transcriptional control of nitrogenase had been observed (Huang and Chou, 1990), with N_2 transcription occurring during the dark periods. Levels of *rbcL* mRNA were found to accumulate in the light and disappear in the dark in this organism, regulated in opposition with the *nif* genes (Chow and Tabita, 1994). In several *Anabeana* strains, *rbcL* transcripts as well as rubisco activase (*rca*) transcripts accumulated in the light and disappeared in the dark (Li and Tabita, 1994).

We are interested in the regulation of rubisco as a means of understanding the control of CO_2 fixation in the oceans. To this end, we have developed methods to quantitatively determine levels of *rbcL* mRNA in natural phytoplankton populations as well as those in culture. A second goal of our studies is to identify the types of transcriptionally active phytoplankton in the water column, though both group-specific hybridization and amplification, cloning and sequencing mRNA from these populations. A portion of these results are presented herein.

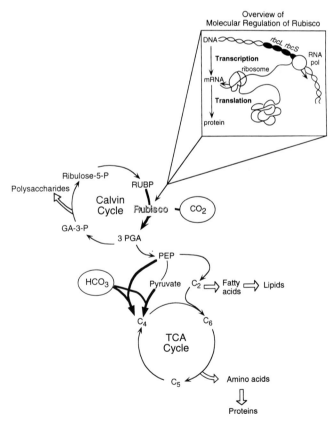

304

Results

Fig. 1 gives a basic overview of the role of Rubisco in the metabolism of autotrophic organisms, and alternate carbon fixation mechanisms (ie. C4 fixation). The former process is linked to the Calvin Cycle, while the latter is used for fixation into C4 acids and amino acids, using several enzymes of the TCA cycle. The inset shows the transcription and translation of *rbcL-rbcS* as would occur in cyanobacteria or in chromophytic algae, where the genes are co-transcribed.

Because of the evolutionary conserve nature of Rubisco, we have used *rbcL* probes to hybridize to natural populations to detect *rbcL* gene expression. We have developed methods to efficiently extract mRNA from natural populations of microorganisms (Pichard and Paul, 1991; 1993; Pichard et al., 1993). The method is based upon filtration of large volumes of seawater through Millipore Durapore filters and extracting the RNA with guanidinium-isothio-cyanate while bead-beating. Once the RNA is extracted, it is dotted and probed with single-stranded RNA probes.

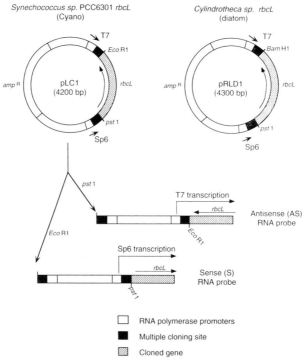

Fig 2 shows the Riboprobe constructs used to make probes. The *Synechococcus* PCC6301 or cyanobacterial probe was subcloned from the plasmid pCS751, while the *Cylindrotheca* or diatom probe was subcloned from the plasmid pVT223, both of which were gifts of F. Robert Tabita at Ohio State University. By use of the Sp6 and T7 RNA polymerase promoters, both sense and antisense

RNA probes can be generated, using ^{35}S-UTP. Because the sense probe can only bind to DNA and not mRNA, it is used as a control to detect contaminating *rbcL* DNA in our RNA preparations. The antisense probe is used to quantitate the levels of *rbcL* mRNA in natural samples. Hybridization is under stringent conditions (55° in 50% formamide) and hybridization signals are compared to diatom and cyanobacterial RNA standards by Ambis radiometric scanning. Thus, the *Synechococcus* probe should detect form I *rbcL* from cyanobacteria, green algae, prochlorophytes and higher plants, while the diatom probe should detect *rbcL* from chromophytic algae, including diatoms, chrysophytes, brown and red algae, cryptomonads, prymnesiophytes, and other unicellular chromophytic phototrophs (see Fig. 5).

In a previous study of the distribution of *rbcL* mRNA and carbon fixation in the water column, we found the *Synechococcus* mRNA to coincide with *Synechococcus* cell counts (orange-fluorescing cells) at a depth of 65 m. However, below this at 80 m there was a peak in chlorophyll *a*, red fluorescing cells, and carbon fixation. This population of autotrophs was not detected by probing with the cyanobacterial probe. We therefore returned to these waters to see if we could detect this autotrophic population with the diatom probe.

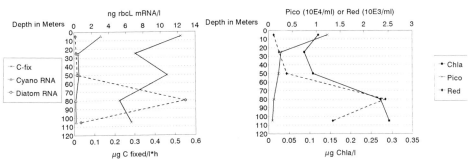

Fig 3. shows the results of the use of both these probes to probe RNA extracts of samples from the water column in the southeastern Gulf of Mexico. Samples were taken with niskin bottles in a rosette attached to a CTD equipped with a fluorometer. The depth of the *in situ* fluorescence maximum (referred to

as the chlorophyll *a* maximum) was 80 m at this station. Also measured were picoplankton direct counts (labelled "pico") which are yellow/orange fluorescing, phycoerythrin-containing *Synechococcus*-like cells and red fluorescing (labelled "red" cells). Chlorophyll *a* was also measured (Fig. 3B), as well as photosynthetic carbon fixation by ^{14}C-incorporation in flasks that had the light attenuated to approximately match *in situ* light conditions. The surface waters (3 m depth) had the greatest amount of cyanobacterial mRNA (Fig. 3A) as well as the greatest number of *Synechococcus*-like cells (Fig. 3B) and the greatest level of carbon fixation. The greatest amounts of chlorophyll *a* and red-fluorescing cells were found at 80 m, which also coincided with the peak in diatom-like *rbcL*. Thus, there is a spatial separation in the water column of these two distinct autotrophic populations. We have not as yet characterized these "red cells", but will be looking at them by TEM for the presence of diatoms or some other morphologically identifiable forms in the future.

Carbon Fixation and rbcL mRNA

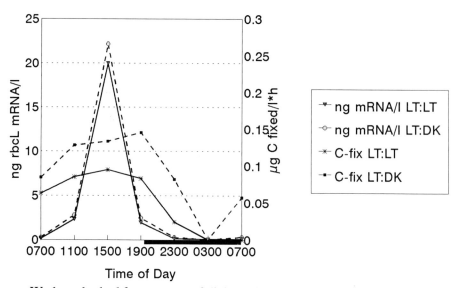

Time of Day

We have looked for patterns of diel regulation of carbon fixation in natural phytoplankton populations by enclosing ~150 l of water in polyethylene tanks on shipboard. In the most recent experiment, two tanks were covered in black plastic, and the temperature maintained by use of a chiller unit. Both had submersible light banks in them, with one on a continuous light regime, and the other on a 12:12 light dark regime. Both tanks showed a broad peak in carbon fixation, from 0700 to 2300, with the highest amounts between 1100 and 1900 (Fig. 4). Both tanks also showed a sharp peak in *rbcL* mRNA, with a maximum at 1500 hr. This suggests that the synthesis of Rubisco is transcritionally regulated on a diel basis, and the cessation of this synthesis is not necessarily contingent upon darkness (ie. mRNA disappeared in the continuously illuminated tank as rapidly as in the one on the 12:12 lt:dk regime). There may be a circadian (entrained) rhythm in *rbcL* transcription in natural phytoplankton populations as observed for nitrogenase in *Synechococcus* RF-1 (Huang and Chow, 1990).

We have also designed primers based upon conserved portions of the *rbcL* sequence in an attempt to amplify *rbcL* from RNA and DNA samples of natural phytoplankton populations. The primers amplify a 483-489 bp fragment of the large subunit. This has been done to identify the phototrophs in the water column in the ocean by sequence analysis and comparison.

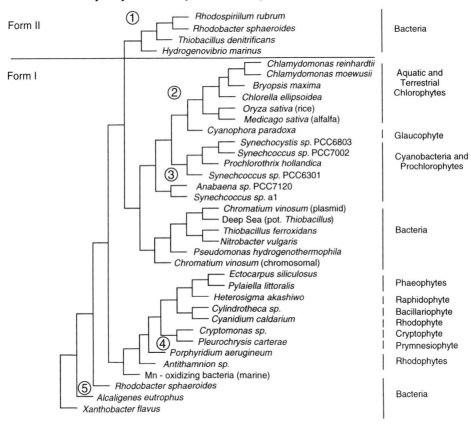

Figure 5 shows a consensus tree (unrooted) based on the UPGMA method done with the CLUSTAL program supplied by Intelligenetic PC Gene Software. All the sequences presented are based upon the full length sequence. Above the long horizontal line are the Form II Rubiscos (L_2) and below the line are the Form I (L_8S_8). The circled 1 indicates the approximate location of a recently

308

discovered eucaryotic sequence from a dinoflagellate, the first form II from a eucaryotic source. The circle 2 indicates the location of recently sequenced marine prasinophytes, which fall in the chlorophytic (green algal) clade. The circled 3 indicates the location of several marine *Synechococcus* and *Prochlorococcus* from which we have sequenced the PCR amplified portion of the *rbcL* gene. The circled 4 indicates the approximate genetic location of an amplified mRNA sequence that we obtained from a surface water sample in the Gulf of Mexico. This sequence was most similar to a prymnesiophyte, a unicellular chromophytic flagellate. The circled 5 indicates the genetic relatedness of an amplified *rbcL* gene which came from the 80 m sample containing the chromophytic mRNA peak shown in Fig. 3. Initial sequence analysis placed this sequence close to the diatom *Cylindrotheca*, but inclusion of other sequences in the tree building placed this sequence was actually most closely related to Mn-oxidizing bacteria isolated from the marine environment. The exact identity of such organisms remains unknown, but the data suggests that unique organisms make up the autotrophic population of the deep euphotic zone.

Conclusions
      We have demonstrated the utility of direct hybridization of mRNA to quantitate the transcription of the *rbcL* gene in ambient phytoplankton populations. Using cyanobacterial and chromophytic probes, we have defined two spatially (and evolutionarily) distinct populations of autotrophs in the water column. Finally, we have found the identity (or nearest relative) of some of these autotrophs by amplifying, cloning, and sequencing a portion of the *rbcL* gene. These studies have helped us understand the molecular regulation of PCF and the identity of the organisms responsible for this process in the oceans.
      This work was supported by the Department of Energy's Ocean Margin Program and the National Science Foundation Biological Oceanography program.

References
Carpenter EJ, Lively JS (1980) In Falkowski PG, ed, Primary productivity in the sea, pp 161-178, Plenum Press, New York.
Chisholm SW et al (1988) Nature 334, 340-343.
Chow TJ, Tabita FR (1994) J Bact (in press)
Falkowski PG et al (1992) 5, 84-91.
Hartman FC, Harpel MR (1994) Annu Rev Biochem 63, 197-234.
Huang T-C, Chow T-J (1990) Curr Microbiol 20, 23-26.
Jannasch HW et al (1985) Science 229, 717-720.
Li L-A, Tabita FR (1994) J Bact 176, 6697-6706.
Pichard SL et al (1993) Mar Ecol Progr Ser 101, 55-65.
Pichard SL, Paul JH (1991) Appl Environ Microbiol 57, 1721-1727.
Pichard SL, Paul JH (1993) Appl Environ Microbiol 59, 451-457.
Shapiro LP et al (1989) J Phycol 25, 794-797.

Steemann Nielsen E (1952) J Cons Cons Int Explor Mer 18, 117-140.
Tabita FR (1994) In Bryant DA, ed, The molecular biology of cyanobacteria, pp 299-329, Kluwer Academic Publishers, Dordrecht, the Netherlands.
Tabita FR (1988) Microbiol Rev 52, 155-189.
Waterbury JB et al (1979) Nature 277, 293-294.

Microbial Cycling of Methyl Bromide

Ronald S. Oremland, U.S. Geological Survey, Menlo Park, CA 94025

INTRODUCTION

Environmental concern about brominated halocarbons like methyl bromide (MeBr) is focused on their potential to destroy stratospheric ozone. Photocatalysis of MeBr and other halocarbons in the stratosphere results in the liberation of reactive Cl and Br atoms. Because Br atoms are perhaps as much as 100-fold more efficient at attacking ozone than are Cl atoms, bromine's lower abundance is partly compensated for by its higher reactivity. Furthermore, the coupling of Br and Cl chemistry can accelerate ozone destruction (McElroy et al., 1986; Solomon et al., 1990; Wahner, Schiller, 1992):

$$Cl + O_3 \quad \text{------------>} \quad ClO + O_2 \qquad (1)$$
$$Br + O_3 \quad \text{------------>} \quad BrO + O_2 \qquad (2)$$
$$ClO + BrO \text{----------->} \quad Cl + Br + O_2 \qquad (3)$$

Hence, even though the tropospheric residence time of a reactive molecule like MeBr is short, its potential to degrade stratospheric ozone is large. In order to better assess the ozone degrading potential of MeBr, more information is needed with regard to its biogeochemical global balance with respect to sources and sinks (Prather, Watson, 1990). This information is required in order to calculate more precise estimates of tropospheric residence times. The longer the tropospheric residence time (τ) of a halocarbon, the greater will be the amount of it transported to the stratosphere. Hence, precise estimates of τ will result in more accurate assessments of the ozone degradation potential of compounds like MeBr.

Methyl bromide is a trace constituent of the lower atmosphere with ambient mixing ratios of about 10-15 parts per trillion (ppt) in the northern hemisphere. An interhemispheric gradient exists with slightly lower levels occuring in the southern hemisphere (Singh et al., 1983, Cicerone et al., 1988). The atmospheric residence

M. E. Lidstrom and F. R. Tabita (eds.), Microbial Growth on C₁ Compounds, 310–317.
© 1996 Kluwer Academic Publishers. Printed in the Netherlands.

time and burden of MeBr have been estimated to be ~ 2 years and ~ 200 Gg, respectively (1 Gg = 1 x 10^9g) (Khalil et al., 1993; Singh, Kanakidou, 1993). More precise estimates have placed the residence time closer to 1.7 years (Mellouki et al., 1992). The residence times were calculated from the kinetics of MeBr oxidation by tropospheric hydroxyl radicals.

Initial estimates of the magnitude of the annual global source term for MeBr placed it at about 100 Gg, with perhaps one third derived from anthropogenic sources like fumigation (Singh, Kanakidou, 1993; Khalil et al, 1993). With a τ of 1.7 years, this translates to an annual atmospheric sink term of ~ 78 Gg. Natural sources of MeBr include 10-50 Gg produced annually from biomass burning (Mano, Andreae, 1994) with the rest thought to be derived from biological production by oceanic phytoplankton and perhaps from terrestrial sources like fungi. The biological formation of such compounds as bromoform and dibromomethane by marine algae and macrophytes has been established (Sturges et al., 1992; Manley et al., 1992), but much less is known about biogenesis of MeBr. Furthermore, because of the high solubility of MeBr in water, the oceans play an important role in regulating atmospheric concentrations (Butler, 1994). The oceans were originally thought to represent an important source of MeBr to the atmosphere (Singh et al., 1983), but recent findings suggest that much of the surface waters of the open ocean are undersaturated with respect to MeBr (Lobert et al., 1995). This oceanic sink would consume the global atmospheric burden of MeBr in 2.9-3.6 years, which translates into an annual consumption term of about 50 Gg. The combined effects of the atmospheric hydroxyl radical sink plus the oceanic sink have lowered estimates of τ to 1.1 years (Lobert et al, 1995).

MeBr is chemically reactive in water and will undergo a number of nucleophilic substitution reactions with other halides (Swain, Scott, 1953; Zafiriou, 1975) as well as hydrolysis to methanol (Elliott, Rowland, 1993). In anoxic environments with free sulfide present, MeBr reacts to form alkyl sulfides like methane thiol and dimethylsulfide (Oremland et al., 1994a). However, the active involvement of bacteria in the degradation of MeBr in nature has not been examined until quite recently (Oremland et al., 1994 a; Oremland et al., 1994 b; Shorter et al., 1995). In this paper I will review some of the recent experimental work which demonstrates the mechanisms by which MeBr undergoes bacterial attack, and the possible global significance of these reactions in the cycling of MeBr in nature.

RESULTS

Degradation of MeBr in anoxic sediments. Sediments with free sulfide present react with MeBr by nucleophilic substitution of sulfide for bromine, resulting in the formation of methane thiol (MeSH) and dimethylsulfide (DMS). These alkylated sulfur gases can subsequently undergo attack by sediment methanogens and/or sulfate-reducers to produce CH_4 and/or CO_2, as well as H_2S (Kiene et al., 1986). Figure 1 illustrates the intermediates and products formed when MeBr was added to anoxic sediment slurries from a sulfide-rich San Francisco Bay saltmarsh. MeBr was rapidly consumed and both DMS and MeSH were observed as transient intermediates

312

as was MeCl. However, all of these gases were removed upon the occurrence of methanogenic activity. No MeBr was consumed in killed controls lacking sulfide (Oremland et al, 1994 a).

Degradation of MeBr in aerobic soils: Soils consumed 10 ppm MeBr rapidly while the rate was much slower in anoxic soils or those inhibited with methyl fluoride (MeF) (Fig 2). This demonstrated that methanotrophs could be of importance in the oxidation of MeBr in nature, however the persistence of activity in MeF-inhibited controls suggested that other types of aerobes also oxidize MeBr. This conclusion was reinforced in experiments with ^{14}C-MeBr (Table 1). MeF inhibited activity by 72 %, and therefore 28 % of MeBr oxidation could be attributed to non-methanotrophic aerobes.

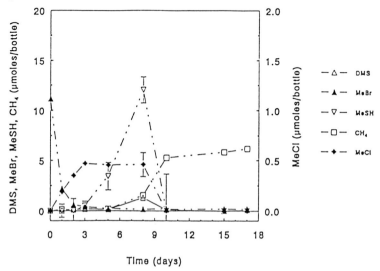

Figure 1: Degradation of MeBr in saltmarsh slurries. Symbols represent the mean of 3 slurries and bars indicate $\pm$ 1 std dev. From Oremland et al. (1994a) reproduced with permission.

Degradation of MeBr during agricultural field fumigations: Release of MeBr to the atmosphere during field fumigation events is dependent upon the type of soil involved. Hence, the capacity for MeBr destruction by chemical and/or biological mechanisms is highly variable, ranging from ~10 % to ~ 70 % (Yagi et al., 1995). Figure 3 shows the results of ^{14}C-MeBr incubation of soil from a strawberry field located on the Monterey Peninsula of California. The major product was ^{14}CO$_2$ and production was near linear for several days (Fig 3 A). However, some ^{14}CH$_4$ was also formed, especially after prolonged incubation (Fig 3 B), presumably because of the occurrence of anoxia within the sample and the subsequent participation of methanogens. Killed controls did not produce ^{14}C-labeled gaseous products (data not shown). Vertical profiles of ^{14}C-MeBr degradation have shown that microbial

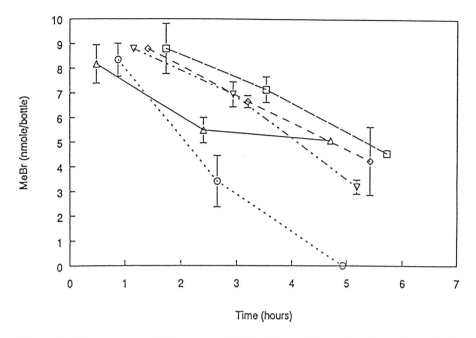

Figure 2: Consumption of headspace MeBr (10 ppm) by soils. Symbols: (o), air only; (Δ), autoclaved; (▽), air + MeF; (◊), N_2; (□), N_2 + MeF. Symbols represent the mean of 3 soils and bars indicate ± 1 std. dev. From Oremland et al. (1994 b) with permission.

Table 1: Oxidation of ^{14}C-MeBr to $^{14}CO_2$ by soils.

Condition	$^{14}CO_2$ (nCi)	% Oxidation[a]
Aerobic	558 ± 30	43.3
Aerobic + MeF	158 ± 41	12.3
Anaerobic	85 ± 15	6.6
Autoclaved	0	0

[a]Based on a 4 h incubation with 1.29 μCi ^{14}C-MeBr. From Oremland et al. (1994b) with permission.

oxidation can account for as much as 12 % of MeBr destruction, with highest activity occurring during actual fumigation events when the fields were tarped (Miller et al., submitted). Fumigation also increased levels of soil CH_4 and N_2O due to the inhibitory effects of MeBr and chloropicrin (also added during fumigation) on methane-oxidation, nitrification and denitrification (Miller et al., 1994).

Degradation of ^{14}C-MeBr in Mono Lake waters: The oceans represent an important sink for atmospheric MeBr, however the biological mechanism(s) by which it is destroyed is not known. Because we have previously characterized the hypersaline and alkaline water column of Mono Lake, California with regard to its chemical and biological properties (Oremland et al., 1993), and have shown the potential link of MeBr degradation to anoxic methanogenic (Oremland et al, 1994 a) and oxic methanotrophic biomes (Oremland et al, 1994 b) we conducted in situ experiments to measure ^{14}C-MeBr oxidation by bacteria in Mono Lake. Figure 4 shows the biological activity evident in a water sample taken from 11m depth during April, 1995. This activity translates to a turnover time of 60 days with the caveat that addition of the radioisotope raised the MeBr pool size to 2.5 M, which is several orders of magnitude greater than in situ concentrations based on equilibrium with the atmosphere. Vertical profiles of activity measured in July, 1994 did not show strong correlation with methane oxidation or nitrification, thereby implying the involvement of other types of bacteria in MeBr oxidation (Connell et al., 1994).

CONCLUSIONS

The bacterial degradation of MeBr occurs in a variety of environments, including anoxic sediments (Oremland et al., 1994 a), aerobic soils (Oremland et al., 1994 b), fumigated agricultural fields (Miller et al., ms submitted), and the water column of a hypersaline lake (Connell et al., 1994). In the case of the fumigation experiments, addition of the ^{14}C-MeBr tracer did not raise soil MeBr levels, and hence the observed oxidation represents a realistic appraisal of in situ bacterial activity. However, in all the other cases manipulations of soils, waters, and sediments raised the MeBr pool size by several orders of magnitude above that achieved by equilibrium with the atmosphere. Hence, these results are constrained by being MeBr degradation "potentials" which are valid only at the concentrations MeBr applied in order to conduct the experiment. It is therefore of great significance that Shorter et al. (1995) have demonstrated rapid bacterial uptake of MeBr by forest topsoil at ambient tropospheric mixing ratios. Thus, soil microbes can degrade MeBr at ~ 10 ppt mixing ratios, a value about five orders of magnitude below that carried out by soil methanotrophs oxidizing atmospheric methane at its ambient mixing ratio of 1.7 ppm (Cicerone and Oremland, 1988). Preliminary results with soils suggest that bacteria other than methanotrophs are responsible for the observed activity (P. Crill, personal communication). In addition, by examining a number of representative soils and extrapolating their measured MeBr-consuming activities to a global scale, these authors estimated an annual terrestrial soil sink for MeBr of ~ 42 Gg, a value which in conjunction with the oceanic sink, diminshes the estimate of τ to roughly 0.9 years. Future research directed at determining the microbes responsible for this activity, the enzymatic mechanism(s) by which it operates, as well as refining global estimates of the soil MeBr sink represent obvious challenges. In addition, the possibility that soil microbes can oxidize other trace constituents of the troposphere which occur at ppt levels, such as methyl chloride, methylchloroform and hydrochlorofluorocarbons needs to be examined. Finally, if the atmospheric global burden estimate of MeBr is correct (~ 200 Gg), and a revised τ of 0.9 years is valid, a doubled annual global flux is implicated (Table 2). Hence, as yet unidentified anthropogenic and natural sources of MeBr must exist, and a search of terrestrial and aquatic environments for biogenic emission of MeBr is justified.

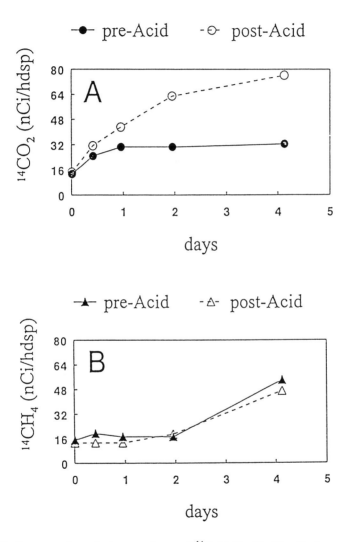

Figure 3: Gaseous degradation products of ^{14}C-MeBr (2 Ci added per soil core) from a California strawberry field (L.G. Miller, unpublished data). Acid was added to liberate dissolved $^{14}CO_2$.

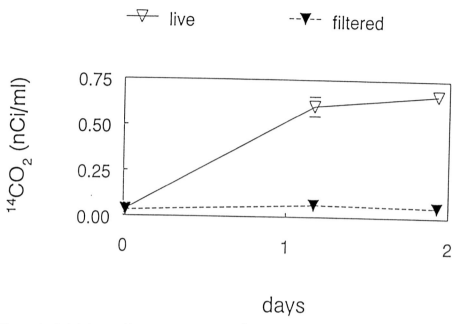

Figure 4: Oxidation of ^{14}C-MeBr (45 nCi ml^{-1} added) by Mono Lake water (T.L Connell, unpublished).

Table 2. A comparison of estimated annual fluxes of MeBr to the atmosphere assuming only a tropospheric OH sink (Khalil et al, 1993; Singh, Kanakidou, 1993), and with a tropospheric OH sink plus oceanic (Lobert et al, 1995) and soil sinks (Shorter et al, 1995) included.

TERM	OH SINK ONLY	OCEAN & SOIL SINKS
ATMOS. BURDEN	200 Gg	200 Gg
RESIDENCE TIME	2.0 Yr	2.0 Yr
ANNUAL GLOBAL FLUX	100 Gg	215 Gg
ANTHROPOGENIC FLUX (% of total)	30 %	14 %

REFERENCES
Butler JL (1994) Geophys. Res. Lett. 21, 185 - 188.
Cicerone RJ et al (1988) J. Geophys. Res. 93, 3745 - 3749.
Cicerone RJ, Oremland RS (1988) Global Biogeochem. Cycles 2, 299 - 327.

Connell TL et al (1994) Eos Abstr. Fall Meet. Am. Geophys. Union, p. 110.

Elliott S, Rowland FS (1992) Geophys. Res. Lett. 19, 1043 - 1046.

Khalil MAK et al (1993)J. Geophys. Res. 98, 2887 - 2896.

Kiene RP et al (1986) Appl. Environ. Microbiol. 52, 1037 - 1045.

Lobert JM et al (1995) Science 267, 1002 - 1005.

Manley SL et al (1992) Limnol. Oceanogr. 37, 1652 - 1659.

Mano S, Andreae MO (1994) Science 263, 1255 - 1258.

McElroy MB et al (1986) Nature 321, 759 - 762.

Mellouki A et al (1992) Geophys. Res. Lett. 19, 2059 - 2062.

Miller LG et al (1994) Eos Abstr. Fall Meet. Am. Geophys. Union, p. 110.

Miller LG et al (1995) Soil Biol. Biochem. (submitted).

Oremland RS et al (1993) In Oremland RS, ed, Biogeochemistry of Global Change:
Radiatively Active Trace Gases, pp. 704 - 741, Chapman and Hall, New York, USA.

Oremland RS et al (1994 a) Environ. Sci. Technol. 28, 514 - 520.

Oremland RS et al (1994 b) Appl. Environ. Microbiol. 60, 3640 - 3646.

Prather MJ, Watson RT (1990) Nature 344, 729 - 733.

Shorter JH et al (1995) Nature (in press).

Singh HB, Kanakidou M (1993) Geophys. Res. Lett. 20, 133 - 136.

Singh HB et al (1983) J. Geophys. Res. 88, 3684 - 3690.

Solomon S et al (1990) J. Geophys. Res. 95, 13,807 - 13,817.

Sturges WT et al (1992) Nature 358, 660 - 662.

Swain CG, Scott CB (1953) J. Am. Chem. Soc. 75, 141 - 147.

Wahner A, Schiller C (1992) J. Geophys. Res. 97, 8047-8055.

Yagi K et al (1995) Science 267, 1979 - 1981.

Zafiriou OC (1975) J. Mar. Res. 33, 75 - 81.

Regulation of methane oxidation: contrasts between anoxic sediments and oxic soils

Gary M. King
Darling Marine Center
University of Maine
Walpole, ME 04573

Although the majority of organic matter mineralization occurs under oxic conditions, significant rates of mineralization occur in the anoxic guts of herbivores, and in the anoxic sediments and water columns of freshwater and marine ecosystems (Henrichs, Reeburgh 1987). Anaerobic metabolism in these systems is often coupled to methane production, with variations in the extent of methanogenesis among systems a function of organic matter, nitrate, ferric iron, and sulfate availability. Controls of anaerobic metabolism and methanogenesis have been reviewed extensively elsewhere (e.g., Henrichs, Reeburgh 1987; Oremland 1988), and will not be considered further here. Instead, I will illustrate the controls of methane oxidation by comparing two very different: anaerobic methane oxidation in marine systems and methane consumption in terrestrial systems.

A large fraction of the methane produced in benthic and water column systems is oxidized to carbon dioxide (King 1992; Reeburgh et al 1992), either anaerobically or aerobically. Anaerobic methane oxidation is primarily limited to marine environments. Though the mechanism is uncertain, and energy yields low, it is a very effective process, consuming virtually all of the methane produced in marine sediments (Reeburgh et al 1992). As a result, the oceans are only a minor atmospheric methane source (Cicerone, Oremland 1988).

In spite of numerous and diverse biogeochemical and microbiological efforts, anaerobic methane oxidation has essentially remained enigmatic. Several lines of evidence have implicated sulfate-reducing bacteria (SRB), but pure cultures of methane-oxidizing SRB have not been documented. Hoehler et al (1994) have proposed that a methanogen-SRB consortium oxidizes methane. The hypothetical pathway involves hydrogen and CO_2 production from methane by methanogens; SRB consume the hydrogen, thereby maintaining hydrogen partial pressures low enough for favorable free energy yields. Hoehler et al (1994) estimate from thermodynamic calculations that hydrogen concentrations < 0.29 nM are required for net methane oxidation, while concentrations > 2.4 nM promote net methane formation from CO_2 reduction.

Evidence from Cape Lookout Bight sediments and a variety of other observations are consistent with the basic hypothesis, if not confirmatory. For example, the absence of net methane oxidation in freshwater sediments might be explained by hydrogen concentrations in excess of the 0.3 nM threshold; lower hydrogen concentrations in sulfidogenic sediments would facilitate oxidation. The localization of anaerobic methane

M. E. Lidstrom and F. R. Tabita (eds.), Microbial Growth on C_1 Compounds, 318–325.

oxidation at the base of the sulfate-reducing zone is also consistent with oxidation by a methanogen-sulfidogen consortium, since the abundance and activity of methanogens is low until sulfate is substantially depleted (Oremland 1988).

A microbiological test of the consortium concept would require use of methanogen-sulfidogen co-cultures instead of the monocultures examined thus far. Additional tests should include assays of hydrogen concentrations in sediments where methane is actively oxidized anaerobically. Lovley and Phillips (1987) have reported 1.4 and 0.2 nM hydrogen concentrations for sulfidogenic and iron-reducing fresh-water sediments, respectively, but did not measure methane oxidation. Hoehler et al (1994) have measured 0.2 nM hydrogen concentrations in sulfidogenic marine sediments where methane oxidation was evident, but the hydrogen database is otherwise too limited for further speculation. However, it is evident that hydrogen concentrations are a major control of anaerobic methane oxidation if the Hoehler et al. hypothesis proves correct.

While hydrogen and possibly sulfate may determine the distribution and rates of anaerobic methane oxidation, a more diverse group of parameters appears to control aerobic methane oxidation. In terrestrial systems, methane oxidation is limited primarily to the consumption of atmospheric methane by forest, grassland and agricultural soils, and to the oxidation of methane generated within landfills (King 1992). Methane oxidation also occurs in association with leakage from natural gas pipelines and storage facilities, but the rates and significance are unclear. For atmospheric methane consumption, the supply of methane is an important regulatory factor (e.g., King, Adamsen 1992). In contrast to other atmospheric trace gases, e.g., hydrogen and carbon monoxide, methane uptake is diffusion limited (Conrad 1988; King 1992 and references therein). Diffusion limitation results from the sub-surface localization of methanotrophic activity, and is substantially modulated by soil water content but not temperature (e.g., Steudler et al 1989; Whalen et al 1990; King, Adamsen 1992; Adamsen, King 1993). The sub-surface zonation of methanotrophic activity in most soils (Fig. 1) remains enigmatic but may be attributed to several factors, including water stress, ammonium inhibition and high rates of grazing by bactivores (King, Schnell 1994).

As a consequence of diffusion limitation, atmospheric methane consumption rates may respond minimally to changes in regional or global mean temperatures. This contrasts markedly with predicted responses of methanogenesis, a process much more sensitive to temperature change (e.g., Hameed, Cess 1983). On the other hand, rates of atmospheric methane consumption are sensitive to regional-scale hydrologic patterns that are in turn linked to global temperature changes. Thus, predictions of future trends in the global methane sink must consider indirect consequences of temperature change on abiological factors rather than direct effects on the methanotrophic microbiota.

Natural and anthropogenic sources of ammonium represent another factor affecting atmospheric methane consumption. With a few notable exceptions, ammonium fertilization of soils results in rapid and sustained inhibition of methane uptake (e.g., Steudler et al 1989; Mosier et al 1991; Adamsen, King 1993; Schnell, King 1994). At a physiological level, the mechanism of inhibition appears to involve competition between ammonia and methane for methane monoxygenase; in addition, ammonia oxidation may deplete intracellular pools of reduced $NADH+ H^+$ (King, Schnell 1994a, b). The products of ammonia oxidation, hydroxylamine and nitrite, also inhibit methane oxidation via several mechanisms. The limited ability of methanotrophs in culture or in situ to grow actively with methane at atmospheric to sub-atmospheric concentrations constrains their recovery from ammonium toxicity (Schnell, King in press), and may account for the longevity of inhibition. This is clearly evident in cultures where elevated methane concentrations (> 200-500 p.p.m.) alleviate completely inhibition by 500-1000 μM ammonium (King, Schnell 1994).

Figure 1. Methane concentrations in the headspace of soils from the Darling Marine Center Forest (see Adamsen, King 1993). Soils from various depths were incubated initially with atmospheric methane; after sealing in jars, methane concentrations were measured at the indicated times (data from A.P.S. Adamsen, 1991.Specialerapport, "Oxidation af atmosfœrisk methan I jord". Institut for Genetik og Økologi, Aarhus Universitet).

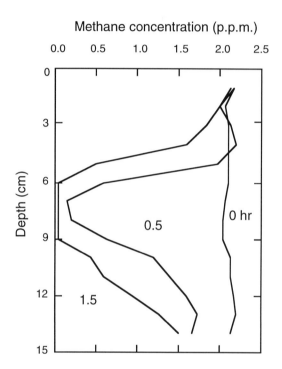

At an ecological level, a variety of factors affect the relationship between ammonium and atmospheric methane oxidation. First, it appears that some methanotrophs may not transport ammonium (Lees et al 1990); these methanotrophs may therefore exhibit limited sensitivity to ammonium, if any. Ammonium insensitive methanotrophs may occur in certain types of soils, especially those subjected to chronic ammonium inputs, and account for reports of negligible inhibition by added ammonium in agricultural soils (e.g., Hütsch et al 1993). Shifts in the relative importance of methanotrophs versus ammonia-oxidizing bacteria (AOB) may also occur in response to ammonia additions. For example, Castro et al (1994) have reported an increase in methane oxidation by AOB in sandy pine forest soils fertilized with urea. Although AOB may not typically dominate atmospheric methane uptake, edaphic parameters in some cases may select against methanotrophs.

The *availability* of ammonium in soils is a second, often overlooked factor. Mosier et al (1991) have indicated that nitrogen mineralization rates rather than extractable ammonium concentrations per se correlate with inhibition of atmospheric methane consumption. In addition to nitrogen mineralization rates, ammonium availability is also determined by the partitioning of ammonium between dissolved and various ionically bound phases. At a given total soil ammonium concentration, high cation exchange capacities decrease the concentration of dissolved ammonium and limit ammonium toxicity. Differences among soils in cation exchange capacity likely contribute to differences in the response to exogenous or endogenous sources of ammonium. Unfortunately, data on cation exchange capacities, extractable and dissolved ammonium, and other ions have not been published for the soils used in most of the existing analyses of methane uptake.

Soil cation exchange capacities also account for the reported inhibition of methane consumption by various inorganic salts, e.g., sodium chloride, sodium nitrate, potassium chloride (Adamsen, King 1993; Nesbit, Breitenbeck 1992; Crill et al 1994). Direct biological effects are unlikely, since inorganic salts at low to moderate concentrations do not adversely affect pure cultures of methanotrophs (King, unpubl. data). However, when added to soils, salts desorb bound ammonium that can obviously inhibit methanotrophic activity. The extent of desorpton varies predictably as a function of the particular alkali metal or alkaline earth cations used (Table 1). For example, rubidium and cesium chlorides desorb more ammonium than lithium and sodium chlorides. This pattern is consistent with the lyotropic series for the affinities of alkali metals for ion exchangers, and is a consequence of the increase in affinity for the exchange sites of clays with decreasing hydration radius (Kinniburgh 1981).

Table 1. Relative adsorption or desorption or of ammonium after addition of 1 ml (g fresh weight soil)$^{-1}$ of solutions containing 1 meq salt l^{-1}. Ammonium desorption is expressed relative to NaCl = 1, and adsorption relative to NH$_4$Cl = 1.

Salt	Relative change
LiCl	0.75
NaCl	1.00
KCl	1.22
RbCl	1.30
CsCl	1.41
NaNO$_3$	0.38
Na$_2$SO$_4$	0.32
MgCl$_2$	1.32
MgSO$_4$	0.09
CaCl$_2$	0.37
K$_2$SO$_4$	0.01
NH4F	1.22
NH$_4$Cl	1.00
NH$_4$Br	0.97
NH$_4$I	1.13
(NH$_4$)$_2$SO$_4$	0.59
(NH$_4$) $_3$PO$_4$	0.54

Though alkali metal and alkaline earth cations clearly play an important role in ammonium adsorption, empirical evidence and theoretical considerations indicate a role for anions as well. In particular, ammonium is desorbed differentially by the chloride, nitrate and sulfate counterions of cations such as sodium or magnesium (Table 1). This phenomenon can also be attributed to the effect of hydration radius on anion adsorption, and to changes in charge distribution on clay surfaces that affect cation adsorption. Differences in adsorption of various ammonium salts illustrate the effect, with phosphate and sulfate salts promoting adsorption relative to halide salts (Table 1).

An important implication of these results is that the form of exogenous ammonium inputs to soils can affect the extent of inhibition of methanotrophic activity, with somewhat stronger inhibition occurring for chloride than nitrate, sulfate or phosphate salts. Schnell and King (in prep.) have observed exactly such sensitivities for a forest soil. The differential response of methane uptake to various ammonium salts precludes direct comparisons among studies that have used different forms of ammonium. It is also evident that desorption of ammonium by inorganic salts precludes the use of inorganic salts as controls for changes in water potential due to additions of ammonium salts. This problem is illustrated well by the fact that sucrose, a non-electrolyte, is less inhibitory for methanotrophic activity in soils than salts added at equivalent finalwater potentials (Schnell and King, in prep.). Thus, without measuring ammonium desorption and changes in water potential attendant with salt additions, it is not possible to conclude, as have Kightley et al (1995), that the effects of added ammonium in various published field and laboratory studies are "non-specific" or otherwise due to changes in ionic strength. Osmotic effects may have played a role in the study of Kightley et al (1995), but they added salts at levels 10-71 fold higher than used by others (e.g., Adamsen, King 1993; Schnell, King 1994). Moreover, Schnell and King (in prep.) have observed that elevated methane concentrations increase the inhibitory effects of salts; this result cannot be explained in the context of osmotic effects, but does agree with the mechanism for ammonium inhibition (King and Schnell, 1994a, b; Schnell and King, 1994).

Methane concentrations also affect the extent of ammonium inhibition of methane uptake in pure cultures and in situ (King, Schnell 1994a, b; Schnell, King, 1994). The mechanism for this effect likely involves the requirement of methane monooxygenase for a reductant, and the depletion of reductant by ammonia oxidation. With increasing methane concentrations from sub-atmospheric levels to > 100 p.p.m., rates of methane oxidation increase as does the supply of reductant. In the presence of ammonium, elevated methane concentrations also increase reductant supply, but at a level less than in the absence of ammonium; since ammonium oxidation is dependent on the supply of reductant, increasing methane concentrations enhance ammonia oxidation and the concomitant production of two inhibitors, hydroxylamine and nitrite. The net result is an increase in the relative extent of inhibition, even though absolute rates of methane oxidation in the presence of ammonium increase with increasing methane concentration. At relatively high methane concentrations (e.g., > 250 p.p.m.), ammonium inhibition is diminished in pure cultures, presumably as a result of the competitive exclusion of ammonia from methane monooxygenase by methane, and by greater availability of $NADH + H^+$ and ATP to offset nitrite toxicity. Since methane concentrations in situ are often low, and in soils generally << 10 p.p.m., reversal of inhibition is not likely. Indeed, King and Schnell (1994a) demonstrated enhanced inhibition of atmospheric methane uptake for concentrations approximating pre-industrial values (about 0.7 p.p.m.) to values expected within the next 50 years (4 p.p.m.). Even without considering changes in land use and enhanced anthropogenic ammonium inputs to soils, these results suggest that the relative significance of the soil methane sink might decrease in the future simply as a result of increased methane concentrations.

Landfill soils also affect global methane budgets (Cicerone, Oremland 1988). A significant fraction of the methane produced within landfills is oxidized during transit throught the overlying soil burden (Whalen et al 1990; Kightley et al 1995). Not surprisingly, the controls of landfill methane oxidation differ somewhat from those in forest and grassland soils. Soil water content (or water potential) is important in landfill soils (e.g., Whalen et al 1990), yet methane oxidation does not appear to be diffusion limited. This is largely due to the relatively high methane concentrations produced within buried waste. One consequence of high methane concentrations and the absence of diffusion limitation is that landfill methane oxidation responds more strongly to temperature changes than forest soils (Whalen et al 1990). In addition, high methane concentrations promote oxidation rates that can substantially deplete oxygen in the soil gas phase. Indeed oxygen is a major limiting factor in some landfill soils (Kightley et al 1995); in contrast, there is no evidence for such a limitation in typical forest and grassland soils. High rates of methanotrophic activity also result in a local enrichment of microbial biomass, potentially depleting nitrogen, phosphorous, trace metals and other nutrients. Again, such a depletion is highly unlikely in association with atmospheric methane consumption. In the presence of high methane concentrations, ammonium inhibition may be less important than in other soils. Kightley et al (1995) report evidence for ammonium inhibition, but the concentrations of ammonium added in their study were high (≥ 10 µmol [g fresh weight soil]-1). A more comprehensive understanding of the controls of landfill methane oxidation should provide a basis for sharply reducing methane emissions from landfills, perhaps partially offsetting increased emissions from other sources.

In addition to methane transport, oxygen availability, ammonium concentrations and water content, a number of other parameters regulate soil methane oxidation. Water potential may limit activity in soils subjected to sustained or periodic drying. Neither methanotrophs in culture nor methanotrophic activity in soils show any adaptation for potentials less than about -0.1 MPa (Schnell, King in prep.). As a result of this intolerance of water stress, the very low water potentials ($<< $ -1 MPa) that develop in surface soils may limit methanotrophic activity to horizons with a more constant and favorable water potential regime. Although low water contents promote methane transport, low water potentials inhibit methane uptake, and can confine activity to sub-surface soils,ultimately resulting in diffusion limitation and temperature insensitivity. While hydrologic regimes are predicted to vary at regional scales in response to global climate change, the effects on regional or global scale atmospheric methane oxidation are not understood. Drying might enhance methane transport, but might also affect the depth distribution of activity, perhaps decreasing local or regional methane uptake. Similarly, increased precipitation and soil water contents could decrease methane transport, but facilitate activity in surface soils where diffusive transport is less significant as a limiting factor than low water potential.

Various organic substrates, copper, and other nutrients affect methane oxidation. However, the role of these factors in soils is unknown. Schnell and King (in press) found no evidence for stimulation or inhibition of methane consumption by additions to a forest soil of methanol, several multi-carbon organics, or copper. Schnell and King (in press) also observed a dramatic loss of methanotrophic activity during incubations of soils with methane-free air. These results strongly suggested that the methanotrophs responsible for atmospheric methane consumption derive little, if any, benefit from endogenous soil organic matter. In addition, the data indicate that atmospheric methane consumption in these soils is likely the province of obligate methanotrophs, rather than mixotrophic or ammonia-oxidizing methanotropths. The susceptibility of methanotrophs to starvation also raises several questions: what physiological mechanisms facilitate survival of methanotrophs during periods when atmospheric methane fluxes are low (e.g., winter)? what levels of atmospheric methane represent a starvation medium? how have the capacity and distribution of the soil methane sink changed during the post-industrial period, when atmospheric methane concentrations have risen from 0.3-0.7 to > 1.7 p.p.m.?

In summary, the controls of methane oxidation are diverse, involving a range of parameters and processes that operate at scales from cells to ecosystems, minutes to millenia. The net result is a system with complex heirarchical interactions that determine in large part methane fluxes from the biosphere to the atmosphere, and the temporal and spatial trends in these fluxes. The complexity of the control system, its susceptibility to anthropogenic perturbations, and the consequences of variations in methane concentration in the atmosphere-biosphere system are becoming more fully understood. However, definitive predictions of future trends in methane and the role of methane oxidation in any changes remain elusive.

Acknowledgements

The author acknowledges support from NSF DEB 9107315, USDA NRI 94-37107-0488, and NASA NAGW-3746. Contribution 287 from the Darling Marine Center.

Literature cited

Adamsen APS, GM King (1993) Appl. Environ. Microbiol. 59, 485-490.

Castro MS et al (1994) Can. J. Forest Res. 24, 9-13.

Cicerone R, RS Oremland (1988) Global Biogeochem. Cyc. 2, 299-327.

Conrad R (1988) Adv. Microb. Ecol. 10, 231-283.

Crill PS et al (1994) Soil Biol. Biochem.

Hameed S, RS Cess (1983) Tellus 35(B), 1-7.

Henrichs SM, WS Reeburgh (1987) Geomicrobiol J 5, 191-237

Hoehler TM et al (1994) Global Biogeochem. Cyc. 8, 451-463.

Hütsch BW et al (1993) Soil Biol. Biochem. 25, 1307-1315.

Kightley D et al (1995) Appl. Environ. Microbiol. 61, 592-601.

King GM (1992) Adv. Microbial Ecol. 12, 431-468.

King GM, APS Adamsen (1992) Appl. Environ. Microbiol. 58, 2758-2763.

King, GM (1993) In, Murrell JC and Kelly DP, eds, Microbial growth on C1 compounds, pp. 303-313, Intercept Scientific Publications, Andover.

King GM, S Schnell (1994a) Nature 370, 282-284.

King GM, S Schnell (1994b) Appl. Environ. Microbiol. 60, 3508-3513.

Kinniburgh DG (1981) In Anderson MA and Rubin AJ, eds, Adsorption of inorganics at solid-liquid interfaces, pp. 91-160, Ann Arbor Science, Ann Arbor, Michigan.

Lees V et al (1991) Arch. Microbiol. 157, 60-65.

Lovley DR, EJ Phillips (1987) Appl. Environ. Microbiol. 53, 2636-2641.

Mosier A et al (1991) Nature 350, 33-332.

Nesbit SP, GA Breitenbeck (1992) Agric. Ecosyst. Environ. 41, 39-54.

Oremland RS (1988) In Zehnder AJB, ed, Biology of anaerobic microorganisms, pp. 641-706, Wiley Interscience, New York, New York.

Reeburgh WS et al (1992) In Murrell JC and Kelley DP, eds, Microbial growth on C1 compounds, pp. 1-14, Intercept Scientific Publications, Andover.

Schnell S, GM King (1994) Appl. Environ. Microbiol. 60, 3514-3521.

Schnell S, GM King (in press) FEMS Microbiol. Ecol.

Steudler PA et al (1989) 341, 314-316.

Whalen SC et al (1990) Appl. Environ. Microbiol. 56, 3405-3411.

Anaerobic methane oxidation by a methanogen-sulfate reducer consortium: geochemical evidence and biochemical considerations

T.M. Hoehler and M.J. Alperin. Curriculum in Marine Sciences, University of North Carolina, Chapel Hill, North Carolina, 27599, USA.

Introduction

During the last twenty years, studies employing geochemical models, radiotracers, naturally occurring stable isotopes, and sample incubations have provided compelling evidence in support of anaerobic methane oxidation (Table 1). The first evidence came from geochemical models which show that "concave up" methane concentration profiles observed in many marine sediments can only be accounted for by consumption of methane in anoxic depth intervals. The models have been corroborated by radiotracer experiments which show that $^{14}CH_4$ is converted to $^{14}CO_2$ in anoxic systems. Distributions of naturally occurring stable isotopes provide further evidence of methane oxidation. The presence of ^{2}H- and ^{13}C-enriched CH_4 is consistent with kinetic fractionation arising from methane consumption, while the presence of ^{13}C-depleted CO_2 implies oxidation of an isotopically "light" substrate (methane). Finally, time-series incubations have shown methane concentration decreases in contained anoxic samples. Given the independent nature of these approaches, the number of investigators, and the wide range of study sites, it is clear that methane is consumed in anoxic environments.

Despite the abundance of evidence supporting the occurrence of anaerobic methane oxidation, the mechanism has yet to be established. Sulfate-reducing (Davis, Yarborough, 1966; Iversen, 1984) and methanogenic (Zehnder, Brock, 1979) bacteria can oxidize small amounts of methane during growth on other substrates, but no anaerobic bacterium has been grown on methane as the sole carbon and energy source. Attempts to identify the organism(s) responsible for environmental methane oxidation through the use specific inhibitors have yielded inconclusive results (Alperin, Reeburgh, 1985; Sandbeck, 1987).

Zehnder and Brock (1979) suggest that methanogens might be capable of net methane oxidation given appropriate environmental conditions . We propose that this mechanism is responsible for methane oxidation in the anoxic sediments of Cape Lookout Bight, North Carolina. Our hypothesis holds that methanogens carry out the actual chemical transformation of the methane, while sulfate reducers provide the appropriate environmental conditions via their influence on pore water H_2 concentrations. A previous paper (Hoehler et al, 1994) provides field and laboratory evidence supporting the methanogen-sulfate reducer consortium hypothesis. The purpose of this paper is to briefly review these arguments and to evaluate the feasibility of this hypothesis within a framework of thermodynamics and methanogen biochemistry.

M. E. Lidstrom and F. R. Tabita (eds.), Microbial Growth on C₁ Compounds, 326–333.
© 1996 Kluwer Academic Publishers. Printed in the Netherlands.

327

Table 1. Environmental Studies of Anaerobic Methane Oxidation

Study Type	Environment	Reference
Geochemical models	Santa Barbara Basin sediments	Barnes, Goldberg, 1976
	Cariaco Trench waters, sediments	Reeburgh, 1976
	Long Island Sound sediments	Martens, Berner, 1977
	Gulf of Mexico sediments	Bernard, 1979
	Eckernförder Bay sediments	Whiticar, 1982
	Skan Bay sediments	Alperin, Reeburgh, 1985
	Cariaco Trench waters	Scranton, 1988
	Eckernförder Bay sediments	Albert, Martens, 1995
	Amazon Shelf sediments	Blair, Aller, 1995
Radiotracer experiments	Santa Barbara Basin sediments	Kosiur, Warford, 1979
	Lake Mendota waters	Panganiban et al, 1979
	Skan Bay sediments	Reeburgh, 1980
	Lake Mendota, Izembeck Lagoon sediments, digested sludge	Zehnder, Brock, 1980
	Kysing Fjord sediments	Iversen, Blackburn, 1981
	Saanich Inlet sediments	Devol, 1983
	Skan Bay sediments	Alperin, Reeburgh, 1985
	Kattegat, Skaggerak sediments	Iversen, Jørgensen, 1985
	Big Soda Lake waters	Iversen et al, 1987
	Saanich Inlet sediments	Sandbeck, 1987
	Cariaco Trench waters	Ward et al, 1987
	Saanich Inlet waters	Ward et al, 1989
	Black Sea waters, sediments	Reeburgh et al, 1991
	Cape Lookout Bight sediments	Hoehler et al, 1994
	Eckernförder Bay sediments	Albert, Martens, 1995
Stable isotope measurements	Santa Barbara Basin sediments	Doose, 1980
	Guinea Basin sediments	Miller, 1980
	Big Soda Lake waters	Oremland, DesMarais, 1983
	Skan Bay sediments	Alperin, Reeburgh, 1984
	Eckernförder Bay sediments	Whiticar, Faber, 1986
	Skan Bay sediments	Alperin et al, 1988
	Eckernförder Bay sediments	Albert, Martens, 1995
	Amazon Shelf sediments	Blair, Aller, 1995
Time-series incubations	Framvaren Fjord waters	Lidstrom, 1983
	Japanese rice paddy sub-soil	Miura et al, 1992
	Japanese rice paddy sub-soil	Murase, Kimura, 1994(a,b,c)

The Consortium Hypothesis

Anaerobic methane oxidation in Cape Lookout sediments appears to occur in two distinct "modes". During the summer, methane oxidation occurs only in actively methanogenic, sulfate-free sediments (below the sulfate depletion depth) at 10-25% of the concurrent methane production rate (Figure 1a). This mode of oxidation is consistent with that attributed to methanogens in freshwater sediments (Zehnder, Brock, 1980). Presumably, water is the electron acceptor for the process: $CH_4 + 2H_2O \longrightarrow CO_2 + 4H_2$. In winter months, when sulfate penetrates to greater depths and impinges on these

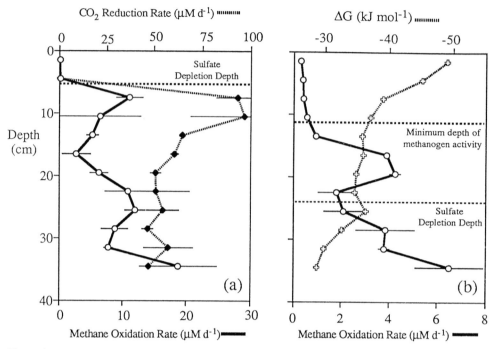

Figure 1. The two "modes" of methane oxidation in Cape Lookout Bight sediments. (a) CO_2 reduction and methane oxidation rates from 16 July, 1990; horizontal dashed line indicates the depth of sulfate depletion; error bars represent one standard deviation about the mean of two samples. (b) Methane oxidation profile and calculated ΔG for methane oxidation with sulfate from 28 January, 1991. Horizontal dashed lines represent the depth of sulfate depletion and the depth above which methanogens have not been active (sulfate always present). Error bars represent one standard deviation about the mean of two samples.

methanogenic sediments, methane production ceases and net oxidation occurs (Figure 1b). It is unlikely that net methane oxidation in Cape Lookout sediments is mediated solely by sulfate reducing bacteria. Methane oxidation with sulfate as electron acceptor would be thermodynamically most favorable in the upper portion of the sediment column (Figure 1b) and this zone is also expected to harbor the highest population of sulfate reducers. However, no oxidation is observed in the upper 10 cm of sediment: the process appears confined to a deeper zone in which methanogens have been active (Figure 1b). We propose that methanogenic bacteria are responsible for the oxidation observed in the winter as well as that in the summer and that the two "modes" represent a response to differing thermodynamic conditions. We hypothesize that flooding methano-genic sediments with sulfate in the winter brings about a change in H_2 concentrations sufficient to change the sign of ΔG and reverse the CO_2 reduction reaction.

Table 2. Effects of Adding Sulfate to Methanogenic Sediments

	Without Sulfate	With Sulfate
H_2 Concentration (nM)	1.82 ± 0.04	0.16 ± 0.02
ΔG, indicated reaction (kJ mol^{-1})	-9.9 (CO_2 Red.)	-13.3 (CH_4 Ox.)
CH_4 Oxidized/CO_2 Reduced	0.08 - 0.15	1.4 - 4.0

The results of a laboratory sediment incubation experiment (Table 2) support this hypothesis. We measured pore water H_2 concentrations and rates of methane oxidation and CO_2 reduction in methanogenic sediments with or without sulfate added. Under sulfate-free conditions, methane oxidation occurred at 8-15% of the concurrent CO_2 reduction rate. Steady-state H_2 concentrations in this treatment would result in $\Delta G = -9.9$ kJ mol[-1] for CO_2 reduction. The addition of sulfate caused a drop in the H_2 concentration to a level at which methane oxidation with water as electron acceptor would be favorable ($\Delta G = -13.3$ kJ mol[-1]). Concurrently, net methane oxidation began to occur in the sediments. These results stress the importance of sulfate in bringing about net oxidation and show that sulfate reducers can create environmental conditions in which methanogen-mediated oxidation would be energetically favorable.

Discussion

For the consortium hypothesis to represent a feasible mechanism for methane oxidation, methanogens must: (1) have the enzymes necessary to convert CH_4 to CO_2; (2) derive sufficient free energy from methane oxidation to allow for cell growth and be able to couple this free energy to ATP formation; and (3) be able to reverse their metabolism in response to environmental changes. In the following sections, these requirements are examined in detail.

Biochemical Considerations. Pure culture studies show that methanogens can transform methane to CO_2 while simultaneously producing methane (Zehnder, Brock, 1979). Though *net* oxidation was not observed, this suggests methanogens possess the enzymes necessary for methane oxidation. An examination of methanogen biochemistry suggests that much of the CO_2 reduction pathway could be used for methane oxidation, though mechanistically, the process may not occur via a simple "back reaction".

The biochemical requirements for a bacterium to oxidize methane anaerobically might best be considered in two parts: (1) binding and activation of methane followed by (2) sequential oxidation of the methane carbon with liberation of molecular hydrogen. It is clear that reactions *a-f* of the CO_2 reduction pathway (Scheme 1) are reversible (Thauer et al, 1993) and methylotrophic methanogens use directly analogous reactions to oxidize a fraction of the methyl substrate to CO_2. Typically, the electrons resulting from this oxidation are used to reduce biological electron carriers such as F_{420}, rather than to produce H_2. However, the maintenance of an equilibrium between H_2 and the electron carriers via hydrogenase enzymes provides a mechanism for liberation of H_2. Phelps et al (1985) showed that when a hydrogen sink is present, methylotrophic methanogens convert a substantial quantity of their "reducing equivalents" to H_2. Hence, methanogens seem capable of the reaction CH_3-H_4MPT + $2H_2O$ ----> CO_2 + $3H_2$ + H_4MPT-H, or, given the reversibility of reaction *f*, the equivalent methyl-CoM oxidation.

Scheme 1:

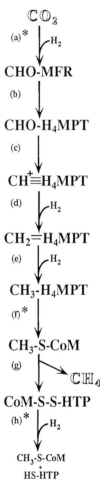

MFR, methanofuran
H_4MPT, tetrahydromethanopterin
CoM, mercaptoethanesulfonate
HTP, heptanoylthreoninephosphate
(∗) = ion gradient energy coupling

330

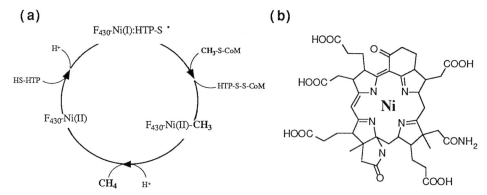

(a)

F$_{430}$-Ni(I):HTP-S •

H$^+$

HS-HTP

F$_{430}$-Ni(II)

F$_{430}$-Ni(II)-CH$_3$

CH$_3$-S-CoM

HTP-S-S-CoM

CH$_4$ H$^+$

(b)

HOOC

O

COOH

HOOC

N   N

Ni

HOOC

N   N

CONH$_2$

N

O

COOH

Figure 2. Aspects of the Methyl-Coenzyme M Reductase (MCR) system. (a) Modified version of the MCR reaction cycle proposed by Berkessel (1991). (b) Structure of the F$_{430}$ prosthetic group (Färber et al., 1991).

The initial activation of methane is a more difficult problem. At present, there is no known biological mechanism for anaerobic cleavage of the methane C-H bond. It is unlikely the activation could occur by a direct reversal of the CO_2 reduction pathway. The final step in the proposed cycle of methyl-CoM reductase (MCR), the enzyme which catalyzes step *g*, involves protonation of methyl nickel (Berkessel, 1991; Fig. 2a). Reversal of this pathway would require deprotonation of methane as the first step and MCR is thus considered irreversible.

However, anaerobic activation of the methane C-H bond is not a chemically intractable problem. Wayland et al (1991) have shown that porphyrin-bound rhodium(II) can homolytically cleave methane to Rh-CH$_3$ and Rh-H. The reaction is thought to proceed through a linear Rh--CH$_3$--H--Rh transition state via a concerted mechanism; this indicates that free radical H• and CH$_3$• are not formed during the course of the reaction. The details of the rhodium system have interesting implications for a would-be biological mechanism of methane activation. Clearly, the metalloradical character of rhodium in the (2+) state is important in bringing about *homolytic* cleavage; the bond geometry or electronic environment provided by the porphyrin system may play a secondary role. The concerted mechanism suggests that the stepwise reaction, Metal + CH$_4$ ---> Metal-CH$_3$ + H• is enthalpically "too expensive" and that an enzyme would require *two* proximal active sites to cleave methane.

The model system has at least two close analogs among methanogenic enzymes. The F$_{430}$ prosthetic group of MCR is a nickel porphyrin (Fig. 2b) and spectroscopic evidence indicates that nickel cycles through the metallo-radical (1+) state (Rospert et al, 1991). Among the proposed intermediates in the MCR cycle is an adduct of F$_{430}$ (in the nickel(I) state) and the radical •S-HTP (Berkessel, 1991; Fig. 2a) which could meet the two-center requirement. Activation of methane by MCR would introduce the methane carbon at the methyl-CoM stage and would make the process unique to methanogens.

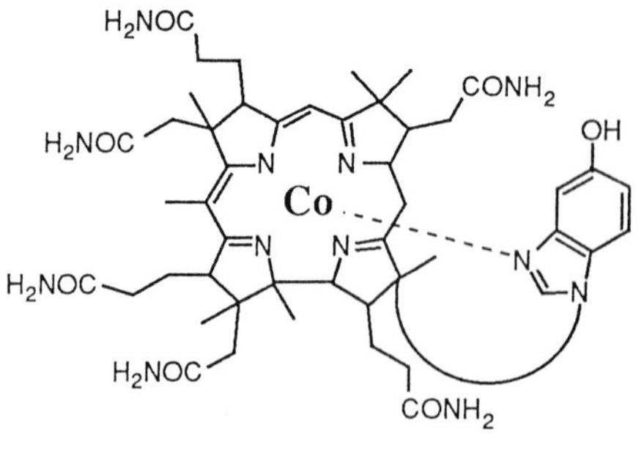

The cobamide group (Fig. 3) of

H$_2$NOC

CONH$_2$

H$_2$NOC

N   N

OH

Co

N   N

N

H$_2$NOC

N

H$_2$NOC

CONH$_2$

Figure 3. 5-hydroxybenzimidazoyl cobamide, the prosthetic group of a methanogen methyltransferase enzyme (Kräutler et al, 1987).

methanogen methyl transferase (MT), the enzyme which catalyzes reaction f, bears a striking analogy to the model system. The porphyrin-bound cobalt(II) state has a valence electron configuration exactly analogous to that of the rhodium compound. The *in vivo* reduction of cobalt in one-electron steps (Vogels et al, 1988) suggests a one-electron transfer agent periodically has proximity to the cobalt center, which might meet the requirement for a second active site. The presence of Fe-S clusters in MT (Thauer et al, 1993) might also meet this requirement. Activation of methane via MT would yield methyl-H_4MPT. It is interesting to consider that corrinoid enzymes are common and activation by MT (or possibly by other common metalloenzymes) might make methane oxidation a more general bacterial reaction.

Based on the foregoing considerations, it seems likely that the methanogenic pathway from CO_2 to methyl-H_4MPT or methyl-CoM could be readily used for methane oxidation. The initial activation of methane almost certainly does not occur through an *exact* reversal of the methanogenic pathway. However, methane *can* be chemically activated without molecular oxygen and the conversion of methane to CO_2 in pure cultures indicates that such a mechanism exists in methanogen biochemistry.

Thermodynamic Considerations. Net oxidation of methane by methanogens requires that the free energy of reaction is sufficient to allow for cell growth and that a means of coupling this energy to the formation of ATP is available. The sediment incubation experiment (Table 2) provides a basis for comparing the energetics of CO_2 reduction and methane oxidation under *in situ* conditions (without sulfate and with sulfate, respectively). CO_2 reducers in Cape Lookout sediments apparently can survive on an energy yield of -9.9 kJ mol^{-1} (as can acetotrophic methanogens (ΔG = -10.0 ± 0.9 kJ mol^{-1}; Hoehler et al, 1994). This suggests that the -13.3 kJ mol^{-1} resulting from methane oxidation would also allow for methanogen growth. Under *in situ* conditions, the initial step of CO_2 reduction (Scheme 1, a) requires at least +50 kJ mol^{-1} and is energetically coupled via a sodium gradient. Methane activation requires a similarly high energy input and would likely also require energetic coupling. It is interesting to note that methane activation via MT could be coupled while that via MCR could not (perhaps arguing for the former). The energy yield of CO_2 reduction is concentrated in the final steps, and two of these (Scheme 1, f and h) allow for energy conservation. In methane oxidation, the final step (reverse a) proceeds with an energy yield of about -60 kJ mol^{-1} and all of this energy could be conserved via the associated sodium gradient. This meets the important requirement of coupling free energy to ATP formation and allowing for cell growth. If the energetics of CO_2 reduction in Cape Lookout sediments are taken to represent minimum requirements, it seems likely that methanogens could survive by carrying out net methane oxidation.

Bacterial Precedents. The final requirement of the consortium hypothesis is that changes in environmental conditions can alter bacterial metabolism to the point of reversing it. Two studies provide useful evidence in this regard. First, a precedent exists regarding the question of "reversible metabolism" in bacteria. Lee and Zinder (1988) isolated a bacterium that converted H_2 and CO_2 to acetate in pure culture. When co-cultured with a methanogen, the acetogen catalyzed the reverse reaction, forming H_2 and CO_2 quantitatively from acetate. The metabolic reversal was likely due to the lower H_2 concentrations maintained by the methanogen, in direct analogy to the methanogen-sulfate reducer consortium. Second, it is clear that an external hydrogen sink can affect methanogen metabolism. Phelps et al (1985) found that a pure culture of *Methanosarcina barkeri* produced 0.26 moles of CO_2 per mole of methanol consumed (in keeping with predicted stoichiometry) and maintained a headspace H_2 partial pressure of ~230 µatm (equivalent to about 130 nM solution concentration). When co-cultured with the sulfate reducer *Desulfovibrio vulgaris*, the headspace H_2 concentration dropped approximately five-fold and the quantity of CO_2 produced increased to 0.59 moles per mole of methanol. If the relative amounts of methanol reduced and oxidized in this experiment bear some analogy to the "forward" and "reverse" components of CO_2 reduction, the influence of a sulfate reducer on H_2 concentrations could serve as a control on the ratio of methane oxidation to methane production. In summary, while there is no direct evidence that a decrease in H_2 concentrations can bring about a reversal of methanogenesis, it is clear that methanogens are influenced by external H_2 concentrations and that such a "metabolic reversal" has a close analog in nature.

Conclusions

The large body of geochemical evidence supporting anaerobic methane oxidation makes it clear that a biological mechanism must exist for the consumption of methane in anoxic environments. The feasibility of the methanogen-sulfate reducer consortium rests on several requirements: (1) methanogens must possess enzymes capable of converting methane to CO_2; (2) sulfate reducers must be capable of maintaining H_2 concentrations at which methane oxidation with water is thermodynamically favorable; (3) methanogens must be able to "reverse" their metabolism to take advantage of prevailing conditions; (4) methanogens must have a means for conserving the free energy of methane oxidation to allow for growth.

Pure culture work shows that methanogenic bacteria can convert methane to carbon dioxide; much of the CO_2 reduction pathway is likely used in this process, though it is not clear how methane is activated. Our experiments show that addition of sulfate to methanogenic sediments lowers H_2 to levels at which methane oxidation is thermo-dynamically favorable. While there is no direct evidence that methanogens can completely reverse their metabolism, it is clear that such a reversal can happen in anaerobic bacteria and that the degree of oxidative metabolism in methanogens is strongly dependent on external H_2 concentrations. The sodium gradient coupled to formyl-methanofuran oxidation (Scheme 1, *a*) provides a mechanism for conserving the free energy available from methane oxidation, which would allow for growth at the expense of this reaction. These factors suggest that a methanogen-sulfate reducer consortium is a feasible mechanism for methane oxidation in Cape Lookout Bight.

Acknowledgments. This work was supported by National Science Foundation grant OCE-9017979. TMH received support from an NDSEG Fellowship (Department of Defense). We are particularly grateful to Dan Albert, who aided in many aspects of these studies, as well as to Chris Martens. Joe Templeton and Maurice Brookhart of UNC's inorganic chemistry faculty provided helpful discussions concerning aspects of transition metal chemistry.

References

Albert DB, Martens CS (1995) Report 22, Forschungsanstalt der Bundeswehr für Wasserschall- und Geophysik.
Alperin MJ, Reeburgh WS (1984) In Crawford RL and Hanson RS, eds, Microbial Growth on C1 Compounds, pp 282-289, American Society for Microbiology, Washington, DC, USA.
Alperin MJ, Reeburgh WS (1985) Appl. Environ. Microbiol. 50, 940-945.
Alperin MJ, et al (1988) Global Biogeochem. Cycles 2, 279-288.
Barnes RO, Goldberg ED (1976) Geology 4, 297-300.
Berkessel A (1991) Bioorg. Chem. 19, 101-115.
Bernard BB (1979) Deep-Sea Res. 26, 429-443.

Blair NE, Aller RC (1995) Geochim. Cosmochim. Acta. 59, 3707-3715.
Davis JB, Yarborough HF (1966) Chem. Geol. 1, 137-144.
Devol AH (1983) Limnol. Oceanogr. 28, 738-742.
Doose PR (1980) Ph. D. Thesis, University of California, Los Angeles, USA.
Färber GW et al (1991) Helv. Chim. Acta 74, 697-716.
Hoehler TM et al (1994) Global Biogeochem. Cycles 8, 451-463.
Iversen N (1984) Ph. D. Thesis, Aarhus University, Aarhus, Denmark.
Iversen N, Blackburn TH (1981) Appl. Environ. Microbiol. 41, 1295-1300.
Iversen N, Jørgensen BB (1985) Limnol. Oceanogr. 30, 944-955, 1985.
Iversen N et al (1987) Limnol. Oceanogr. 32, 804-814.
Kaesler B, Schönheit, P (1988) Eur. J. Biochem. 174, 189-197.
Kosiur DR, Warford AL (1979) Estuarine Coastal Mar. Sci. 8, 379-385.
Kräutler B et al (1987) Eur. J. Biochem. 162, 275-278.
Lee MJ, Zinder SH (1988) Appl. Environ. Microbiol. 54, 124-129.
Lidstrom ME (1983) Limnol. Oceanogr. 28, 1247-1251.
Martens CS, Berner RA (1977) Limnol. Oceanogr. 22, 10-25.
Miller LG (1980) M. S. Thesis, University of Southern California, USA.
Miura Y et al (1992) Soil Sci. Plant Nutr. 38, 673-679.
Murase J, Kimura M (1994) Soil Sci. Plant Nutr. 40, (a) 57-61; (b) 505-514; (c) 647-654.
Oremland RS, DesMarais DJ (1983) Geochim. Cosmochim. Acta 47, 2107-2114.
Panganiban AT et al (1979) Appl. Environ. Microbiol. 37, 303-309.
Phelps TJ et al (1985) Appl. Environ. Microbiol. 50, 589-594.
Reeburgh WS (1976) Earth Planet. Sci. Lett. 28, 337-344.
Reeburgh WS (1980) Earth Planet. Sci. Lett. 47, 345-352.
Reeburgh WS et al (1991) Deep-Sea Res. 38, 1189-1210.
Rospert S et al (1991) FEBS Lett. 291, 371-375.
Sandbeck KA (1987) Ph.D. Dissertation, University of Washington, USA.
Scranton MI (1988) Deep-Sea Res. 35, 1511-1523.
Thauer RK et al (1993) In Ferry JG, ed, Methanogenesis, pp 209-252, Chapman & Hall, New York, USA.
Vogels GD et al (1988) In Zehnder AJB, ed, Biology of Anaerobic Microorganisms, Wiley-Interscience, New York, USA.
Ward BB et al (1987) Nature 327, 226-229.
Ward BB et al (1989) Continental Shelf Res. 9, 65-75.
Wayland BB et al (1991) J. Am. Chem. Soc. 113, 5305-5311.
Whiticar MJ (1982) In Fanning KA and Manheim FT, eds, The Dynamic Environment of the Ocean Floor, pp 219-235, Heath, Lexington, USA.
Whiticar MJ, Faber E (1986) Org. Geochem. 10, 759-768.
Zehnder AJB, Brock TD (1979) J. Bacteriol. 137, 420-432.
Zehnder AJB, Brock TD (1980) Appl. Environ. Microbiol. 39, 194-204.

"SOFT SPOTS" IN THE GLOBAL METHANE BUDGET

William S. Reeburgh
Earth System Science
University of California, Irvine
Irvine, CA 92717-3100

ABSTRACT
Direct rate measurements under *in situ* conditions are needed to assess the balance between methane production and consumption in the global methane budget. Measurements of potential CH_4 production and oxidation have advanced our understanding and provide a start at predicting the role of CH_4 oxidation under altered climate conditions in a range of important CH_4 source terms. *In situ* rate measurements or methods for assessing the fraction of CH_4 oxidized have not been possible in wetlands, rice paddies, and landfills. This paper summarizes recent work on these uncertain budget terms and points out approaches needed for *in situ* measurements

INTRODUCTION
King's 1992 review of ecological aspects of CH_4 oxidation showed that methanotrophy occurs in a diverse array of aquatic and soil systems, and serves as a global-scale control on the atmospheric CH_4 budget. Understanding the global CH_4 budget under present conditions and predicting changes in various terms under altered climate requires information on processes that affect CH_4 prior to emission to the atmosphere. Reeburgh et al (1993) attempted to quantify the role of methanotrophy in the global CH_4 budget, highlight information gaps, and focus future research directions. This study used the global atmospheric CH_4 budget as a framework (Cicerone, Oremland, 1988; Fung et al, 1991) and employed available rate measurements and estimates of the fraction of CH_4 oxidized to estimate the importance of methanotrophy in the global CH_4 budget. The compilation included rate measurements made under *in situ* conditions as well as potential CH_4 oxidation rates. Table 1 summarizes this work with columns representing net emission (E, based on the Fung et al, 1991 CH_4 budget), consumption (C), and their sum, gross CH_4 production (P). Estimates of CH_4 oxidation cannot be constrained by burdens, turnover times and isotope composition as in the atmospheric budget. However, conservative estimates of microbially-mediated CH_4 oxidation in major atmospheric CH_4 budget source terms resulted in an oxidation estimate of 688 Tg CH_4 yr^{-1}, 188 Tg CH_4 yr^{-1} larger than net emission to the atmosphere.

This area is advancing rapidly, and recent studies in environments representing key terms in the global CH_4 budget, such as wetlands, rice paddies, landfills, oceans, and soils, have clarified processes and emphasized uncertainties or "soft spots" in these estimates. The "soft spots", which are also important terms in the global CH_4 budget, are italicized and underlined in Table 1 for emphasis. This work proposes no revisions to the atmospheric CH_4 budget, but points out areas where production and consumption based on limited data or potential rates may be larger than reported in

M. E. Lidstrom and F. R. Tabita (eds.), Microbial Growth on C₁ Compounds, 334–342.

Reeburgh et al (1993). This paper summarizes recent work on these uncertain budget terms and points out approaches that may be useful in obtaining measurements needed for better CH_4 budget estimates.

Table 1. Global Net CH_4 Emission, Consumption, and Gross Production, Tg CH_4 yr[-1]

	E	+	C	=	P
Source/Sink term	Net Emission[a]		Consumption[b]		Gross Production
Animals	80		0		80
Wetlands	*115*		*27*		*142*
bogs/tundra (boreal)		35		15	50
swamps/alluvial		80		12	92
Rice Production	*100*		*477*		*577*
Biomass burning	55		0		55
Termites	20		24		44
Landfills	*40*		*22*		*62*
Oceans, freshwaters	*10*		*75.3*		*85.3*
Hydrates	*5?*		*5*		*10*
Coal production	35		0		35
Gas production	40		18		58
venting, flaring		10		0	10
distribution leaks[c]		30		18	48
Total Sources	500[d]				
Chemical destruction	-450				
Soil consumption	*-10*		*40*		*40*[e]
Total Sinks	-460[d]		688.3		
			Total Production		1188.3

a - Scenario 7, Fung et al, 1991; b - From Table 1, Reeburgh et al, 1993;
c - Should be considered P; d - 500 - 460 = 40 Tg CH_4 yr[-1] = annual atmospheric (0.9% yr[-1]) increment; e - Soil consumption of atmospheric CH_4 added to the gross budget as an equivalent production term.

___

RECENT STUDIES

    Wetlands and Rice Paddies. Processes affecting CH_4 production and oxidation in wetlands and rice paddies are similar, so that the two environments, which account for about half of CH_4 emission to the atmosphere, can be considered together. Conditions in rice paddies are more homogeneous than wetlands and provide better opportunities for experimental manipulation. Methane oxidation in wetlands and rice paddies is a complex process that depends on the availability of oxygen and occurs in different locations depending on environmental conditions. King (1990a, 1990b) demonstrated that wetland CH_4 fluxes responded rapidly to light and dark conditions and differed by approximately a factor of 2. The difference in fluxes was attributed to the penetration of photosynthetic O_2 and subsequent oxidation by a community with high CH_4 oxidation potential. Methane oxidation in rice paddy and swamp floodwaters (Conrad, Rothfuss, 1991; Happell et al, 1993, 1994) is a process capable of effectively oxidizing 80-90% CH_4 diffusing from flooded anoxic soils.

    The rhizosphere of wetland and rice systems is also an important locus of methanogenesis and CH_4 oxidation. Although early work in rice systems (de Bont et al,

1978) suggested that CH_4 oxidation in the rhizosphere was of minor importance, methanotrophs with high CH_4 oxidation potential have been observed in a range of rhizospheres. The methanotrophs are associated primarily with fine root material (King, 1993; Gerard, Chanton, 1993). The ratio of maximum potential CH_4 oxidation (MO) to CH_4 emission (ME) determined in incubation experiments ranged from 0.7 to 1.7 in rice to 5.6 to 51 in *Typha* and *Cladium* (Gerard, Chanton, 1993). However, CH_4 oxidation is oxygen-limited and these environments are typically net CH_4 emitters. The CH_3F inhibition technique (Oremland, Culbertson, 1992) was used by Epp, Chanton (1993) in greenhouse and field studies, and showed that 23 to 90% and 10 to 47% of the CH_4 produced (emission in the absence of oxidation) was oxidized. Care was taken to avoid inhibition of methanogenesis in the Epp, Chanton measurements, so these likely represent *in situ* rates. Tracer experiments on CH_4 oxidation with $^{14}CH_4$ have not been performed in wetland rhizospheres, and should be attempted to confirm the inhibition experiments. The role of the rhizosphere in CH_4 production is unclear. A correlation between net ecosystem exchange and CH_4 flux has been observed (Whiting, Chanton, 1993), and based on ^{14}C measurements in emitted CH_4, a significant fraction of emitted CH_4 must originate from recently-fixed carbon (Chanton et al, 1995). These results suggest close coupling of CH_4 production with plant growth. The mechanism supporting this correlation and the methanogenic substrate are unknown. Rapid turnover of root exudates, less rapid turnover of fine roots, or longer term turnover of detritus or soil organic matter to CH_4 are possibilities that cannot be resolved with existing measurements. Pulse labelling experiments focusing on the turnover of recently fixed $^{14}CO_2$ to CH_4 are needed to determine the time scale, mechanism, and degree of coupling between the rhizosphere and CH_4 emitted by vascular wetland plants.

Field and laboratory studies have identified water table level as an important control on CH_4 flux from wetlands and rice systems. Lowered water tables are believed to control CH_4 fluxes by enhancing aerobic CH_4 oxidation. Reeburgh, Whalen (1992) used potential oxidation rate measurements to suggest that water table lowering of 20-30 cm could convert tundra systems from a CH_4 source to a sink. Laboratory core incubations (Funk et al, 1994, Moore, Knowles, 1989; Moore, Roulet, 1993), field experiments in drained peatlands (Dise, Gorham, 1993; Lein et al, 1993, Roulet et al, 1993), and measurements in systems experiencing seasonal water table variations (Happell, Chanton, 1993; Whalen, Reeburgh, 1992) all show consistent CH_4 flux decreases with lowered water table, but no general relationship has emerged from this work. Field water table manipulation experiments at a variety of high latitude sites are in progress (Reeburgh et al, unpublished). The focus of most of the above work has been on CH_4 oxidation, but changes in CH_4 production rates can also affect the balance between production and consumption. It appears that most of the CH_4 oxidizing activity is confined to a zone adjacent to the water table (Roulet et al, 1993; Whalen, Reeburgh, unpublished). Soils from this zone are able to oxidize and produce CH_4 with no lag when incubation conditions are changed from oxic to anoxic (Whalen, Reeburgh, unpublished).

Landfills. Landfills are important contributors to the global CH_4 budget, emitting <10 to >50 Tg CH_4 yr^{-1} to the atmosphere (Bingemer, Crutzen, 1987; Augenstein, 1992). Since landfilling rather than dumping or burning is becoming a common practice in both the non-industrialized and industrialized worlds, the landfill CH_4 term is expected to increase in importance in the next century. Surprisingly few direct CH_4 flux measurements have been made at landfill surfaces, and the available measurements indicate high spatial variability. An active engineering literature deals with model estimates of landfill CH_4 production rates and lifetimes, and focuses on recovery and utilization of landfill gas for heating or electrical generation. The approach taken in estimating landfill CH_4 emission involves inventories of the amounts of refuse landfilled and laboratory measurements of the CH_4 generating potential of the materials. Because the waste composition is poorly known and conditions for CH_4 production are not optimum, these models are typically tested against gas recovery from existing landfills by considering a material balance as follows:

$$CH_4 \text{ generated} = CH_4 \text{ recovered} + \text{migration} + \text{emission} + \text{oxidation} \pm \text{storage}$$

The amount of CH_4 recovered is often the most reliable term in landfill CH_4 budgets. There are over 100 commercial gas recovery sites in the U.S., recovering (consuming) an estimated 40-90% of the potential production. The amount of CH_4 oxidized prior to release from landfills has been regarded as small term in the above material balances. Mancinelli, McKay (1985) observed that oxidation removed only 10% of the CH_4 supplied in laboratory soil microcosm experiments with landfill core columns arranged to mimic natural fluxes and soil gradients. Based on small changes in $CO_2:CH_4$ ratios in landfill depth profiles cited in previous work, Bingemer and Crutzen (1987) considered that only small amounts of CH_4 are oxidized in landfill cover soils.

Whalen et al (1990) measured CH_4 oxidation in a landfill cover soil using jar experiments. The potential rate of 45 g CH_4 m^{-2} day^{-1} was applied in the Reeburgh et al (1993) budget. Jones, Nedwell (1990) measured CH_4 and O_2 depth distributions in a landfill cover soil and observed a variety of profiles; some indicated complete oxidation of CH_4 below the the soil surface, and others showed overlap of the CH_4 and oxygen profiles at depth, a zone termed the "methane:oxygen crossover". Soil core microcosm studies by Kightley et al (1995) duplicated natural fluxes and conditions, and reported the highest CH_4 oxidation rates measured to date (166 g CH_4 m^{-2} day^{-1}) at the "methane:oxygen crossover".

The role of CH_4 oxidation in landfills remains uncertain because of the lack of *in situ* rate measurements. Landfill cover soils are typically 1-3 m thick, and are frequently low-porosity clay. Methane oxidation in landfills occurs within the cover soil at depths too great for direct measurements using inhibitors or oxygen replacement techniques. A technique that measures the cumulative effect of CH_4 oxidation in emitted CH_4 would allow more measurements and better estimates of spatial variability. It should be possible to use stable isotope fractionation associated with CH_4 oxidation to quantify the fraction of CH_4 oxidized.

Ocean. The ocean remains a small term in the net CH_4 emission budget because oxidation processes are so effective at consuming CH_4 within the ocean. The bulk of ocean methanogenesis occurs in high deposition rate shelf and near-shore sediments, which cover about 3.5% of the ocean area, and is effectively oxidized, largely by anaerobic oxidation in a subsurface sediment zone before it reaches the water column. This process is recognized as an important CH_4 sink, but the responsible organism and mechanism have not been elucidated. Recent work by Hoehler et al (1994) Hoehler, Alperin (this volume) present a plausible mechanism involving a methanogen-sulfate reducer consortium. This work shows that sulfate reducers are able to maintain porewater H_2 concentrations low enough to allow methanogens to accomplish net oxidation of CH_4 and derive sufficient energy for growth. This mechanism is consistent with previous work and neatly explains anaerobic CH_4 oxidation in anoxic sediments. It should be extended and tested in other environments supporting large-scale anaerobic CH_4 oxidation, such as anoxic water columns.

Sources of CH_4 in the upper water column have been reported recently. Karl, Tilbrook (1994) found that CH_4 is associated with sinking particles, possibly as a dissolved constituent of particle interstitial fluids, and that it exchanges with adjacent waters as the particles sink. Particle:water CH_4 fluxes measured on trapped particulate matter are high enough to result in an upper water column CH_4 replacement time of 50 days. Marty (1993) observed production of CH_4 from suspended particles, zooplankton, and fresh fecal pellets under anoxic conditions and concluded that methanogens originate in the digestive tracts of zooplankton and survive in fecal pellets, producing CH_4 with competitive and non-competitive substrates. No production rate estimates resulted from this work. Laboratory studies on zooplankton grazing on phytoplankton (de Angelis, Lee, 1994) showed that CH_4 production dependent on zooplankton species occurred at rates of 4 - 20 nmol copepod^{-1} d^{-1}. Anaerobic microniches within zooplankton digestive tracts were considered the likely site for methanogenesis, which occurs at rates sufficient to contribute to the formation and maintenance of oceanic subsurface CH_4 maxima. Trimethylamine was present in sufficient concentrations to account for the CH_4 produced; conversion of only 4 -12 % of the phytoplankton methylamines available to the grazing zooplankton could account for the observed CH_4 production in the experiments.

Pockmarks in continental shelf sediments, which are believed to result from gas discharges, have been documented by Hovland, Judd (1988), but only a few measurements of CH_4 flux from these features are available. Hovland et al (1993) estimated submarine CH_4 fluxes and applied them to areal estimates of high seepage potential sediments to obtain a global emission estimate of 8 - 65 Tg CH_4 yr^{-1}. The impact of CH_4 from submarine seeps on the atmospheric budget is unclear. In addition to questions about the flux of CH_4, other questions such as whether it is emitted in fluids or as bubbles, the extent of dissolution of rising bubbles in the water column, and the effectiveness of water column oxidation processes must be addressed before reliable assessments can be made. Isotopically light carbonate cements, which must result from CH_4 oxidation, have been observed in a number of ocean locations (see Jensen et al, 1992), and point to extensive CH_4 emission and oxidation not considered by Reeburgh et al (1993). It is clear that CH_4 oxidation and emission at submarine seeps has been underestimated, and that water column oxidation in oxic water columns must be more important than previously believed. There are very few direct water column CH_4 oxidation rate measurements, and most have been performed in anoxic systems (Ward et al, 1987; Reeburgh et al, 1991). These measurements, combined with recently developed mass spectrometric techniques capable of stable isotope ratio measurements on small quantities of CH_4 (Merritt, et al, 1995, Popp et al, 1995) offer approaches to understanding the role of CH_4 oxidation adjacent seeps and the balance between CH_4 production and consumption in the near-surface oceanic CH_4 maximum.

Hydrates. Methane clathrate decomposition appears as a placeholder term in most CH_4 budgets and has been frequently cited as a possible positive feedback to global warming. A recent modeling study by Harvey, Huang (1995) considers that marine sediments contain the bulk of CH_4 hydrates and concludes that the impact of global warming on clathrate destabilization is smaller than the differences between a range of CO_2 emission scenarios. This study did not consider CH_4 dissolution or oxidation, so the actual impact of clathrate destabilization is expected to be much smaller.

Soils . Soils have been recognized relatively recently as a sink for atmospheric CH_4, and oxidation has been observed in a wide variety of soils that do not produce CH_4. This process is diffusion-limited and occurs in a subsurface zone (Whalen et al, 1992; Adamsen, King 1993). Soils are capable of oxidizing CH_4 to concentrations well below atmospheric levels and do so over a relatively narrow range of rates which appear to depend heavily on soil porosity. One puzzling aspect of soil CH_4 oxidation is the observed low concentration thresholds. The organisms mediating soil oxidation are unknown, but kinetic work (Bender, Conrad, 1992; Koschorrek, Conrad, 1993) indicates that organisms with high and low affinities and different kinetic parameters are responsible.

FUTURE APPROACHES

As pointed out by King (1992) the importance of analyzing the behavior of CH_4 oxidizing bacteria under *in situ* conditions cannot be over-emphasized. Measurements of oxidation potential are essential in understanding how systems function, but cannot substitute for direct rate measurements under *in situ* conditions to assess the balance between CH_4 production and consumption and how these affect CH_4 emission.

Methane oxidation in several important environments occurs at depths below the surface great enough to preclude application of inhibitors like CH_3F (Oremland, Culbertson 1992) or oxygen replacement (Happell et al, 1994) to obtain net flux and gross flux measurements. These environments include wetlands with lowered water tables and landfill cover soils. Stable isotope techniques offer a means of obtaining an integrated measure of the fraction of CH_4 oxidized in these systems, provided that appropriate isotope fractionation factors applicable to the systems are available. Table 2 summarizes isotope fractionation factors associated with CH_4 oxidation. These measurements result from a wide range of experimental approaches in a variety of field and laboratory studies. Table entries include open and closed system measurements on pure cultures and enrichments, isotope changes in flux chamber headspaces, model studies, and model fits to rate and stable isotope data. These fractionation factors vary too widely for universal application and should be determined under conditions appropriate for the environments under study. Soil microcosms (Kightley et al, 1995) offer a means of determining stable isotope fraction factors appropriate for the study of CH_4 oxidation in soil environments. Measurements of source CH_4 collected with soil probes and emitted CH_4 collected under large canopies, combined with appropriate isotope fractionation factors could be used to estimate the fraction of CH_4 oxidized.

Pulse-labelling experiments with $^{14}CO_2$ have been performed on commercially important plant species grown in moist soils to determine plant carbon allocation. One study (Minoda, Kimura, 1994) describes pulse-labelling rice plants with $^{13}CO_2$ and indicates emission of labelled CH_4 with a time scale of hours to days. Experiments focusing on the turnover of recently-fixed $^{14}CO_2$ to CH_4 in other wetland plants grown in waterlogged soils are needed to determine the time scale, mechanism, and degree of coupling between the rhizosphere and CH_4 emitted by vascular plants. Longer-term studies will provide information on turnover of detritus and dissolved organic carbon. If a means of distributing $^{14}CH_4$ and $^{14}CO_2$ in wetland rhizospheres can be developed, below-ground pulse labelling experiments could provide additional information on the transport and fate of CH_4 in the rhizosphere. These experiments in waterlogged systems will require development of equilibration samplers suitable for introducing labeled tracers and sampling dissolved gases in the rhizosphere. Rate measurements involving $^{14}CH_4$ should also attempted in wetland and rice system rhizospheres to determine whether inhibition and oxygen replacement techniques lead to increased CH_4 production.

Questions regarding the importance of submarine seeps and the processes that maintain the near surface ocean CH_4 maximum should be addressed with tracer techniques and stable isotope measurements. Tritium-labelled methane (C^3H_4, Reeburgh et al, 1991) can be prepared with specific activities high enough to permit true tracer measurements in the ocean water column, but has only been applied in only one study. Newly introduced techniques for stable carbon isotope measurements on CH_4 dramatically reduce sample preparation time, permitting analysis of large numbers of small samples. These techniques should be particularly useful in studies of CH_4 oxidation in the ocean water column and soils. Isotope-ratio-monitoring gas chromatography/mass spectrometry (irm GC/MS, Merritt, et al, 1995, Popp et al, 1995) uses a gas chromatographic separation, on-line combustion, and measurement on an

340

Table 2. Kinetic Isotope Fractionation Factors Associated with Methane Oxidation

Study	αC	αH	Comment
Silverman, Oyama, 1967	1.011	n.d.	Calculated by Whiticar, Faber, 1986.
Coleman et al, 1981	1.013-1.025	1.103-1.325	Closed system enrichment culture experiment, 18-30% CH_4.
Barker, Fritz, 1981	1.005-1.103	n.d.	Closed system enrichment culture experiment.
Whiticar, Faber. 1986	1.002-1.014	n.d.	Calculated using models (% residual methane, higher hydrocarbon enrichment, CO_2-CH_4 coexisting pairs) and field data.
Zyakun et al, 1987	1.011-1.033	n.d.	Open system (continuous culture) with *Methylomonas methanica* at range of temperatures and growth rates.
King et al, 1989	1.027 (14°C) 1.016 (4°C)	n.d.	Estimated from change in chamber headspace $\delta^{13}C$-CH_4 above CH_4 consuming tundra.
Alperin et al, 1988	1.0088 1.0250	1.157 1.177	Anaerobic CH_4 oxidation, modeled with and w/o diffusion effect.
Happell et al, 1994	1.003-1.021	1.050-1.129	Calculated using models and field data from Florida swamp floodwaters.
Tyler et al, 1994	1.017-1.022	n.d.	Estimated from measured change in chamber headspace $\delta^{13}C$-CH_4 above CH_4 consuming temperate forest soil; suggested diffusion controls KIE.

Kinetic isotope fractionation factor = α_K = (k_L/k_H), where k_L and k_H are the first order rate constants of the light and heavy isotopes.

isotope ratio monitoring mass spectrometer. The technique (Merritt et al, 1995) is capable of analyses precise to 0.5‰ on 5 cc air samples containing 1.7 ppm CH_4. Popp et al (1995) report similar precision and a detection limit of 10 nM on 10 ml seawater samples. Analyses are reported to take 15 min for air samples and 30 min for water samples. These techniques should permit high resolution studies of CH_4 isotopes in water column, soil, and sediment environments.

ACKNOWLEDGMENTS

This work was supported by grants from NSF, EPA and DOE (WESTGEC).

REFERENCES
Adamsen APS, King GM (1993) Appl. Environ. Microbiol. 59, 485-490.
Augenstein D (1992) Global Environ. Change 2, 311-328.
Barker F, Fritz P (1981) Nature 293, 289-291.
Bender M, Conrad R (1992) FEMS Microbiol. Ecol. 101, 261-270.
Bingemer HJ, Crutzen PJ (1987) J. Geophys. Res.92, 2181-2187.
Chanton JP et al (1995) Geochim. Cosmochim. Acta 59:3663-3668.
Cicerone RJ, Oremland RS (1988) Global Biogeochem Cycles 2, 299-327.
Coleman DD et al (1981) Geochem. Cosmochim. Acta 45, 1003-1037.
Conrad R, Rothfuss F (1991) Biol. Fertil. Soils 12, 28-32.
De Angelis MA, Lee C (1994) Limnol. Oceanogr. 39, 1298-1308.
De Bont JAM et al (1978) Ecol. Bull. (Stockholm) 26, 91-96.
Dise NB, Gorham E (1993) J. Geophys. Res. 98, 10,583-10,594.
Epp MA, Chanton JP (1993) J. Geophys. Res. 98, 18413-18,422.
Fung I et al (1991) J. Geophys. Res. 96, 13,033-13,065.
Funk DW et al (1994) Global Biogeochem. Cycles 8, 271-278.
Gerard G, Chanton J (1993) Biogeochemistry 23, 79-97.
Happell J D, Chanton JP (1993) Global Biogeochem. Cycles 7, 475-490.
Happell JD. et al (1993) J. Geophys. Res. 98(D8), 14,771-14,782.
Happell JD. et al (1994) Geochim. Cosmochim. Acta 58, 4377-4388.
Harvey LDD, Huang Z (1995) J. Geophys. Res. 100, 2905-2926.
Hoehler TM et al (1994) Global Biogeochem. Cycles 8, 451-463.
Hoehler TM, Alperin M (this volume).
Hovland M, Judd AG (1988) Seabed Pockmarks and Seepages: Impact on Geology, Biology and the Marine Environment, Graham & Trotman, London.
Hovland M et al (1993) Chemosphere 26, 559-578.
Jensen P et al (1992) Mar. Ecol. Prog. Ser. 83, 103-112
Jones HA, Nedwell DM (1990) Waste Mgt. Res. 8, 21-31.
Karl DM, Tilbrook BD (1994) Nature 368, 732-734.
King GM (1990a) Nature 345, 513-515.
King GM (1990b) FEMS Microbiol. Ecol. 74, 309-324.
King GM (1992) Adv. Microb. Ecol. 12, 431-468. Plenum Press, New York.
King GM (1993) In Murrell JC and Kelley DP, eds, Microbial Growth on C-1 Compounds, pp 303-313, Intercept Ltd., UK.
King SL et al (1989) J. Geophys. Res. 94(D15), 18,273-18,277.
Koschorrek M, Conrad R (1993) Global Biogeochem. Cycles 7, 109-121.
Kightley D et al (1995) Appl. Environ. Microbiol. 61, 592-601.
Lein T et al (1993) Suo 43:267-269.
Mancinelli RL, McKay CP (1985) In AA Antonopoulos, ed, 1st Symposium on Biotechnological Advances in Processing Municipal Wasters for Fuels and Chemicals, Argonne National Laboratory Report ANL/CNSV-TM-167.

Marty DG (1993) Limnol. Oceanogr. 38, 452-456.
Merritt DA et al (1995) J. Geophys. Res., 100, 1317-1326.
Minoda T, Kimura M (1994) Geophys. Res. Lett. 21, 2007-2010.
Moore T, Knowles R (1989) Can. J. Soil Sci. 69, 33-38.
Moore T, Roulet NT (1993) Geophys. Res. Lett. 7, 587-590.
Oremland RS, Culbertson CW (1992) Nature 356, 421-423.
Popp BN et al (1995) Anal. Chem., 67, 405-411.
Reeburgh WS et al (1991) Deep-Sea Res. 38, S1189-S1210.
Reeburgh WS, Whalen SC (1992) Ecol. Bull. (Copenhagen) 42, 62-70.
Reeburgh WS et al (1993) In Murrell JC and Kelley DP, eds Microbial Growth on C-1 Compounds, pp 1-14, Intercept Ltd, UK.
Roulet NT et al (1993) Global Biogeochem. Cycles 7, 749-769.
Silverman MP, Oyama VI (1968) Anal. Chem. 40, 1833-1877.
Tyler SC et al (1994) Geochim. Cosmochim. Acta 58, 1625-1633.
Ward BB et al (1987) Nature 327, 226-229.
Whalen SC et al (1990) Appl. Environ. Microbiol. 56, 3405-3411.
Whalen SC et al (1992) Biogeochemistry 16, 181-211.
Whalen SC, Reeburgh WS (1992) Global Biogeochem. Cycles 6, 139-159.
Whiticar MJ, Faber E (1986) Org. Geochem. 10, 759-768.
Whiting GJ, Chanton JP (1993) Nature 364, 794-795.
Zyakun AM et al (1987) Geokhimiya 7, 1007-1003.

Cycling of Methanol between Plants, Methylotrophs and the Atmosphere

RAY FALL
Department of Chemistry and Biochemistry, University of
Colorado, Boulder, Colorado, 80309-0215, U.S.A.

Introduction

Several recent developments have kindled new interest in volatile organic compounds (VOCs) of the C1 class (i.e., methanol, formaldehyde and formic acid) that are found in the troposphere. First, Paul Goldan and his colleagues at the NOAA Aeronomy Laboratory (Boulder) have discovered that methanol, is one of the most abundant VOCs in forest air, and exhibits a diurnal cycle indicative of major methanol sources and sinks in forest canopies (Fehsenfeld et al., 1992; Goldan et al., 1995). Second, we have discovered that all plants tested emit methanol, some at relatively high emission rates (MacDonald, Fall, 1993; Fall, 1994; Nemecek-Marshall et al., 1995), and thus, are probably major regional and global sources of atmospheric methanol (Guenther et al., 1995). Third, there has been renewed recognition that a group of bacteria that consume methanol, called methylotrophs, are widespread and numerous on leaves. This raises the issue that methylotrophs on plant surfaces could be a major sink for methanol, and perhaps other C1-VOCs (Fall, 1994).

These developments, discussed below in more detail, have led us to the conclusion that there is natural cycling of methanol between the biosphere and atmosphere, and that if large releases of anthropogenic methanol occur, there could be alterations of this natural balance. The main thrust of this presentation is that we should endeavor to understand the biogenic methanol cycle, and develop the knowledge base that will allow assessment of current and future impacts of human activities on the C1-VOC balance in the atmosphere.

Methanol in the atmosphere

Prior to 1992 there were only a few measurements of atmospheric methanol. For example, Snider and Dawson (1985) used condensation sampling and measured methanol in air in Tucson, Arizona and two rural sites; mean methanol concentrations were 7.9 ppb (city) and 2.6 ppb (rural). The earlier literature contains only a few scattered reports of measurements of methanol and other alcohols in air samples. Then, during field measurements at the ROSE I and ROSE II field site in rural Alabama, Paul Goldan and colleagues determined that methanol is the major VOC in forest air, averaging about 6 ppbv at night and 11 ppbv during the day. Of the more than 50 VOCs detected in these experiments, methanol typically represented 40-46% of the total volatile organic carbon detected (Fehsenfeld et al., 1992; Goldan et al., 1995).

M. E. Lidstrom and F. R. Tabita (eds.), Microbial Growth on C₁ Compounds, 343–350.
© 1996 Kluwer Academic Publishers. Printed in the Netherlands.

Some typical data from the ROSE II study are shown in Figure 1; the hourly atmospheric concentrations of methanol for 23 consecutive days in June, 1992 are illustrated. As can be seen, the air concentrations of methanol show distinct diurnal cycles with large sources during the day, maximum concentrations occurring in the late afternoon, followed by large declines indicative of major sinks for this compound. This pattern parallels that for isoprene, a biogenic hydrocarbon known to be produced by various plant species at this forest site. For isoprene, which is photochemically reactive, the major sink is chemical. Methanol, on the other hand, is much less reactive with an estimated half-life in the free troposphere of 69 hours (daylight) and >1000 hours (night) (Jacob et al., 1989). The strong sink for methanol each afternoon and night (Figure 1) is uncharacterized, but may be due to dry and wet surface deposition to the forest canopy and soil like that seen for other polar organics such as formic and acetic acid (Talbot et al., 1990). This surface sink may also include uptake by leaf bacteria as discussed below.

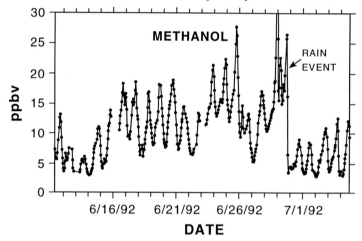

Figure 1. Diurnal concentrations of methanol above a pine forest canopy in rural Alabama in summer 1992. This unpublished data (with 5-point smoothing) was kindly provided by P. Goldan, NOAA Aeronomy Laboratory, from a data set presented in Goldan et al. (1995). The pattern of daily sources and sinks for methanol parallel those for biogenic isoprene; air concentrations for anthropogenic compounds (e.g., benzene) do not show such regular diurnal cycles.

The similarity in diurnal peaks for isoprene and methanol concentration around midday suggests that the production of both VOCs is linked to photosynthetic processes.

It is also clear that some tropospheric methanol must be anthropogenic, since it is a widely used industrial chemical; in 1993 estimated world wide production of methanol was about 20 million tons (Crocco, 1994); of this total about 4.8 million tons were produced in the United States (C&EN, 1995). Because of its chemical stability, methanol could be transported long distances from pollution sources. The amount of tropospheric methanol that is anthropogenic is unknown.

Methanol emission from plants

Paul Goldan's work stimulated us to investigate the possibility that vegetation is a source of tropospheric methanol. For these experiments we used leaf and plant enclosure methods as diagrammed in Figure 2. Methanol emissions have been routinely analyzed by gas chromatography or, more recently, by enzymatic analysis of gas phase methanol using methanol oxidase (Nemecek-Marshall et al., 1995). The clamp-on leaf cuvette shown in Figure 2 has the advantage that methanol emissions can be monitored from either leaf surface, but involves direct contact of the cuvette with the leaf. The flux chamber method allows measurement of methanol fluxes from a large number of leaves without physical contact with leaf surfaces, but has the disadvantages that temperature is hard to control and light levels vary in the canopy. With either sampling method leaf photosynthesis and stomatal conductance can also be monitored.

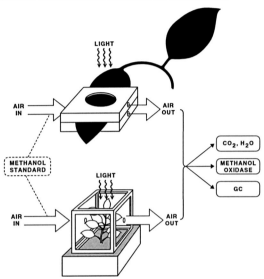

Figure 2. The experimental design for measuring leaf methanol fluxes with either a clamp-on leaf cuvette (upper) or a flux chamber (lower). The use of each of these devices for measurement of leaf or canopy gas exchange has been described in detail elsewhere (Fall et al., 1988; Fall, Monson, 1992). The flux chamber is lined with PFA Teflon film, and usually the plants to be tested are grown through Teflon film allowing methanol emission measurements free of the complication of exchange with the underlying soil. Methanol recovery in both of these systems is assessed with a gas phase methanol standard. Analysis of the emerging gas stream includes: CO_2 (photosynthetic CO_2 assimilation), H_2O (stomatal conductance), methanol oxidase (enzymatic determination of methanol), or GC (gas chromatographic analysis of methanol).

Using these methods we have found substantial emissions of methanol from leaves of all plants tested, including 19 different C3 species (MacDonald, Fall, 1993; Fall 1994; Nemecek-Marshall et al., 1995). Some typical methanol emission rates that we have measured for fully expanded leaves are compared to those for two other biogenic hydrocarbons, isoprene and monoterpene, in Table 1. For most plants the magnitude of these emissions, 0.3 to 17 µg (methanol C) h^{-1} gdw^{-1}, is lower than isoprene emission (from isoprene emitting species), but comparable to or greater than that for monoterpene emission from plants. It is noteworthy that while most plants do not emit significant isoprene or monoterpenes, methanol emission may be ubiquitous in plants. As mentioned below even higher methanol fluxes are seen with young leaves.

Table 1. Typical leaf emission rates for methanol and other major biogenic VOCs (units of μg C gdw^{-1} h^{-1})[a]

Plant type	Methanol Emission Rate	Isoprene Emission Rate	Monoterpene Emission Rate
deciduous trees:			
aspen	6	70	<0.1
maple	4	<0.1	1.6
coniferous trees:			
fir	0.6	<0.1	3
pine	0.3	<0.1	3
understory species:			
blackberry	10	<0.1	0.2
sassafras	3	<0.1	<0.1
crops:			
corn	17	<0.1	<0.1
soybean	7	0.1	<0.1

[a] Methanol data is from MacDonald and Fall (1993), Fall (1994), and Nemecek-Marshall et al. (1995); data for isoprene and monoterpenes is from Guenther et al. (1994). Emission rates are typical for mature, fully expanded leaves measured at 30°C and PAR=1000 μE (stomata open).

This comparison suggests that vegetation could be a significant regional and global source of atmospheric methanol. As part of a recent International Global Atmospheric Chemistry core project a large group of scientists estimated global natural organic compound emissions from oceans and plant foliage (Guenther et al., 1995). The global annual VOC flux was estimated to be 1150 Tg C composed of 44% isoprene, 11% monoterpenes, and 45% of other VOCs. It is noteworthy that methanol could account for a large fraction of these other VOC emissions, possibly more than 100 Tg C. Many more field measurements of methanol emission fluxes are needed to establish the global methanol flux to the atmosphere.

If these methanol flux estimates are validated, and methanol is deposited to plants and soils each evening as suggested in Figure 1, one can imagine atmospheric methanol as a carbon source for methylotrophic microorganisms that inhabit leaves, soils, and waters.

Methanol exchange at leaf surfaces

Several lines of evidence support the view that methanol is produced inside leaves and emitted from stomata (MacDonald, Fall, 1993; Nemecek-Marshall et al., 1995). Although other factors, such as leaf temperature, may also affect methanol emission rate, short-term changes in methanol emission rate are probably controlled by stomatal conductance. Figure 3A illustrates our current model for methanol emission when stomata are open. Free methanol contained in the leaf is released via stomatal pores along with transpired water vapor. The figure also indicates a smaller methanol flux from the leaf cuticle; fluxes of methanol from leaf surfaces, prior to stomatal opening, have been seen in laboratory and field experiments. Surface methanol may be the result of atmosphericic deposition to leaves at night, a process indicated in Figure 3B, as well as diffusion of internal methanol through the cuticle to the leaf surface. We have noted that for hypostomatous leaves, only a small fraction of the methanol flux (i.e., 3-6%) occurs from the non-stomatal surface (Nemecek-Marshall et al., 1995), suggesting that if methanol diffuses through the cuticle this process is small in magnitude compared to release via stomata. The methanol emission and deposition processes shown in Figure 3 might account in part for the diurnal behavior of methanol noted in Figure 1.

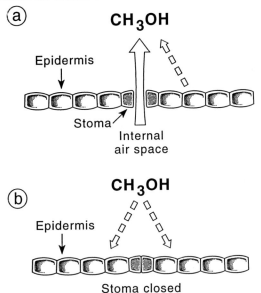

Figure 3. Models for emission (A) and deposition (B) of methanol at leaf surfaces.

Do methylotrophs on leaves consume leaf methanol?

As seen in Figure 4, methanol fluxes from bean leaves decline substantially with increasing leaf age. This has been observed in several plant species including bush bean, soybean and cottonwood (Nemecek-Marshall et al., 1995). This is also a time when leaf methylotroph populations might be increasing. For example, Hirano and Upper (1991) have shown that the population of a particular group of methylotrophs, pink pigmented facultative methylotrophs (PPFMs), on snap bean leaves climbs throughout the growing season; in mature leaves PPFMs are the dominant bacterial group. PPFMs have been found on most plants (Corpe, Rheem, 1989; Holland, Polacco, 1994). We wondered if the methanol flux of a leaf is affected by PPFM populations. It seems possible that these bacteria or other leaf methylotrophs might consume enough methanol to decrease the net flux from a leaf to the atmosphere.

Three types of preliminary experiments were conducted to explore the relationship of PPFM populations and leaf methanol emissions. First, we analyzed PPFM populations of young and old leaves of field and greenhouse-grown bush bean and cottonwood plants. The results for greenhouse-grown bush beans are shown on Figure 4. Essentially, the PPFM populations on young and old leaves were indistinguishable, around 10^3 cfu/g, despite that fact that the methanol emission rate decreased by more than 8-fold in the older leaves. Similar results were seen with field-grown beans and cottonwood, and greenhouse-grown cottonwood. In these experiments the PPFM populations were several orders of magnitude lower than those reported for beans grown in Wisconsin (Hirano, Upper, 1991), possibly because of the hot, dry conditions in Colorado in summer 1995.

Figure 4. Methanol emissions decrease with increasing leaf age at a time when populations of leaf methylotrophic bacteria are unchanging. Methanol emissions were measured with a clamp-on cuvette, and viable PPFM colonies in leaf homogenates were determined essentially as described by Corpe and Rheem (1989); the PPFM data represent the ranges for four different plants.

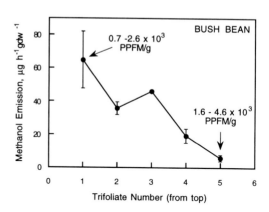

A second type of experiment we have pursued involved attempts to prepare methylotroph-free plants by heat treatment of bush bean or soybean seed (50°C for 48 h before planting). This method has been used with success by Holland and Polacco (1994). The resulting plants and control plants were analyzed for alterations in PPFM population levels as well as methanol fluxes. Although in some experiments PPFM populations in leaves from heat treated seed were 100-fold lower than in controls, we were unable to reproducibly lower PPFMs in soybean or bush bean leaves when seeds were treated this way. The resulting fluxes of methanol from control plants and plants derived from heat treated seed were indistinguishable.

A third approach was to analyze adaxial and abaxial leaf surfaces for PPFM populations, and relate these populations to the distribution of stomata. The rationale was that if PPFMs consume leaf methanol, they might be enriched on the leaf surface(s) bearing stomates; as mentioned above most leaf methanol is thought to be released from stomata. Five different plant species were analyzed: two hypostomatous species, aspen and velvet bean; and three amphistomatous species, bush bean, soybean and cottonwood. PPFM populations on leaf surfaces were determined by leaf blotting on a minimal methanol medium (with cycloheximide to inhibit fungal growth; Corpe and Rheem, 1989). The results, while difficult to quantitate, indicated that for all five plant species, 5 to 20 times more PPFM colonies were detected on the upper (adaxial) surface, despite the fact that for aspen and velvet bean no stomata occur on that surface.

Although these experiments need to repeated more rigorously, it does not appear that PPFMs play a large role in attenuating the flux of methanol from leaves, at least in the plants used in our experiments. The results may be different in leaves growing in humid environments with much larger populations of PPFMs. It seems more likely that leaf methylotrophic bacteria might consume methanol deposited to surfaces from the atmosphere or methanol that diffuses through the leaf cuticle. Experiments in progress are directed at measuring methanol deposition to leaves and methanol levels on leaf surfaces. An interesting experiment would be the direct demonstration of the rates of metabolism of plant produced methanol by leaf PPFMs. This is a technically difficult experiment, complicated by the fact that plants also appear to metabolize methanol (Cossins, 1964).

Biogeochemical cycling of methanol

The work summarized above suggests that methanol is a natural product that cycles between the biosphere and the atmosphere. A model for methanol cycling is presented in Figure 5. We propose that a major source of atmospheric methanol is emission from leaf surfaces; another significant source is likely to be release of industrial and fuel methanol as a result of leaks, spills and incomplete combustion. Major sinks for tropospheric methanol are probably a) reaction with OH radical, and b) wet and dry deposition to surfaces. Chemical reaction of methanol with OH is relatively slow, leading to the production of formaldehyde and HO_2 radical. Surface deposition may be enhanced by methanol's miscibility in water and the presence of methylotrophs in soils, water and vegetation. The proposed methanol cycle shares many similarities to other natural atmosphere-biosphere cycles. For example, like biogeochemical cycling of methane (Cicerone, Oremland, 1988), the biosphere includes both sources and sinks of methanol, and its atmospheric fate is determined in part by reaction with OH radical. Unlike methane, which is insoluble in water, a large fraction of methanol may be removed from the troposphere in rain, fog and dew.

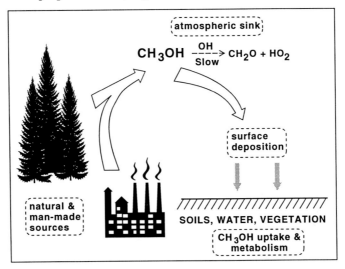

Figure 5. Model for the methanol cycling between the biosphere and atmosphere.

Anthropogenic uses of methanol might increase in the near future. Methanol fuels for vehicles have been introduced in some parts of the U.S. (Short, 1994), and the use of methanol sprays (10-50% methanol solutions) for crop growth stimulation is under field testing in many areas (Nonomura et al., 1994). What is the likelihood that increasing anthropogenic uses of methanol could affect the proposed methanol cycle regionally or globally? This is a relevant question when one considers that virtually all known biosphere-atmosphere cycles have been altered by man's activities. For methanol, increased anthropogenic releases would have a relatively small predicted impact on atmospheric chemistry due to methanol's low chemical reactivity. A more likely impact, in the case of large scale use of methanol fuels, would be on atmospheric formaldehyde levels, due to incomplete combustion of methanol and inefficient trapping of exhaust formaldehyde. Increased atmospheric methanol is also likely to impact regional biological processes by a "fertilization effect" analogous to the effect of increasing atmospheric CO_2 on land plants and phytoplankton (Bazzaz, Fajer, 1992). Methanol removal from the atmosphere by wet and dry deposition almost certainly leads to daily increases in the methanol concentration on soil and leaf surfaces. One result of this might be fertilization of methylotrophs on leaves; most bacteria that inhabit leaf surfaces are thought to be nutrient limited. Enhanced growth of methylotrophs might have some secondary effects on plant growth or viability. At present we know so little of the role of these abundant bacteria that we can't predict effects of methanol fertilization. This may be an important issue if widespread spraying of crops with methanol is adopted.

It is almost certain that man's activities will alter the natural methanol cycle. These alterations could have important effects on regional atmosphere or biosphere processes, perhaps in ways that we can not predict now. We should learn more about the methanol cycle before such changes occur.

Acknowledgments

The contributions of the following colleagues are gratefully acknowledged: Jennifer Franzen, Paul Goldan, Alex Guenther, Bob Holland, Bob MacDonald, Barbara Monday, Michele Nemecek-Marshall, and Cheryl Wojciechowski. Supported by grants ATM-9206621 and ATM-9418073 from the National Science Foundation.

References

Bazazz FA, Fajer ED (1992) Sci. Amer. January, 68-74.
Chemical & Engineering News (1994) July 4, 28-74.
Cicerone RJ, Oremland RS (1988) Global Biogeochem. Cycles 2, 299-327.
Corpe WA, Rheem S (1989) FEMS Microbiol. Ecol. 62, 243-250.
Cossins EA (1964) Can. J. Biochem. 42, 1793-1802.
Crocco J. (1994) In Cheng W-H and Kung HH, eds, Methanol Production and Use, pp 283-317, Marcel Dekker, New York.
Fall R et al. (1988) J. Atmos. Chem. 6, 341-362.
Fall R, Monson RK (1992) Plant Physiol. 100, 987-992.
Fall R (1994) Thirteenth Ann. Symp. Curr. Top. Plant Biochem., Physiol. Mol. Biol., Columbia, Missouri, Apr. 14-16, p. 9-10.
Fehsenfeld FC et al. (1992) Global Biogeochem. Cycles 6, 389-430.
Goldan P et al (1995) J. Geophys. Res., in press.
Guenther A et al. (1994) Atmos. Environ. 28, 1197-1210.
Guenther A et al. (1995) J. Geophys. Res. 100, 8873-8892.
Hirano SS, Upper C.D. (1991) In Andrews JH and Hirano SS, eds, Microbial Ecology of Leaves, pp 271-294, Springer-Verlag, New York.
Holland MA, Polacco JC (1994) Annu. Rev. Plant Physiol. Plant Mol. Biol. 45, 197-209.
Jacob DJ et al. (1989) J. Geophys. Res.94, 8497-8509.
MacDonald RC, Fall R (1993) Atmos. Environ. 27A 1709-1713.
Nemecek-Marshall M et al. (1995) Plant Physiol. 108, 1359-1368.
Nonomura AM, Benson AA (1992) Proc. Natl. Acad. Sci. USA 89, 9794-9798.
Nonomura AM et al. (1994) In Cheng W-H and Kung HH, eds, Methanol Production and Use, pp 253-260, Marcel Dekker, New York.
Short G.D. (1994) In Cheng W-H and Kung HH, eds, Methanol Production and Use, pp 215-252, Marcel Dekker, New York.
Snider J., Dawson G. (1985) J. Geophys. Res. 90, 3797-3805.
Talbot RW et al. (1990) J. Geophys. Res. 95, 16799-16811.

The role of autotrophs in global CO_2 cycling

J.A. Raven
Department of Biological Sciences
University of Dundee
Dundee DD1 4HN
UK

INTRODUCTION

Autotrophs in the narrow sense are photolithotrophs or chemolithotrophs, using respectively photons and exergonic inorganic reactions to convert inorganic C, N, S and P into biomass. This definition includes methanogenesis from CO_2 and H_2. In the broader sense all methanogens and organisms consuming C_1 compounds are included as autotrophs while, concentrating on autotrophs in the strict sense, we shall also consider some other 'autotrophs' and production of C_1 compounds (CO, CH_3OH, CH_3Cl, CH_3Br, CH_3I, $CHBr_3$ and by slightly stretching the definition, $(CH_3)_2S$ and $((CH_2)_2Se)$ by photolithotrophs.

The obvious role of autotrophs in the global CO_2 cycle is to fix CO_2. Globally by far the major CO_2 fixation process is by O_2-evolving photolithotrophs. This is especially the case if we consider CO_2 fluxes associated with atmospheric CO_2 rather than the cycles occurring within predominantly chemoorganotrophic soil and deep ocean (chemolithotrophs) and illuminated oxic/anoxic interfaces (non-O_2-evolving phototrophs, e.g. in the Black Sea). CO_2 fluxes between the atmosphere and the ocean and terrestrial ecosystems have recently received much attention in relation to the regulation of atmospheric CO_2 as a function of anthropogenic CO_2 release. Figure 1 shows that the most significant pools and fluxes involved in the global C cycle.

Essentially all of the inorganic C fixation processes involved in the production of organic C in Figure 1 (~ 100 Pg C each year on land and ~ 40 Pg in the oceans) involve Rubisco as either the first carboxylase encountered by inorganic C (C_3 metabolism) or in refixation after a (C_3 + C_1) carboxylation followed by a (C_4-C_1) decarboxylation (C_4, CAM). The latter processes account for about ¼ of CO_2 fixed on land but a negligibly small fraction of that fixed in the oceans (Raven et al., 1992). In addition to C_4 and CAM involving a carboxylation-decarboxylation cycle in series with Rubisco, all photolithotrophs employ many other carboxylases in parallel with Rubisco which account for < 5% of the net inorganic C fixation but which are essentially for biosyntheses (Raven, 1995a).

351

M. E. Lidstrom and F. R. Tabita (eds.), Microbial Growth on C₁ Compounds, 351–358.
© 1996 Kluwer Academic Publishers. Printed in the Netherlands.

352

Figure 1

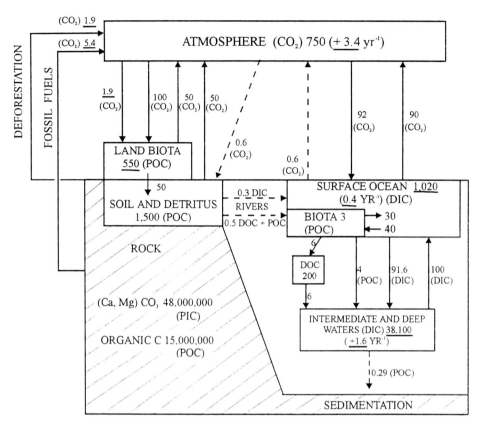

Reservoirs (in boxes) in Pg. Fluxes (beside arrows) in Pg yr. DIC dissolved inorganic C.

PIC = particulate inorganic C. DOC = dissolved organic C. POC = particulate organic C.

– – – –▶ = Fluxes down rivers, and in marine sedimentation, and balancing fluxes from ocean to atmosphere and atmosphere to biota.

Numbers underlined have been changed by human activities

Modified from Siegenthaler & Sarmiento (1993);Meybeck (1993)

Table 1
Inputs of C to the atmosphere from methane and its sulfur and halogen derivatives. For comparison terrestrial biota exchange $\geq 60,000$ Tg C y^{-1} as CO_2 with the atmosphere and marine biota exchange $\geq 35,000$ Tg C y^{-1} as CO_2 with the surrounding seawater (Raven et al., 1992, 1993).

Input to atmosphere (Tg C y^{-1})

Compound	Biogenic	Anthropogenic + Non-Biological	Total	Reference
CH_4	276	118	394	Houghton et al. (1990)
$(CH_3)_2S$	16-43	0	16-43	Kelly et al. (1994)
CH_3Cl	0.7-1.9	0.15	0.85-2.05	Harper (1994)
$CHCl_3$	0.036	0	0.036	Khalil et al. (1983)
CH_3Br	0.0013-0.038	0.003-0.0085	0.004-0.065	Harper (1994); Lobert et al. (1994)
$CHBr_3$	0.031-0.037	0	0.031-0.037	Harper (1994)
CH_3I	0.0253-0.110	0	0.0253-0.110	Harper (1994)
Total halo-methanes	0.79-2.12	0.15-0.17	0.95-2.28	sum of above

What roles does this leave for non-Rubisco-based C_1 metabolism in the global CO_2 cycle? Before considering the quantitative contribution of NH_4^+-oxidising chemolithotrophs to CO_2 fixation, and how this can influence the fixation of CO_2 by photolithotrophs which assimilate the NH_4^+ substrate or the NO_2^- and NO_3^- products of nitrifying chemolithotrophs as their N source, we consider other direct and indirect effect of C_1 metabolism on the C cycle.

1. *Methanogens* Table 1 show that biologically produced CH_4 accounts for over half of the total CH_4-C entering the atmosphere each year. The fate of most atmospheric CH_4 is oxidation to (ultimately) CO_2 by ·OH. While having a minimal effect on the CO_2 flux to the atmosphere, CH_4 has a significant greenhouse effect (Houghton *et al.*, 1990) with potential effects on autotrophic CO_2 fixation.

2. *Production of methylated S and Se compounds* The major source of the atmospheric burden of these gases is the production of $(CH_3)_2S$ (DMS; dimethylsulfide) as a breakdown product of DMSP (dimethylsulphonium propionate), a compatible solute in many eukaryotic marine primary producers. The atmospheric oxidation of $(CH_3)_2S$ yields, ultimately, CO_2 and SO_2, some of which is further oxidised to H_2SO_4. This CO_2, as for that arising from CH_4, contributes a very small fraction of the total CO_2 flux to the atmosphere (Table 1). However, DMS provides almost all of the biogenic S flux to the atmosphere, thus accounting for much of the total S flux to the atmosphere. DMS is thus climatically important in that its oxidation H_2SO_4 droplets which are major cloud condensation nuclei (CCN) and have significant effects on cloud formation, temperature and rainfall Mitchell *et al.*, 1995), thus influencing at least terrestrial primary productivity. The S inputs to terrestrial ecosystems as volatile S marine biota can increase productivity in S-deficient soils (Lovelock, Margulis, 1974; Raven, 1995b).

3. *Production of C_1 halocarbons* The major biogenic source of volatile halocarbons, most of which are C_1 compounds, is marine photolithotrophic O_2-evolvers (Table 1; Collén, 1994). These constituted the major flux of halogens to the atmosphere before the anthropogenic inputs of CFCs and agricultural use of CH_3Br, the latter converting the ocean into a net sink for, rather than a source of, CH_3Br (Butler, 1995; Lobert *et al.*, 1995; Yagi *et al.*, 1995). Their oxidation in the atmosphere, as for CH_4 and DMS, yields only a small fraction of the total input of CO_2 to the atmosphere but does supply a significant fraction ($\sim 20\%$) of the total (biogenic + anthropogenic) organic halogen input to the stratosphere. These biogenic compounds would have been major regulators of stratospheric O_3 levels, and hence of UV-B screening (Wever, 1988; Raven, 1995; Raven, Sprent, 1989). This provides another feedback of C_1 metabolism on CO_2 fixation, this time *via* UV-B effects on species composition and overall photosynthetic potential of primary producers. Biogenic C_1 iodocarbons are also important in supplying I to I-deficient terrestrial habitats (Lovelock, Margulis, 1974), thus influencing growth of I-requiring animal components of the food chain, with implications for primary productivity *via* herbivory.

4.   *Biogenic CO and CH₃OH*   These C_1 compounds come inter alia from primary producers (Arzee *et al.*, 1985; Fischer, Lüttge, 1979; MacDonald, Fall, 1993; Mlot, 1995), contribute little to total atmospheric CO_2 inputs upon their oxidation, and have (at their present partial pressure) no known direct or indirect effects on the functioning of other primary producers (Beevers, 1961; Hemming *et al.*, 1995).

Having considered these <u>indirect</u> effects of C_1 metabolism by autotrophs on the CO_2-fixing limb of the global C cycle, we now turn to an analysis of the contribution of chemolithotrophs to the global C cycle, both *via* their own CO_2 fixation and their impact on CO_2 fixation by photolithotrophs. The particular case analysed is that of chemolithotrophs which obtain their energy for growth by converting the NH_4^+, which is the immediate product of biological mineralisation of photolithotrophs, to NO_2^- and, mainly, NO_3^-. Both NH_4^+ and NO_3^- are N sources for photolithotrophs, with NH_4^+ having (theoretically at least) the potential for a more efficient use of light and (on land) water in producing plant biomass, as well as permitting a higher specific growth rate under high-resource conditions. NO_3^- is, furthermore, the substrate for denitrification in anoxic habitats and is prone to leaching on land thus leading to a loss of combined N from the ecosystem. These considerations suggest that the effects of chemolithotrophic NH_4^+ oxidation in limiting photolithotrophic CO_2 fixation may exceed the additional CO_2 fixation due to the chemolithotrophs.

RESULTS

Here we compare CO_2 fixation by photolithotrophs using the NH_4^+ from chemoorganotrophic mineralisation with CO_2 fixation by a combination of photolithotrophs consuming NO_3^- and chemolithotrophs regenerating NO_3^- from the product (NH_4^+) of chemoorganotrophic mineralization.

The chemolithotrophic oxidation of NH_4^+ to NO_2^- is brought about by such organisms as *Nitrosomononas*, *Nitrosolobus*, *Nitrosovibrio* and *Nitrosococcus* (Prosser, 1986):

$$NH_4^+ + 1.5\ O_2 \rightarrow NO_2^- + 2\ H^+ + H_2O$$

The second step, the chemolithotrophic oxidation NO_2^- to NO_3^- i.e. brought about by *Nitrobacter*, *Nitrospira*, *Nitrospina* and *Nitrococcus* (Prosser, 1986):

$$NO_2^- + 0.5\ O_2 \rightarrow NO_3^-$$

The data of Belser (1984) and Glover (1985) give a mean mol CO_2 fixed per mol NH_4^+ oxidized to NO_2^- of 0.048, with the highest value of 0.086, while the mean mol CO_2 fixed per mol NO_2^- oxidized to NO_3^- of 0.031, with the highest value of 0.087. This means that the increment of CO_2 fixation due to chemolithotrophic conversion of 1 NH_4^+ to 1 NO_3^- is (maximally) 0.09 for each of the steps $NH_4^+ \rightarrow NO_2^- \rightarrow NO_3^-$, i.e. ~0.18 CO_2 per NH_4^+ converted to NO_3^-.

The global CO_2 fixation by chemolithotrophic nitrification in the oceans can be computed by using the estimate of global CO_2 fixation of 35 Pg C per year in the oceans, a C:N atomic ratio of 6.625 (the Redfield Ratio), the knowledge that 20% of oceanic primary productivity is supported by NO_3^- produced by nitrification (from Raven et al., 1992) and the ratio of 0.18 mol CO_2 fixed per mol NH_4^+ converted to NO_3^- obtained above. This yields a global value of 0.19 Pg C fixed as CO_2 each year by marine chemolithotrophic nitrifiers.

This can be compared with any decrement of CO_2 fixation as a result of the use of NO_3^- rather than NH_4^+ as N source. In each of the cases considered the decrement of CO_2 fixation is referred to 1 N being assimilated. When all resources (including photons and N) are available in saturating amounts the specific growth rate at a given temperature is maximal. This may be lower with NO_3^- due to the need to allocate resources to catalysts of NO_3^- uptake, NO_3^- reduction and NO_2^- reduction, a need which is absent with NH_4^+ as N source. The allocation of N to such catalysts may be 0.05 of the total plant N, so that the allocation elsewhere to other essential catalysts is reduced by 0.05. The simplest analysis suggests that the presence of only 0.95 as much N etc. in non-NO_3^--related catalysts in the NO_3^--fed organisms would reduce growth rate of NO_3^- to 0.95 that of NH_4^+ (see Raven, 1984).

When photons are limiting, and again assuming optimal allocation and maximal efficiencies, we may conclude that the growth rate with NO_3^- is reduced to 0.805 of the growth rate with NH_4^+. This is because, if 10 photons are needed for growth equivalent to assimilation of 1 CO_2 into phytoplankton cell material when NH_4^+ is the N source, then an additional 2.415 photons are needed for NO_3^- reduction (16 photons to reduce 1 N, and 16 N per 106 C, i.e. the Redfield Ratio). This gives a ratio of photon costs with NO_3^- as N source to those with NH_4^+ as N source of 1.24 so that, when photons are limiting, growth with NO_3^- should proceed at 1/1.24 or 0.805 of the rate with NH_4^+. Such differences are not always realised (see Raven et al., 1992; 1993).

In terms of decrements of CO_2 fixation per N assimilated, we assume the Redfield Ratio of 6.625 C per N throughout. For resource-saturated growth or N-limited growth, the restriction of the rate of growth with NO_3^- to 0.95 that with NH_4^+ means a reduction in CO_2 fixed per unit time per N supplied as NH_4^+ from 6.625 (NH_4^+ as N source) to 6.294 (NO_3^- as N source), i.e. 0.331 CO_2 less with NO_3^-. This compares with the 0.18 CO_2 fixed when 1 NO_3^- is produced from 1 NH_4^+ by the most efficient chemolithotrophic nitrifiers. For photon-limited growth, 6.625 mol C are fixed per N with NH_4^+ as N source, while only 6.625 x 0.805 or 5.336 mol C are fixed in the same time with NO_3^- as N source under the same conditions of photon supply. The decrement of CO_2 fixation as a result of use of NO_3^- rather than NH_4^+ is thus (6.625-5.336) or 1.289 CO_2 per unit time, referred to 1 NH_4^+-N assimilated. This is substantially greater than the largest reported increment of CO_2 fixation in converting 1 NH_4^+ to 1 NO_3^-, i.e. 0.18 CO_2 per N (Table 2).

Table 2
  Comparison of CO_2 fixation at resource saturation by optimally allocating marine phytoplankton using NH_4^+ or NO_3^- as N source and CO_2 fixation by marine chemolithotrophs converting NH_4^+ to NO_3^-.

N source for phytoplankton	Phytoplankton	Nitrifiers	Total
	mol CO_2 fixed per mol N assimilated by photoplankton		
NH_4^+	6.625	-	6.625
NO_3^-	6.294	0.18	6.474

These calculations suggest that CO_2 fixation in a N cycle involving chemolithotrophic nitrification may, with the most efficient, optimally allocating participants, be less than that in a cycle lacking nitrification by chemolithotrophs.

  However, this is not always the case, since the resource-saturated, N-limited or photon-limited specific growth rate of photolithotrophs with NH_4^+ as N source may on occasion be lower than that with NO_3^- as N source (Raven et al., 1992, 1993).

  Further effects of nitrification on CO_2 fixation by photolithotrophs can be seen in terrestrial ecosystems when water is a limiting resource. While theory suggests that NH_4^+ as N source should permit more CO_2 to be fixed per unit of water lost in transpiration than is the case for NO_3^- as N source, observation generally does not bear this out (Raven et al., 1992, 1993). A final influence of nitrification on the potential for photolithotrophic CO_2 fixation comes from the capacity for loss of combined N when NO_3^- is present as a result of throughflow in terrestrial ecosystems, removing NO_3^- to a much greater extent than NH_4^+, and of denitrification in anoxic micro-habitats supplied with organic C and NO_3^-. Both of these factors mean that nitrification may reduce the capacity for CO_2 fixation.

CONCLUSIONS
  Photolithotrophs with Rubisco as their core carboxylase and which evolve O_2 are the main (>99%) global fixers of CO_2.

  The influence which other autotrophs and C_1 production by O_2-evolvers, has on the CO_2-fixing part of the C cycle is mainly via secondary effects on the metabolic activities of Rubisco-based O_2-evolving photolithotrophs rather than via any CO_2 which they themselves fix. Examples are greenhouse effects of methanogens (and indirectly, via denitrification to N_2O, nitrifiers), producers of DMS via their effects on cloudiness, temperature and rainfall after oxidation to

H_2SO_4, as well as their influence in S fertilization of terrestrial ecosystems, producers of C_1 halocarbons *via* their effect on stratospheric O_3 and hence on UV-B flux as well as in influencing herbivory in I-deficient terrestrial habitats, and nitrifiers in converting a (theoretically) more energy and water-efficient N source (NH_4^+) into one (NO_3^-) which is, moreover, more prone to loss by denitrification or (on land) leaching by throughflow.

REFERENCES

Arzee et al. (1985) Bot. Gaz. 146, 365-374.

Beevers H (1961) Respiratory Metabolism in Plants.   Row-Peterson & Co., Evanston, Illinois.

Belser LW (1984) Appl. Env. Microbiol. 48, 1100-1104.

Butler JH (1995) Nature 376, 469-470.

Collén J (1994) Ph.D. Thesis, Uppsala.

Fischer K, Lüttge U (1979) Flora 168, 121-137.

Glover HE (1985) Arch. Microb. 142, 45-50.

Harper DB (1994) Biochem. Soc. Trans. 22, 1007-1011.

Henning et al. (1995) J. Plant Physiol. 146, 193-198.

Houghton JT et al. (1990) Climate Change.   The IPCC Scientific Assessment. Cambridge University Press.

Kelly DP et al. (1994) Biochem. Soc. Trans. 22, 1011-1015.

Khalil MAK et al. (1983) Tellus 35B, 266-274.

Lobert JM et al. (1995) Science 267, 1002-1005.

Lovelock JE, Margulis L (1974) Tellus 26, 2-9.

MacDonald R C, Fall R (1993) Atm. Envir. 27A, 1709-1713.

Meybeck M (1993) Water Air Soil Polln. 70, 443-463.

Mitchell JBF et al. (1995) Nature 376, 501-504.

Mlot C (1995) Science 268, 641-642.

Prosser JI ed (1995) Nitrification.   IRL Press, Oxford.

Raven JA (1985) New Phytol. 101, 25-77.

Raven JA (1995a) Phycologia 34, 93-101.

Raven JA (1995b) Bot. J. Scotland 47, 151-175.

Raven JA, Sprent JI (1989) J. Geol. Soc. Lond. 146, 161-170.

Raven JA et al. (1992) New Phytol. 121, 5-18.

Raven JA et al. (1993) Physiol. Plant 89, 512-518.

Siegenthaler U, Sarmiento JL (1993) Nature 365, 119-125.

Wever R (1988) Nature 335, 501.

Yagi K et al. (1995) Science 267, 1979-1981.

Index